Ceramic Processing Before Firing

CERAMIC PROCESSING BEFORE FIRING

Edited by
George Y. Onoda, Jr.

and

Larry L. Hench

Department of Materials Science and Engineering
University of Florida, Gainesville

A Wiley-Interscience Publication
JOHN WILEY & SONS, New York • Chichester • Brisbane • Toronto

Library of Congress Cataloging in Publication Data:

Main entry under title:
Ceramic processing before firing.

"A Wiley-Interscience publication."
"Edited from the proceedings of the conference . . . held at the University of Florida from January 27 to 30, 1975."
Includes index.
1. Ceramics. I. Onoda, George Y., 1938–
II. Hench, L. L., 1938– III. Florida.
University, Gainesville.

TP807.C45 666 77-10553
ISBN 0-471-65410-8

Printed in the United States of America

10 9 8

To Karl Schwartzwalder and Frederick H. Norton
for their pioneering contributions to the science
and technology of modern ceramic processing

Contributors

R. A. Alleigro, Norton Company, Worcester, Massachusetts
R. B. Bennett, Battelle's Columbus Laboratories, Columbus, Ohio
M. Berg, General Motors, Flint, Michigan
J. E. Burke, General Electric Company, Schenectady, New York
A. R. Cooper, Case Western Reserve University, Cleveland, Ohio
W. B. Crandall, Alfred University, Alfred, New York
I. B. Cutler, University of Utah, Salt Lake City, Utah
R. E. Farris, Kaiser Refractories, Pleasonton, California
W. M. Flock, Plessey Central Development Laboratories, Los Angeles, California
D. W. Fuerstenau, University of California, Berkeley, California
P. K. Gallagher, Bell Laboratories, Murray Hill, New Jersey
D. W. Gates, NASA-Marshall Space Flight Center, Alabama
R. S. Gordon, University of Utah, Salt Lake City, Utah
Y. Harada, IIT Research Institute, Chicago, Illinois
T. M. Hare, North Carolina State University, Raleigh, North Carolina
D. P. H. Hasselman, Lehigh University, Bethlehem, Pennsylvania
L. L. Hench, University of Florida, Gainesville, Florida
H. Heystek, U.S. Bureau of Mines, Tuscaloosa, Alabama
E. J. Jenkins, University of Florida, Gainesville, Florida
D. W. Johnson, Jr., Bell Laboratories, Murray Hill, New Jersey
W. D. Kingery, Massachusetts Institute of Technology, Cambridge, Massachusetts
W. G. Lawrence, Alfred University, Alfred, New York
M. G. McLaren, Rutgers University, New Brunswick, New Jersey
J. S. Masaryk, Kaiser Refractories, Pleasonton, California
E. A. Metzbower, Naval Research Laboratories, Washington, D.C.
R. E. Mistler, Western Electric, Princeton, New Jersey
D. E. Niesz, Battelle's Columbus Laboratories, Columbus, Ohio
F. H. Norton, Gloucester, Massachusetts
G. Y. Onoda, Jr., University of Florida, Gainesville, Florida
C. Orr, Jr., Georgia Institute of Technology, Atlanta, Georgia

J. P. Page, Western Electric Company, Allentown, Pennsylvania
H. Palmour, III, North Carolina State University, Raleigh, North Carolina
J. A. Pask, University of California, Berkeley, California
G. W. Phelps, Rutgers University, New Brunswick, New Jersey
A. G. Pincus, Rutgers University, New Brunswick, New Jersey
F. N. Rhines, University of Florida, Gainesville, Florida
A. Roberts, Sel-Rex Company, Santa Ana, California
G. C. Robinson, Clemson University, Clemson, South Carolina
H. Rumpf, Universitat of Karlsruhe, Karlsruhe, Germany
R. B. Runk, Western Electric Company, Princeton, New Jersey
R. Russell, Jr., Ohio State University, Columbus, Ohio
K. Schwartzwalder, Holly, Michigan
D. J. Shanefield, Western Electric Company, Princeton, New Jersey
H. Schubert, Universitat of Karlsruhe, Karlsruhe, Germany
P. Somasundaran, Columbia University, New York, New York
K. Sommer, Universitat of Karlsruhe, Karlsruhe, Germany
R. T. Tremper, General Electric Company, Cleveland, Ohio
K. K. Verma, Sel-Rex Company, Santa Ana, California
O. J. Whittemore, Jr., University of Washington, Seattle, Washington
W. O. Williamson, Pennsylvania State University, University Park, Pennsylvania

Preface

This book has been edited from the proceedings of the conference on "The Science of Ceramic Processing Before Firing" held at the University of Florida from January 27 to 30, 1975. The conference, the tenth in a series of university conferences on ceramic science, was attended by approximately 200 persons from industry, government, and universities.

It was the first conference in several decades that concentrated on subjects relevant to raw materials, batch processing, and forming of ceramics. Scientists and engineers were invited from throughout the United States to cover topics that were judged to be of particular importance to the theme of the conference. In addition, we were particularly fortunate in having Dr. H. Rumpf and Dr. K. Sommer of Karlsruhe, Germany, with us. A number of the contributors were not ceramists but were experts in areas that are directly related to problems in ceramic processing. Outside areas included fine-particle technology, surface chemistry, mineral dressing, and metallurgy.

At first we planned to have the proceedings published rapidly as a compiled book with a minimum of editing. However, contributors and conference participants encouraged us to prepare a book as close as possible to a textbook, since none exists that covers the topics of the conference. Therefore, many of the chapters were revised, shortened, and sometimes rearranged in an attempt to provide a style more closely approaching a single-author contribution. There were distinct limits in the editing, because in many cases we did not wish to lose the logic and intent of a paper that depended on the style of presentation.

The book is not a descriptive discourse on how ceramics are processed. Other books are available for this purpose. We concentrate on the principles of processing in rather detailed fashion, with the intention of bringing about a better understanding of the science of processing. Students in ceramics should find this book useful as a textbook in ceramic processing, particularly if their course stresses principles. From the response at the conference, where more than three-fourths of the registrants were from industry, we believe that this book will also be useful for ceramic engineers throughout industry.

We are grateful to the U.S. Bureau of Mines (Tuscaloosa, Alabama and Washington, D.C.), the U.S. Army Research Office (Durham, N.C.), and the Department of Materials Science and Engineering, University of Florida, for their partial sponsorship of the conference that made this book possible. Finally, the editors would like to thank Arleen Weintraub for her diligent, competent, and cheerful preparation of the entire written manuscript and Becky McEldowney for her valuable assistance in the administration of the conference.

G. Y. Onoda, Jr.
L. L. Hench
Editors

Gainesville, Florida
October 1977

Honoring KARL SCHWARTZWALDER

Dr. Karl Schwartzwalder
(May 5, 1907–May 2, 1975)

Karl Schwartzwalder was born on May 5, 1907 in Pomeroy, Ohio. As a youth he learned to work and play vigorously and with purpose. He was employed in the backroom of the Pomeroy Daily News, delivered newspapers covering a long route on foot, and did odd jobs to provide supplementary family income. His family included his fine parents, Frances and Frank, and three sisters, with whose families he has remained very close through the years. While not large physically, he was always naturally competitive and courageous, whether in swimming the Ohio River, participating in contact sports, or even as we attempted to pick winners at the racetrack in later years. He has never been one to back away from a cause in which he believes or to waiver in an encounter in support of right. Yet he has always been considerate of others and concerned with their personal welfare. He genuinely enjoys people and helping others is his greatest pleasure.

While Karl's fine intellect and creative genius enabled him to realize great professional success and recognition, he has remained completely unchanged in demeanor or true character from our first professional association some 40 years ago.

Following graduation from Pomeroy High School in 1925, he worked for a year before enrolling at Ohio State University in the fall of 1926. In the depths of the Great Depression he received the Bachelor of Ceramic Engineering degree in 1930 and the Master of Science degree in 1931. He was active in Tau Beta Pi, Texnikoi, Sigma Xi, Sigma Gamma Epsilon, Keramos, and the Student Branch of the American Ceramic Society while on campus.

Those years at Ohio State were the beginning of a long and close friendship with the late Dr. Authur S. Watts, who for over 60 years contributed so much to the advancement of our American ceramic industry. It was Dr. Watts who encouraged Karl in 1931 to become associated with AC Spark Plug, now a division of the General Motors Corporation, an association that lasted until his retirement in 1972. Karl rose through various positions of responsibility in the Research and Development Laboratory to become Chief Ceramic Engineer in 1945 and then Director of Research and Development for all of AC in 1954. His ingenuity is evidenced by an imposing list of over 50 patents in the field of ceramics and metals. He has been a pioneer and vital force in the continuing development of technical ceramics in the United States, especially the refractory oxide types such as alumina ceramics, where his contributions to compositional and processing technology are well documented. Less well known is his role in the ultrasecret Manhattan Project of World War II, where his council and ceramic expertise were widely employed.

While Karl's technical contributions are generally well recognized, perhaps less known is his influence in advancing the careers of the many people who worked under him through the years. His guidance and selfless concern for them have resulted in an imposing alumni body, whose influence will remain far-reaching.

Karl has always enjoyed travel, having visited many foreign countries, inspected countless research and manufacturing operations, and become acquainted with scientific and technical personnel worldwide. He has a peculiar ability to combine business and pleasure in the best sense. His insatiable curiosity has pervaded whatever he encountered in these extensive travels, benefiting both his company and those he contacted along the way.

Karl has been an active member of many technical and professional organizations, including, of course, the American Ceramic Society, for which he served as President from 1956 to 1957 and which has designated him both a Fellow and an Honorary Member. He has also been an active contributor in many other organizations, such as the National Institute of Ceramic Engineers, the American Society for Metals, the Society for Automotive Engineers, and the National Academy of Engineering. His technical affiliations and many committee services through the years are patently too extensive to allow further review here but are indeed imposing. Karl somehow has also found time to be active in civic affairs in such organizations as the Flint, Michigan Chamber of Commerce, the Flint Science Fair, the Five Talents Program for talented high school students, which he founded and sponsored, the Industrial Executives Club, and Saint Rita's Catholic Church, which he long served as a member of the Board of Trustees, and in which Ruth and Karl Schwartzwalder were both uniquely active.

Perhaps among the least widely known facets of Karl's life and also that of his late, wonderful wife, Ruth, are their abiding faith in young people and their concerned interest in nature and wildlife. One must have been exposed to their country home, "their" animals, and the constant influx of young friends to realize the basic goodness of Ruth and Karl. They truly represent the best of human instincts.

It is only natural that many honors, awards, and accompanying responsibilities have come to Karl, including recognition as a Distinguished Alumnus and the Award of an Honorary Doctorate by his Alma Mater, Ohio State University, which he has tangibly supported in many ways throughout his career; the Arthur F. Greaves-Walker Award of the National Institute of Ceramic Engineers; the Bleininger Memorial Award and John Jeppson Medal of the American Ceramic Society; the Man of the Year Award from *Ceramic Age;* the Liberty Bell Award of the Michigan State Bar; the Outstanding Michigan Inventor Award of the Michigan Patent Law Association; the Earl R. Wilson, Sr. Memorial Award of the Society of Automotive Engineers; and special citations from the National Association of Manufacturers and various governmental agencies that have sought and received his counsel.

Karl Schwartzwalder is truly a giant among men. I will ever be most grateful for a warm and enduring friendship with Karl accompanied by genuine feelings of brotherly love and mutual respect. Paraphrasing our Alma Mater, "Time and change will surely show, how firm our friendship." Thank you, Karl.

Ralston Russell, Jr.

On May 2, 1975, Karl Schwartzwalder died after an extended illness. This tribute by Ralston Russell Jr., was delivered at the conference three months before the untimely passing of Dr. Schwartzwalder.

Ed.

Honoring FREDERICK HAREWOOD NORTON

Frederick Harewood Norton
(October 23, 1896–   )

Professor F. H. Norton was born on October 23, 1896, in Manchester, Massachusetts, and obtained a Bachelor of Science degree in Industrial Physics from the Massachusetts Institute of Technology in 1918.

His connections with M.I.T. have been close throughout his life. His father was head of the M.I.T. Physics Department, and his brother, John T. Norton, was a professor of metallurgy at M.I.T. for many years.

Soon after his graduation from M.I.T., he started work at Langley Field for NACA, the group that was largely responsible for early airframe development in this country. Dr. Norton worked in several areas, including the creep rates of heated steel. Before he left in 1923 to join Babcock and Wilcox as a ceramist, he was chief physicist for NACA. (NACA went on to develop supersonic aircraft and eventually turned into NASA.) Dr. Norton later became an expert on refractories.

After about 3 years at Babcock and Wilcox, he went to M.I.T. as a Babcock and Wilcox Research Fellow, and later joined the M.I.T. faculty to start a Division of Ceramics.

As a teacher he was one of the first to recognize the interdisciplinary nature of ceramic science. He required that his students study colloid chemistry in the chemistry department, take X-ray crystallography from Warren, and geology and petrography from Martin Buerger. Considering the research interest both of these professors have had in the ceramics area, I suspect their interaction with Professor Norton was much more than just as teachers of his students.

Norton's success as a teacher is well known. At least two of his students are now deans in major universities and I shall not attempt to enumerate the leading ceramists in the country who have been directly or indirectly strongly influenced by him.

As a ceramist he has had the widest interests. His book on refractories was perhaps the first scientifically based book on ceramics to be published. It first appeared in the 1930s and has gone through four editions. It reflects his continuing close association with Babcock and Wilcox, but also his intense interest in the mechanical techniques of ceramic processing, as well as the origin of the properties in the precurser and product materials.

His is not theoretical knowledge only. He is responsible for the series of insulating firebrick that permitted the inner wall of furnaces to be insulated instead of requiring an inner course of solid firebrick.

He was one of the early workers to investigate the colloidal chemistry nature of the plasticity of clays and the methods of controlling it by dialysis and evaluating it by measuring base-exchange capacity.

Less well known, perhaps, is his seminal work on fiber optics with the American Optical Company. He may be considered the father of this technology, and certainly did a great deal of the early development work.

Finally, I must mention his invaluable work for the Manhattan District in the making of ceramics at the M.I.T. Branch of the project during the war. I was on the receiving end of the project at Los Alamos and recall the exquisite magnesia and beryllia crucibles we received for melting uranium and later plutonium.

Earlier, there had been a similar development on crucibles made of what was called "Brass," which was actually cerium sulfide. Leo Brewer, I believe, had guessed that the melting point of plutonium would be somewhere above 1600°C, where the vapor pressure of MgO would have been too high, so, Professor Norton and his group developed methods for fabricating this esoteric material. Luckily, the melting point of plutonium turned out to be lower than 700°C, so this material was not needed and MgO crucibles were used.

One could go on endlessly. Professor Norton is a Fellow and Honorary Life Member of the American Ceramic Society and the British Ceramic Society. He has honorary doctorates from Alfred and the University of Toledo. He is a most skillful sculptor and artist potter and has written several books in this area.

The most striking aspect that I repeatedly observed in conversations with many of his former students and coworkers is the very high regard in which Professor Norton was and is held. In addition, there were many expressions of gratitude to him for having provided the discipline, the insight, and, indeed, the demand that they understand what they were doing. Many feel that his precepts have guided them throughout their professional careers.

Joseph E. Burke

Contents

PART FOUR GREEN-BODY FORMATION AND MICROSTRUCTURE

PART FIVE PROCESSES AND APPLICATIONS

PART ONE

INTRODUCTORY

It is appropriate that the two introductory chapters are written by the scientist and the engineer who probably contributed most to modern concepts in ceramic processing. The history described in these chapters serves to highlight the changes in approach and philosophy of thinkers and doers in this industry. Before this century ceramics processing was purely an art, but the rapid technological developments in our society required that new materials be developed rapidly. Leisure was no longer an available luxury, and inventiveness and understanding of why processes did or did not work were crucial to meet the new demands of technology.

1

Cycles in Ceramic History

Frederick H. Norton

Anyone studying the world history of ceramics must be aware of the alternating periods of activity and quiescence. I have found the causes of these cycles of great interest and am therefore setting down some thoughts on this subject.

BEFORE THE CHRISTIAN ERA

Ceramic ware during this period was made largely from native clays, which produced a rather soft earthenware. Hand molding and wheel throwing was usual, but some ware, such as the Tanagra figurines, was formed in fired clay molds. Terra cotta was made in rather large pieces. Glazes containing lead and tin were commonly used with a variety of colors. An original surface finish for earthenware, called Terra Sigrillata, was used in Greece and Italy. The finish consisted of a fine fraction of clay separated by suspension in water. When applied as a glaze it could be dried and fired to a smooth dense layer on the ware.

In this period ceramics were used largely in the household, but art ware, such as painted Greek vases, was also produced. The ancient potter did not seem to have the social standing of sculptors and painters.

THE FIRST 17 CENTURIES OF THE CHRISTIAN ERA

In this period a great variety of pottery was made nearly everywhere, but most of it was earthenware. However, one of the most stupendous ceramic developments of all time occurred in China. The earthenware produced at the beginning of the period was gradually refined to a stoneware and finally to the pure white translucent porcelain. This ware reached such heights of perfection that fine specimens were treasured by the wealthy Chinese and by the Ruling Family as well. When specimens of this porcelain trickled into Europe in Marco Polo's days, they were viewed with wonder and awe.

At once the nobility of the Continent recruited potters to search for the secret of making this fine ware. Some progress was made with a frit porcelain, but for many years the secret of making hard porcelain was not discovered in the West.

Today old Chinese porcelains are exhibited in museums all over the world and there are many private collectors who pay enormous sums for fine specimens. I once held in my hand a little vase that had just been purchased for $60,000, and extraordinary pieces have brought much higher prices than this at auctions.

These fine specimens can still be made today, as attested by the very fine forgeries that frequently come on the market. Museums such as the Boston Museum of Fine Arts have laboratories with the most modern equipment to evaluate Chinese pottery and detect forgeries.

What are the factors that permitted this remarkable ceramic development in China? One, undoubtedly, was the availability of a natural porcelain body, a partially decomposed feldspar. This raw body was found in the area of Ching-Te Chen, the center of fine porcelain making. Then there was the development of high-fire kilns capable of reaching a temperature of 1400°C, which is 200 or 300°C higher than could be reached in previous kilns. The Chinese potters used continuous chamber kilns in the form of a chain of chambers running up the hillside both to gain high temperatures and to conserve the scarce wood fuel.

Another factor encouraging fine porcelain manufacture was the intense desire of the Chinese people for beautiful objects. They cherished a cabinet of fine vases more than the architecture of their houses. The Emperor was intensely interested in porcelain, which caused him to subsidize many of the finest potters.

Sometime ago I had a student who was raised in Ching-Te Chen and whose ancestors had lived there for generations. He told a story about a potter who was sentenced to death for some political offence; however, my student's grandfather in some way was able to save his life. In return this potter made a pair of exquisite vases as a present, which took him 5 years. They were small vases and the surfaces were covered with tiny dragons modeled in the finest detail. The eyes of the dragons twinkled when the vases were moved as the black and white eyeballs were free in their sockets.

It was not unusual for a master potter, subsidized by the Emperor, to spend a lifetime in producing just one superb vase for the palace collection. It would seem that the quality of the potter's ware was influenced by the unhurried life-style at this time and the uninterrupted apprenticeship of the young potter to the masters.

Why did the quality of the porcelain ware decline after the Ming Dynasty? Some believe it was due to the exhaustion of the natural body material before the Chinese had learned to make a controlled body of kaolin, feldspar, and quartz. Probably the main cause of deterioration was the artistic fussiness that sets in towards the end of all great art cycles.

EIGHTEENTH CENTURY EUROPE

In the Western World, at the start of the eighteenth century, there was continued production of generally undistinguished earthenware. On the other hand, strenuous efforts were being made on the Continent to duplicate Chinese porcelain. The early attempts produced a type of frit porcelain in which glass was added to the earthenware body. An excellent translucency was achieved, but the ware could not compare with hard porcelain. Then in 1709 a German chemist, Graph von Tschivnhaus, finally discovered the formula for fine porcelain. Attempts were made to keep this valuable knowledge a secret, but soon potters were enticed away from von Tschivnhaus, and many small factories started up all over the Continent, usually supported by wealthy nobles. Most of these enterprises lasted only a few years, but some are still known as producers of hard porcelain, such as Sèvres, Copenhagen, and Meissen.

Thus vitreous pottery produced on the Continent continues to be largely a hard porcelain body made of a kaolin–feldspar–quartz mixture, bisque fired at low temperature and glost fired at a high temperature. Thus a pattern was set for both tableware and artware in the continental area that is still going on.

THE NINETEENTH CENTURY

The earthenware potteries saw few changes during this period except for greater use of coal. However, Hoffman in Germany invented the continuous-chamber kiln, quite unaware that this principle had been used in China many years earlier. Again, England was the original home of the steam engine, so this source of power became available at the end of the century and must have been a factor in modernizing the potteries. However, in 1935 when I visited a large English pottery, almost all work, even wedging, was done by hand.

Toward the end of the nineteenth century electric power pointed to exciting new possibilities of working at very much higher temperatures than had been possible before. Thus electrochemistry began in France and led rapidly to giant steps at the start of the next century.

THE TWENTIETH CENTURY TO 1930

The manufacture of ceramics in this period showed very few significant changes except for the powerful electric furnaces for making silicon carbide, fused alumina, and fused magnesia. Other changes were the increased use of gas and oil for kiln fuel, the start of tunnel kilns, and the use of electric motors for more powerful forming machines.

However, there was an increasing understanding of the various ceramic materials and their forming and firing, and improvements were made in treating clays and other minerals. All these things were building a foundation for the enormous leap in the next period.

A very significant step occurred near the end of the nineteenth century: the initiation of ceramic societies where ceramists could get together and discuss their problems. This did much to break up the years when each pottery was cloaked in a tight veil of secrecy. These societies also began to publish journals to disperse specialized knowledge. In England the British Ceramic Society was started and was carried on by Mellor and later by Green. In Germany an association was started by Seger and carried on actively, and in the United States the American Ceramic Society was conceived by Orton and carried on by Purdy. These organizations and many more that started later were instrumental in uniting the field of ceramics.

THE PERIOD FROM 1925 TO 1974

In 1926 I was asked to start a Ceramic Department at the Massachussetts Institute of Technology by Dr. Stratten, formerly chief of the United States Bureau of Standards. In preparation for this project, I visited a number of existing schools for ceramic engineering. My reaction after these visits was that the schools were doing a fine job in preparing students for manufacturing plants, but the courses seemed rather narrow and inbred. The comments from these schools was general agreement that my lack of a degree in ceramic engineering doomed my project to failure.

The situation seemed to call for integration of other disciplines into the ceramic course to broaden it. This was already being done in Germany to a considerable extent by applying physical chemistry and thermodynamics to ceramic problems. At this time Dr. Bleininger, then at the Homer Laughlin Pottery, was a great influence in steering me toward ceramic science.

The petrographic microscope was extensively used by the mineralogist but only to a limited extent by the ceramist, as it was not adapted to examining the finer particles in clay. Crystal structure determination by X-ray diffraction, started by Bragg in England and carried on in this country by Warren at M.I.T., seemed to be pertinent to many ceramic problems. Also, differential thermal analysis methods seemed useful, so at this time equipment was built in our laboratory for research on clays. The electron microscope was found to be important in many problems, especially morphology of clay particles, so Warren designed and built one in the ceramic laboratory that was very useful in our research.

The precise measurement of temperature is of the greatest importance in the testing or firing of ceramic products. In early times the eye of the skilled fireman was the only judge of temperature. Excavations of old kiln sites indicate much overfired ware, so this method was far from precise. Wedgewood's shrinkage cylinders were a great step ahead and the pyrometric cones went even further. After World War I the optical pyrometer and the thermocouple began to be extensively used and provided better temperature control. Now there are excellent recording and controlling instruments available to hold temperatures to the 1°C tolerances needed for some of the electronic wares.

The colloidal chemist was becoming interested in clay suspensions, from which many aspects of casting slips were made clear. Hauser at M.I.T. was instrumental in sparking increased interest in clay colloids.

During this period great progress was made in the production of equilibrium diagrams applying to ceramics. The United States Bureau of Standards, United States Bureau of Mines, and various universities were instrumental in carrying out this fundamental work so essential to ceramic research and production. The American Society for Testing materials was making progress in setting standards and organizing definitions.

At this time great progress was made in the ceramic field through the study of crystal physics, for the atomic structure of many crystals became of interest to the ceramist.

During the World War II work was active on electrical ceramic bodies, especially in Germany. For high-frequency use steatite porcelains were made in large quantities. At the same time research was going on in the field of titanates. In this country developments also occurred rapidly with ferroelectrics and later with ferromagnetics, largely under the leadership of Von Hipple at M.I.T. These materials were put into use at once and soon became one of the larger branches of ceramics. The miniature printed circuit, first developed by the United States Bureau of Standards for the proximity fuse, has found ever expanding use for printed-on sintered-alumina substrates. The spark plugs with triaxial or mica cores were not suitable for high-powered internal combustion engines, so the sintered-alumina core came into use and now is universally employed for this type of engine.

Around 1930 a new type of conference was started at M.I.T., the first of its kind there and, as far as I know, anywhere else. The idea was to set out a rather specific branch of the ceramic field and other disciplines that could reflect on it. The lecturers were given a reasonable honorarium and those attending paid an entrance fee, hopefully to cover expenses. When first suggested, the plan met with much skepticism from the administration, but finally I got permission to go ahead because it employed unused summer facilities of the school. We included not only lectures and discussions, but also a carefully prepared laboratory course so that those registered could become acquainted with apparatus not generally available at that time, such as the long-arm centrifuge and differential thermal analysis apparatus. Also, attention was given to laboratory note keeping, precision of measurements, patent records, temperature measurements, and many other details pertaining to the ceramic research laboratory.

This type of conference proved so successful that others like it were given annually until my retirement. Other departments began to develop these summer conferences until now there are hundreds every year. Looking back, it seems certain that these conferences have done much to bring up to date the plant worker who often feels rather alone in his domain. Also, the associations made in these groups lead to an excellent interchange of ideas. A further dividend is the opportunity for our graduate students to become acquainted with people in industry; a number of times this has led to excellent positions for the students.

Great progress was made in the refractories field after World War II. Higher-fired and high-alumina silica–alumina brick were commonly used for severe service. Basic brick, for the steel industry of $MgO–Cr_2O_3$ was greatly improved. Fused refractory blocks for the glass industry of alumina and zirconia became common. Perhaps one of the most interesting developments was the insulating refractory which allowed the insulating layer to be on the inside of the furnaces instead of outside of a layer of heavy firebrick. This not only saved fuel, but allowed much quicker heating.

In the field of manufacturing processes, we may mention isostatic pressing used for many products in the last few years, such as spark plug porcelains and tank blocks. The vacuum auger has taken over the field of extrusion, and automated hydraulic presses have found considerable use. Slip casting under pressure is on the way. The roller former has now largely displaced the jigger for making flat and hollow ware. The kilns have been improved to give more even temperature distribution.

In the field of dinner ware, silk screen decoration has been mechanized and decals have been improved. The color maker, applying crystal physics methods, has devised new strains, such as those containing vanadia and zirconia.

Abrasives of the usual silicon carbide and fused alumina are now being made much as they have been for many years. However, the shape and brittleness of the alumina grains is better controlled. Among new materials are the man-made diamonds produced on the basis of the high-pressure carbon-phase diagram. Other hard materials, such as boron nitride, are finding new uses.

It is interesting to speculate on the reason for the tremendous advances of ceramics since World War II. The most important factor was the infiltration of science into the previously rather narrow and empirical ceramic sphere. This caused older processes to be better understood and mysteries to be unraveled and opened the door for original approaches to solve problems. Another factor was the precise determination of material properties, which opened up new uses for them. Lastly, we may look to the much greater interchange of ideas from our ceramic schools, ceramic societies, and technical publications digested by the increasing number of well-trained ceramists.

Has this present expansion in the ceramics field reached a peak or will it continue? I think anyone who reads this book must conclude that the field of ceramics is ready for a further climb.

2

Processing Controls in Technical Ceramics

Karl Schwartzwalder

It was the best of times: it was the worst of times
It was the age of wisdom: it was the age of foolishness
It was an epoch of disbelief: it was an epoch of incredulity
It was the season of hope: it was the winter of despair.

Charles Dickens

These times of challenge in ceramic processing are much like those that Dickens wrote about in the *Tale of Two Cities*, for in many plants therein lies the cause of nonuniformity, unreliability, nonreproducibility, and size limitation of products. In these times of rising material costs and nonavailability, soaring energy costs and shortages, and drops in productivity and profits, a challenge to use creative processing faces us.

In this paper I review (1) processing controls in technical ceramics and (2) major improvements resulting from such controls. Also, I point out the most fertile areas for future development.

CHANGING ATTITUDES IN THE 1930s

Years ago *Punch* had a cartoon captioned "The greatest of all research problems is the people who do research." I have not found this to be true as we were blessed with very capable people. As the problems unfold, I give the names of your peers who solved or contributed to the solution of these problems. The innovations developed in testing resulted in (1) a change from defensive to offensive tactics in processing controls and (2) a simplification in processing.

Let me take you back some 40 years. In those days under the guise of quality control, controls were more for defensive purposes should losses arise than for the improvement of processing or quality. The engineer was called upon to defend himself against management. Following the tactics of the metallurgist, the as-received materials were made into bars and processed (including firing). Test results on these pieces provided process control. There was no apparent reason for change from normal.

My first experience with the company for which I worked over 40 years was as a summer student. My impression as a student was that our plant was the most obsolete that I had ever seen. Professor Watts later convinced me that it was a golden opportunity. I am sure that many summer students feel the same way today.

The composition or body of our insulator was made by blunging, leaf-filter pressing, mixing, and shaping to a roll on French kneading machines (ever hear of them?). It was then extruded from a cylinder into a block with an internal hole and counterbore and finally turned on a lathe into an insulator shape. Moisture content was measured by weighing and by feel, with feel determining the final decision. Modulus of rupture bars, tensile bars, thermal expansion, thermal diffusivity, T_e value discs, and breakdown voltage pieces were made daily. As many people were employed in this area as in development. The insulators were fired in a continuous kiln that Ross Purdy once called a toaster. Lindbergh flew the Atlantic with spark plugs containing insulators made by such methods!

From this method of processing in production, changes were made. Better materials were used in the composition; that is, a switch was made from mullite to alumina. In processing, a change was made to continuous filtration with vacuum drums and continuous extrusion by vacuum pug mills. The lathe continued to be the method for contouring. The salvage material from the operations was reprocessed. While the new materials were tested, they were usually passed as received. A closer control on moisture was maintained since this influenced shrinkage and hence size, which has to be maintained. The only pride in the process was cost, and the process was used at least in part up to the 1950s.

ROLE OF RESEARCH AND DEVELOPMENT

Several unique processing procedures were developed for use in production. Siemens-Halske in Germany developed a method to make a sintered-alumina composition. The patent rights were purchased. The processing was to grind 20% fused alumina and 80% of a low-soda calcined alumina in steel mills to the desired fineness. This mix was then acid treated to remove the iron. The fineness and the pH were controlled by methods developed by Dr. Ralston Russell. The insulators were made by slip casting in plastic molds using chrome-plated copper core pins. The mold life was short. The insulators' shapes were finished on a lathe using a ceramic cutting tool. The processing was costly (as I recall, the company dropped about 500,000 in good depression dollars). But the experience did lead to other processes for making an insulator with alumina as the base material.

The first new processing methods developed in the thirties were the manufacture of insulators using thermosetting and thermoplastic resins as binders, and the application of new techniques developed by the plastics industry to form the insulators. No longer was processing dependent on the plasticity of clays. No longer was wet processing required. From this point on all compositions were prepared by dry grinding, the plastics being added as part of the mill charge. In the case of thermosetting resins, the ground dry powder was used as screened as the mill discharged. The powder was formed into preforms, the several preforms being used to simulate the shape of the formed insulators. Later the preformed system was abandoned and the granulated powder was volumetrically fed into hot 40-cavity dies and cured under pressure to form the insulators.

In the thermoplastic system the resins were part of the dry-milled batch, and the plasticizer was admixed to this material in a hot sigma mixer. The insulators were injection molded in insulator shapes in 40-cavity dies. The plasticizer was removed in a dryer, and the insulators were then fired in continuous tunnel kilns.

During this period of development and manufacture, new methods of control and testing were devised, the former based on resin systems and the latter on sections taken from insulators. Dr. Robert W. Smith, who had worked on the development of the spectrograph for our company at the University of Michigan, devised methods of analysis for the ingredients in raw materials and the insulator, except of course the alkalies. He also devised methods for determining physical characteristics such as thermal expansion, thermal diffusivity, and electrical resistance from sections cut from the insulator. Dr. Helen Barlett cut thin bands from the insulators to determine modulus of rupture. Dr. Morris Berg, while at RCA, worked to improve the method and, as I recall, had it adopted by the American

Society for Testing Materials. Dr. Barlett also sought to determine the tensile strength by measuring internal pressure by closing the hole through the insulator and bursting it with fluid pressure. It was partially successful.

The plastic method of manufacture, utilized because it was a simplified process, required that the insulator be free from ribs on its exposed portion. Functionally this was permissible, since in operation the exposed portion of the spark plug was covered by a protective boot to avoid any flashover due to dirt accumulation. Madison Avenue and national competitors tested the spark plugs for the general public without the boots to show the increased voltage necessary for flashover on ribbed plugs. To this day, American spark plugs have ribs, while Bosch in Germany does not.

The secondary operation required to put on the ribs necessitated a further change in processing. The isostatic method of molding was adopted around 1950. By the way, Dr. Ralston Russell did development work on this process in the early thirties. Mr. Joseph Coors was very helpful in not only making the decision to go in this direction, but in giving us his ideas on processing. The plant facilities were completely changed, and new continuous kilns were built.

The body continued to be dry ground. The binders were added in liquid form in the wet mixing equipment. The blanks were pressed from spray-dried powder and were contoured on a resin-bonded abrasive wheel.

Mr. John Quirk, who at this time was working for Battelle, called our attention to work Battelle was doing for one of the carbon companies that had initiated a new process for making reactive alumina. This was an exciting subject, since at the same time we were also working with Monsanto on a similar matter and had failed. We discussed the potential with the alumina companies and, after several years of hard work, all were able to produce a satisfactory reactive product that has found good usage in a variety of markets. The future was reactive aluminas.

The objective then was to be able to dry grind the entire body and, after screening, feed the material directly to the presses. We also wanted to accelerate grinding. Dr. Walter Gitzen of Alcoa called our attention to the work that had been done on mineral aids, mentioning that napthenic acid had been found to be an excellent aid in ball-milling Bayer aluminas, except that it was hydrophobic and hence was not suitable for alumina to be used in aqueous suspension. Subsequently, both companies undertook development programs. Dr. Gitzen, A. Pearson, and George MacZura worked on Bayer alumina, and at our plant Michael Fenerty in manufacturing, Dr. Morris Berg, and Arthur V. Somers in Research and Development worked on body compositions. Several patents on this manner of processing were issued to both companies.

The attempt to reduce the process to manufacturing practice was not successful. The material tended to bridge in the feeder, and when this happened, the molds were destroyed and replacement was costly, especially in time.

The mill grinding was successful, the output per mill increasing markedly. A system had been devised by Mr. Fenerty so that molding pressure that would yield a given firing shrinkage could be predicted. Losses now were more in handling in the firing operation than in the processing to that point.

MICROSTRUCTURE AND PROPERTIES

During the period of 1959 to 1965, Dr. William Shulhof and Dr. Dave Hinckley undertook to relate physical properties to the interdependent elements of microstructure. Examples were composition relating to phases, size, shape, and orientation and packing of grains. Modulus of rupture was predicted by a few simple measurements on ceramics on which there was good background, such as a production body.

They were also the first, at least in our company, to use designed experiments and statistical analysis to control composition and process variables and to design a ceramic to meet our specific property requirements. Robert Vernetti was the engineer under Mr. Fenerty, and with a few studies he was able to control the kilns to give maximum properties as required for our product.

Dr. Shulhof and Dr. Hinckley developed preparation techniques for sections that permitted better definition of various phases in the body. They may have been the first to use electron microscopy to look at the morphological changes in various phases caused by trace elements.

They also developed X-ray diffraction techniques that measured minor-phase constituents of ceramic bodies. The minor phases were often indications of process conditions; therefore, these techniques led to better process control. X-Ray measurements were found to be a method that predicted "reactivity" of alumina.

Dr. Hinckley developed techniques so that the amount and quality of clays could be measured accurately on production-type materials. For example, X-ray diffraction techniques gave complete analysis of the clay phases in a sample. Percent of bentonite was measured by an ethylene glycol expansion in conjunction with X-ray diffraction. He used base exchange techniques to characterize the "activity" of the clay, for instance, in comparing Florida kaolin with Georgia kaolin. Incidently, Dr. Russell worked on similar base exchange techniques during the thirties.

Dr. William Flock simplified the control aspect of this analysis by correlating the X-ray data with simple optical examination of alumina aggregates. I recall one incident when we were visiting one of the alumina companies and were asked if we had any questions. Bill said, "Yes, why did you increase your rate of precipitation of alumina hydrate?" They were stunned that Bill knew this, and so he discussed his methods of analysis with them.

At the time of my retirement, Dr. Flock's insulator composition was the best that had been developed for our product.

Dr. Charles Ondrick simplified the processing procedures, which has reduced costs and improved quality.

You may wonder why so many worked on processing for relatively short periods of time. The top engineers or scientists move on to other areas for the simple reason that very capable men are always in demand whether we be in periods of depression, recession, or prosperity. Their capabilities saved the company millions of dollars and, at the same time, produced a superior product.

Process control is no longer a defensive measure. Management looks at process control as (1) cost control, (2) a means of lower processing costs and improving productivity through simplification of processing and decreasing salvage operations, and (3) a means of improving the quality and reproducibility of the product.

Our objective was to achieve this through the development of new techniques for the determination of the microstructure of the materials utilized and its relation to properties and microstructure of the final fired product. Like any method, it can and should be improved, and will be in the future. New techniques and instrumentation to cover organic and inorganic distribution are required.

FUTURE ADVANCES IN PROCESSING

To me, the most important advancement that will be made in the next few years is in new processing techniques. Some innovative changes are taking place in ceramics that could lead to new processing techniques in technical ceramics.

One innovation is a process developed by Howard Sunman of 3M that does not require melting and high-temperature firing for making continuous ceramic fibers. These have a visible appearance much like that of fiber glass. Viscous concentrates of salts solutions, sols, or mixture of these are spun into continuous filaments by extruding and drawing through multiple-holed spinnerettes at low temperatures. The drawn fibers can be collected as a strand and fired continuously to decompose the compounds, the fugitives can be removed, and the material can be consolidated into strong continuous fibers. Numerous materials can be formed into fibers by this process. In addition to fiber products, this method should open a vast field for ceramic composites requiring process controls completely different than those now utilized.

About 20 years ago 3M developed a monolithic ceramic honeycomb that could operate at high temperature and radiate or transmit very large heat fluxes. These honeycombs are now being used in automotive catalytic converters made by several companies and, I assume, by different processes. I do not know who actually developed the extrusion process for making honeycombs. Many, I am sure, thought of this process. Corning was apparently the first to reduce the process to commercial practice. I am sure this process required a completely different set of controls and will open new fields in thin-walled ceramics.

One more example is a process developed by Kelsey-Hayes for hydrostatic molding of powders at high temperatures by using a glass or ceramic envelope in the shape of the product desired. The part produced from these powders is fired in an atmosphere best suited for the powder. Here again, a whole new set of process controls will be required. While presently this process is used for powdered metals, it can be used for ceramics.

PART TWO

POWDERS

The first requirement of a successful ceramic process is the availability of a good powder. Not only should the powder have the required purity, it should also have certain desirable characteristics, such as high packing density and good sinterability. Many existing processes have been improved through the use of better raw materials, and the past 20 years have been particularly fruitful in this regard. Methods for fine grinding have been significantly improved and active powders from solution are now providing a greater variety of starting materials. Available also are better powder-characterization techniques, and the practice of characterization is becoming more common. What to characterize is better appreciated, particularly in terms of how it relates to actual processing steps.

3

Active Powders

I. B. Cutler

The goal in processing ceramic powders is to obtain a product with the most desirable properties commensurate with an acceptable cost. Properties of a sintered product reflect the nature of the particle compact. A review of sintering theory and practice can help the processor of ceramic powders and powder compacts get the most active powders for the least effort.

Unfortunately, the term "active" powder may mean different things to different people. In terms of ceramic powders it generally means a sinterable powder. What makes a powder particularly active or responsive to temperature and promotes sintering is at least partially known at present. Among the voluminous literature are some excellent reviews.[1-4] An oversimplified approach clearly reflecting the bias of the author is given here.

DENSIFICATION THEORY

In the majority of sintering operations, densification accompanies sintering. Densification almost always requires shrinkage. The shrinkage takes place as a result of materials being transported by one or more of several diffusion processes. This may involve a liquid or reactive liquid or diffusion at grain boundaries or through the volume of particles. Cearly, any vapor transport or surface diffusion will not contribute to shrinkage and densification. On the contrary, any vapor transport or surface diffusion will interfere with densification processes by dissipating the driving force for the densification and shrinkage processes. Although plastic deformation through dislocation motion has not been entirely eliminated as a mechanism of material transport, the number of instances where dislocation motion makes a contribution is thought to be small.

Considerable effort has been expended in attempting to delineate mechanisms of sintering. There appears to be little, if any, practical justification for much of the effort that has been expended in determining the particular equation(s) or the particular diffusion path(s) that best describes the sintering of a particular material. The theory has been of great practical importance in describing what can be done to change the sintering of many materials either to promote or inhibit densification.

PARTICLE SIZE

The theory makes it clear that particle size is a very important variable in the densification process. The rate of densification, regardless of which theory or equations and mechanisms are employed, is inversely proportional to particle size for a simple viscous system at any temperature and inversely proportional to particle size to the 1.5 power for volume diffusion. This means that the smaller the particle size of the material, the more active the powder will be toward sintering.

To put this in perspective, sufficient data on various Bayer-process alumina powders are available to give us some information on the sintering temperature for various particle sizes. A tenfold reduction in particle size reduces the sintering temperature by approximately 200°C. This, of course, assumes that all the other variables are held constant and that we are only altering the particle size and sintering temperature.

Most everyone agrees that particle size is an extremely important parameter. Confusion arises, however, because many small particles are held in rather tight agglomerates. The problem of agglomeration and the confusion over which is really the particle to consider in sintering are best illustrated in the case of Bayer alumina. The early work of Ryshkewitch[5] emphasized the importance of grinding Bayer alumina to make it into an active powder. Actually, the crystal size was not altered during the grinding operation, but the particle size was altered as it was reduced from an agglomerate to a particle near the size of the crystals contained in the agglomerate. With the aid of scanning electron microscopy, we can visualize this much better today than when the principle was first discovered. Figure 3.1 shows an agglomerate of alpha alumina derived from

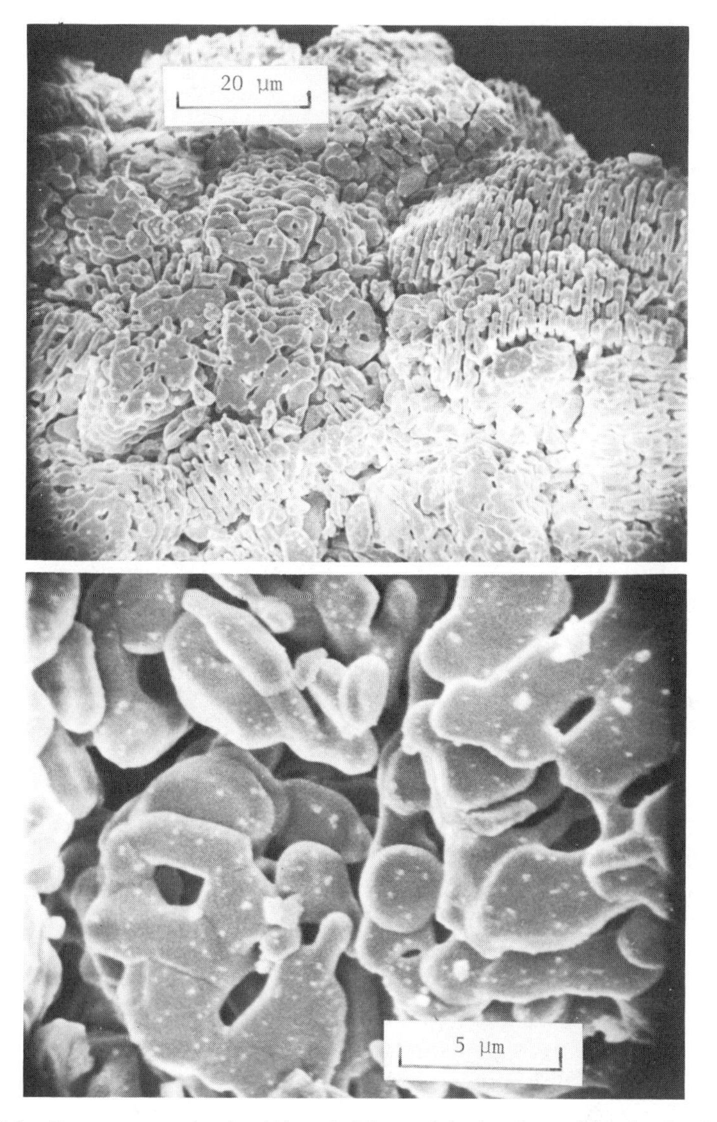

Figure 3.1. Bayer-process alumina (Alcoa A-14) as calcined to form alpha alumina from aluminum hydroxide. The particles retain the morphology of the original hydrate.

aluminum hydroxide by calcination. The particle still shows the morphology of the original aluminum hydroxide. Like so many powders derived by decomposition, there is some preferred orientation of the alpha alumina in the agglomerate. Each agglomerate acts as a spongy particle of very large size and, consequently, it is nearly impossible to sinter below the melt-

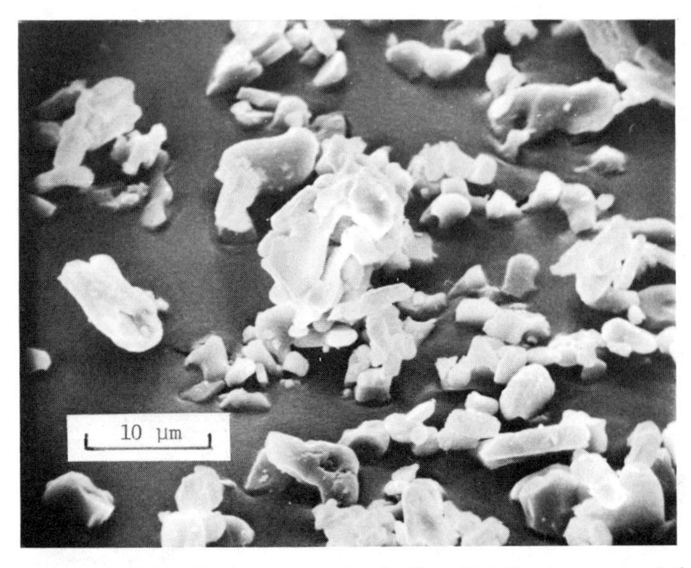

Figure 3.2 Ground alpha alumina (Alcoa A-14). Note that the fragments of the original agglomerate are only partially reduced to their ultimate crystalline size.

ing point. Upon grinding to release the individual crystals, an active powder is obtained. As shown in Figure 3.2, grinding does not always yield individual crystals but rather smaller agglomerates, depending on the grinding efficiencies.

Even alpha alumina derived from alum and other sulfates produces agglomerates. Possibly the agglomerates from alum have lower preferred orientation of the individual crystallites than in the case of the Bayer alumina. As active as the aluminas derived from alum are, they can still be increased in activity by grinding.[6] More important is the increase of green density resulting from the grinding of these spongy agglomerates.

The activation of alumina by grinding has, at times, been related to energy stored in alumina particles through the incorporation of defects. It is certainly possible to increase the defect content of materials through deformation, but it is doubtful that this contributes significantly to the sintering process. Such nonequilibrium vacancies and interstitials will rapidly anneal out at the sintering temperature during the initial stages of sintering and make only a very minor contribution to the sintering process.

PARTICLE SHAPE AND SIZE DISTRIBUTION

Because theory describes particle size in terms of spherical particles, realistic systems must incorporate a measure of the particles in terms of their radii of curvature. As can be seen in Figure 3.2, the grinding of alumina produces angular particles. Judging from experience with angular glass particles in comparison with spherical glass particles this angularity may increase the activity of a powder by as much as four or five times the activity of the comparable spherical powder.[7] This is significant, but, unfortunately, submicron-size particles in agglomerates are very difficult to effectively grind to their ultimate crystals. Thus it is not always possible to utilize the angularity with the more desirable sizes.

Size distribution is important in practical systems where shrinkage is to be minimized. Mostly large particles and sufficient small particles to fill in the interstices will give a particle compact of highest green density. For viscous systems that cannot maintain stress during sintering, the effective particle size in a distribution is simply the average of all particle sizes in the particle compact. It is more complicated for nonviscous systems, as discussed by Coble,[8] and the effective particle size for up to something like 10% coarse particles appears to be essentially that of the fine-size fraction. It rapidly changes to the particle size of the coarse fraction for a distribution with over half coarse particles. From the point of view of minimizing shrinkage and maximizing density of the green compact, it would be nice to have particle compacts of mostly coarse particles. If this choice is made, it must be remembered that the particle compact with the distribution of mostly coarse particles will sinter as though it were made up entirely of coarse particles, and the sintering temperature must be increased accordingly.

PACKING OF PARTICLES

One of the chief arguments against using fine particle sizes ($<0.1\ \mu$) is the decrease in green density of particle compacts of very fine powders and the resultant increase in sintering shrinkage. At least part of this problem is agglomeration. Close-packed spheres have the same density regardless of sphere size. The idea that physical forces between small particles begin to exert an influence on particle packing as particle sizes below 1 μ are utilized is one explanation for poor packing of very fine particles. Ramsey and Avery[9] have some interesting information regarding this factor in their work with ultrafine oxide powders prepared by electron-beam evaporation. They have compared individual particles produced by electron-beam evaporation with agglomerates of a similar surface area obtained by decomposition. The individual particles of MgO were compacted to 50% theoretical density at 66.7 kpsi and 64% of theoretical density at 133 kpsi. The MgO obtained by decomposition of nesquehonite (magnesium carbonate, $MgCO_3 \cdot 3H_2O$) gave significantly lower pressed densities. Sintering behavior of MgO parti-

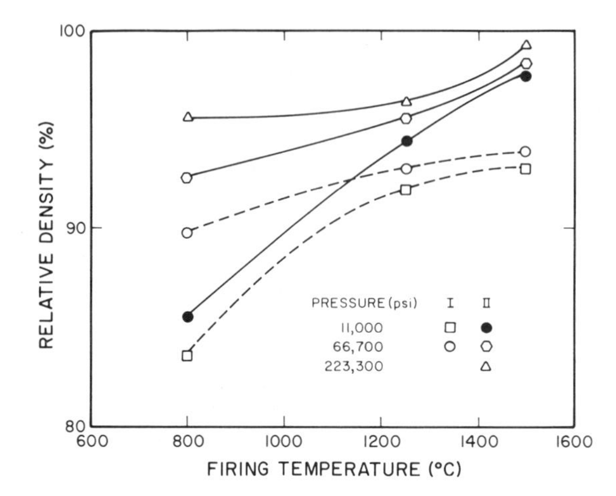

Figure 3.3. Data by Ramsey and Avery for (I) MgO from decomposition of $MgCO_3 \cdot 3H_2O$ *in vacuo* and (II) MgO from electron-beam evaporation, both of about the same surface area (319 and 360 m^2/g, respectively.

cles produced individually is compared with the agglomerated MgO obtained by decomposition in Figure 3.3. The conclusion reached from these and the data of Binns et al.[6] is that the most important deterrent to the use of submicron particles is the difficulty of the packing of agglomerates rather than individual particles. Thus we need to be more clever in devising methods of obtaining unagglomerated crystals of submicron size by techniques other than grinding.

There appears to be some minimum density of packing of particles or minimum number of contacts per particle necessary to produce uniform shrinkage in a particle compact. Bruch[10] showed that a lower limit of density of a particle compact had to be exceeded to produce full densification of alumina during sintering. Exner et al.[11] have noted that particle rearrangement can take place during the initial stages of sintering and result in pore growth, that is, when a limited number of particle–particle contacts are involved in the compact. Ordinarily, this threshold of green density is exceeded in many practical cases; however, we should be aware of the importance of low density in situations where minimum densification is desired.

ADDITIVES AND IMPURITIES

The sintering of stoichiometric oxides, such as aluminum oxide and magnesium oxide, may be dramatically affected by an additive that will enter solid solution and disturb the stoichiometry of the oxide. This is shown in Figure 3.4 for a rather impure commercial aluminum oxide sintered at various temperatures compared with the same aluminum oxide for which the sintering temperature is held steady but the addition of TiO_2 is varied. It is easily seen that temperature and additions of TiO_2 have comparable effects. Both change the diffusion process by about equal amounts. A small addition of TiO_2 may alter the sintering temperature by as much as 200°C, which is about the same effect as reducing the particle size to one-tenth of the original size.

Not all additives that enter solid solution increase the rate of sintering, as has been discussed in the literature.[14,15] Various solutes have different abilities for producing nonstoichiometry, some of which may increase while others decrease the rate of sintering. Finding additives that are grain-growth inhibitors and that promote elimination of porosity is similarly difficult and time consuming. In no field of endeavor is it more difficult to predict the effect that an additive will produce than in the area of sintering.

By exceeding the solubility limit of TiO_2 in aluminum oxide, the kinetics of densification are decreased. This appears to be typical of the sintering

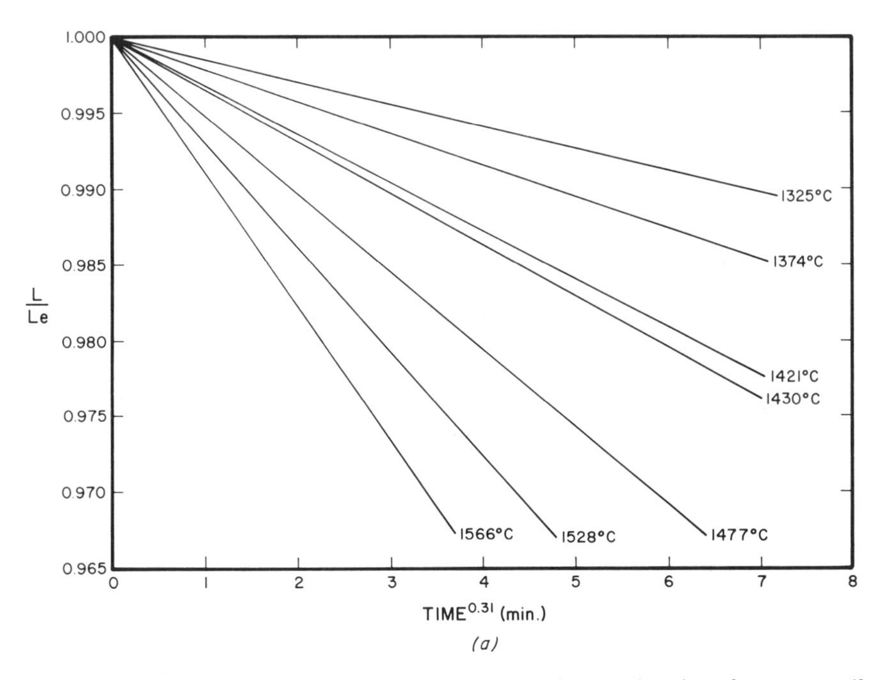

Figure 3.4. Isothermal shrinkage data for A-14 alumina (*a*) as a function of temperature[12] and (*b*) as a function of TiO_2 content at $T = 1520°C$.[13] Additions of TiO_2 are expressed in terms of atomic percent of Ti.

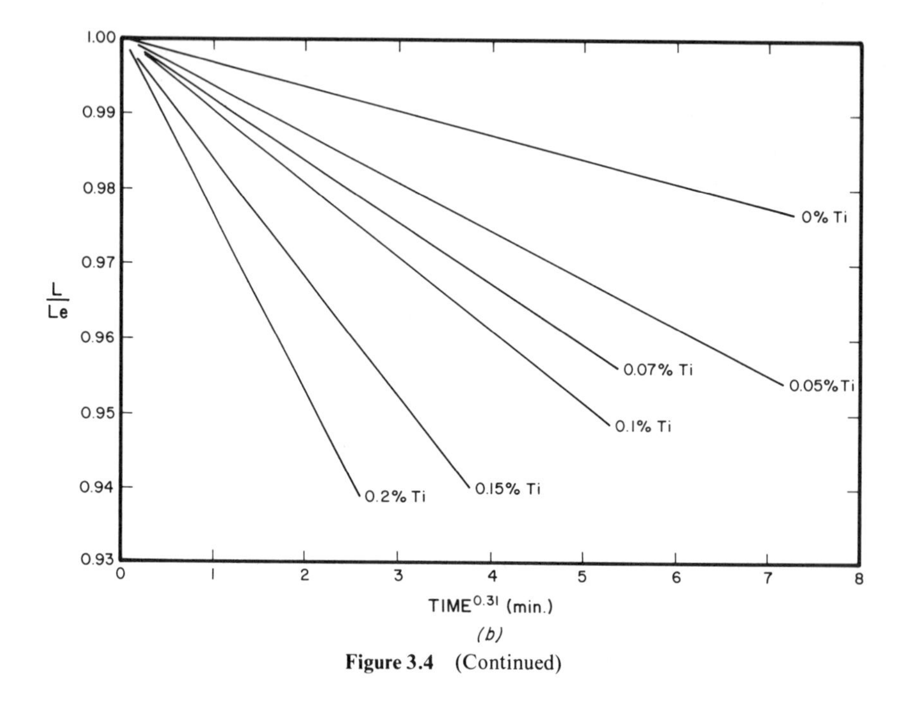

(b)

Figure 3.4 (Continued)

process. Additives that form a second phase without forming a liquid apparently inhibit the densification process. On the other hand, the liquids appear to increase densification. No example of this is more familiar to the ceramist than the case of porcelain where the feldspar contributes a liquid phase that promotes densification.

Reactive sintering in the presence of a liquid phase constitutes still another special case of active sintering. For example, forsterite (Mg_2SiO_4) may be sintered utilizing a eutectic liquid of magnesia and about 65% silica, along with periclase (MgO) at temperatures on the order of 1560 to 1600°C. By the time the liquid has reacted with the magnesium oxide the material will be well sintered and will yield a forsterite composition.

ATMOSPHERE

Although the sintering atmosphere has little to do with ceramic powder processing, it does have a great deal to do with the sintering process. For completeness, it is mentioned here that we should be cognizant of the atmosphere surrounding the particle compact as it is being sintered. Defect concentration can be altered by the atmosphere, and the atmosphere contributes to pore removal in the final stages of sintering.

CONCLUSION

Great progress has been made in understanding the sintering process of particle compacts. Particle size, agglomeration, particle shape, size distribution of particles, and packing in the compact all have a great influence on sintering and densification. Additives and atmosphere contribute to the sintering process as well. During processing most of these parameters may be changed to meet the needs of the ceramist and his sintered product.

REFERENCES

1. F. Thummler and W. Thomma, "The Sintering Process," *Metall. Rev.*, **12** (115), 69 (1967).
2. A. L. Stuijts, "Synthesis of Materials from Powders by Sintering," *Ann. Rev. Mater. Sci.*, **3**, 363–395 (1973).
3. A. L. Stuijts, "Sintering Theories and Industrial Practice," *Sintering and Related Phenomena* (Materials Science Research, Vol. 6), G. C. Kuczynski, ed., Plenum Press, New York, 1973, pp. 331–350.
4. I. B. Cutler, "Sintered Alumina and Magnesia," *Refractory Materials*, Vols. 5–111, A. M. Alper, ed., Academic, New York, 1970, pp. 129–179.
5. E. Ryshkewitch, *Oxydkeramik*, Springer-Verlag, Berlin 1948.
6. D. B. Binns, P. Engel, and P. Popper, "Methods for Increasing the Compactibility of Powder," *Proc. Int. Conf. Compaction Consolidation Particulate Matter, 1st Brighton, England*, **1972**.
7. I. B. Cutler and R. E. Henrichsen, "Effect of Particle Shape on the Kinetics of the Sintering of Glass," *J. Amer. Ceram. Soc.*, **51** (10), 604–605 (1968).
8. R. L. Coble, "Effects of Particle-Size Distribution in Initial-Stage Sintering, *J. Amer. Ceram. Soc.*, **56**, 461–466 (1973).
9. J. D. F. Ramsey and R. G. Avery, "Ultrafine Oxide Powders Prepared by Electron Beam Evaporation," *J. Mater. Sci.*, **9**, 1681–1695 (1974).
10. C. A. Bruch, "Sintering Kinetics for the High-Density Alumina Process," *Bull. Amer. Ceram. Soc.*, **41** (12), 799–806 (1962).
11. H. E. Exner, G. Petzow, and P. Wellner, "Problems in the Extension of Sintering Theories to Real Systems," *Sintering and Related Phenomean*, Material Science Research, Vol. 6, G. C. Kuczynski, ed., Plenum Press, New York, 1973, pp. 351–362.
12. D. L. Johnson and I. B. Cutler, "Diffusion Sintering II: Initial Sintering Kinetics of Alumina," *J. Amer. Ceram. Soc.*, **46** (11), 545–550 (1963).
13. R. D. Bagley, I. B. Cutler and D. L. Johnson, "Effect of TiO_2 on the Initial Sintering of Al_2O_3," *J. Amer. Ceram. Soc.*, **53**, 136–141 (1970).
14. P. Reijner, "Non-Stoichiometry and Sintering of Ionic Solids," *Proc. Int. Symp., Reactivity Solids, 6th, New York*, **1969**, Wiley-Interscience, New York pp. 99–114.
15. D. W. Ready, "Mass Transport and Sintering of Impure Ionic Solids," *J. Amer. Cer. Soc.*, **47**, 366–369 (1969).

4

Characterization and Process Interactions

W. M. Flock

The process engineer must decide when, where, how, and what powder characterization parameters to measure. Furthermore, he is faced with the fact that they must be measured within a reasonable time, at a reasonable cost, and by average operators. Hence material characterization is often the most frustrating tasks facing the ceramic scientist.

Probably the first question that should be posed is, why characterize? Answers to this question are as variable as those to the question of what to characterize. A common answer is to ensure quality or obtain information for future references. Another more negative response is to reject defective material; if these are the only reasons for characterization, then an inactive process-control model results and data are collected and stored until a problem occurs.

If characterization is carried out to predict and adjust, an active process-control model is being employed. By predicting the material-process response and ceramic-product character, adjustments in processing can be made if these predictions are not in line with the desired performance or properties. It is the material characterization required for this active model that is discussed in this chapter. For an active process model, material characterization can be defined as the means by which the ceramic engineer interacts or communicates with the process on a real-time basis.

THE PROCESS-CONTROL-MODEL CONCEPT

To interact with a process on a real-time basis, a process model is required. The model shown in Figure 4.1 is used in principle for nearly all high-volume ceramic-manufacturing processes. Its use may be intentional or unintentional, and it may not be as rigorously developed as shown in Figure 4.1, but economics dictate that variation from this model must be slight.

If we assume that the process objectives are to develop consistent predictable products, then this model requires that processing conditions and raw materials be consistent and predictable. If variations occur in one, then adjustments must be made in the other, that is, raw material plus process equals predictable ceramic character. However, economic conditions require that if process adjustments are made, they must be inexpensive and rapid. In large-volume operations, process parameters such as tooling, body composition, organic composition, and firing profile must be held constant, while parameters such as mill time, forming pressure, and surfactants (grinding aids, viscosity aids, and lubricants) may be changed.

In summary, for this model to be operative, two conditions are required: (1) that constant raw materials or materials that have been sufficiently characterized to predict their material-process response be used and (2) that the process is sufficiently understood to make the required adjustments.

Since high-volume ceramic processing has been successfully carried out for over 50 years, and since the material scientists are in agreement that material characterization is, by in large, inadequate, then the scientific community is out of touch with reality and no problem exists, ceramic articles are not being made with consistent properties, or the above proposed model is incorrect and material variations have no effect on ceramic character, and some other unknown process is active. In actual practice, it is found that all the above are partially true, for in lieu of the capability to

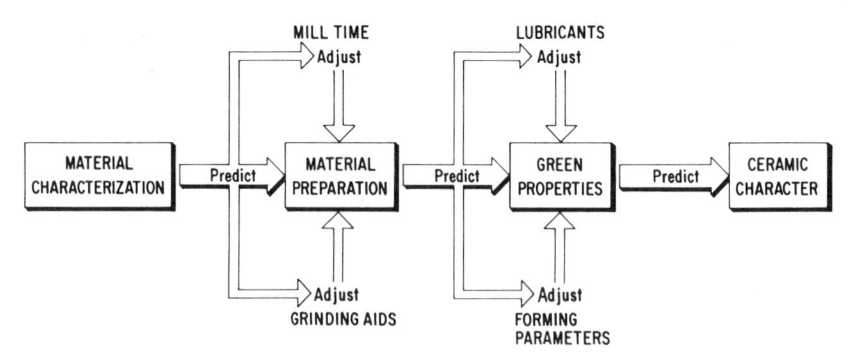

Figure 4.1. Active material process control model.

provide complete material characterization, the process engineer has survived by holding constant as many of the material variables as possible.

MATERIAL CHARACTERIZATION CONCEPT

What material variables must be held constant? Or stated differently, what parameters are required to define the material character? It has been found by a number of investigators that material character can be expressed by the following function:

$$C = f(s, sh, o, p \text{ and } c)$$

where C = character
s = size
sh = shape
o = orientation
p = packing
c = composition

The process model dictates that these variables either be held constant or be measured, but since the capability to measure or characterize such variables as shape, packing, and orientation is sorely lacking, the problem was solved, and continues to be solved for many processes, by the practice of single-source purchasing. The ceramic industry, and for that matter, most industries, has used, and continues to use, single-source raw materials made by a particular manufacturing process and specific manufacturing equipment. These sole source suppliers have learned from loud and quick customer response that process changes are forbidden. It is common to hear statements that only kiln A or pit B can be used for a customer and kiln B or pit C for another customer. The failure of the material scientist to provide meaningful characterization tools has fostered sole-source purchasing. This concept, based on human experience and not scientific concepts, has created misconceptions and doubts that will strew the road to progress with obstacles for years to come, that is, it has created an environment in which change is forbidden.

Another less detrimental but nevertheless costly technique is to negate the need for material characterization by eliminating the previous raw material history, that is, beginning the process over within the users control. This approach transfers the problem of material characterization to process control, which fortunately is much better developed. A typical example of this approach is the destruction of the Bayer-process characteristics by high-temperature sintering (tabular alumina) followed by steel-ball milling for size reduction and an acid wash for iron removal; however, energy costs are making this concept prohibitive.

In summary, increasing costs dictate that material characterization must be advanced to the stage where total characterization can be made regardless of the previous process used and that it can be done sufficiently to predict the material-process response. This is an ambitious step, but the concept presented in Chapter 9 for Bayer-processed alumina is believed to be one approach that practical experience has shown to be in the right direction.

5

Physical Characterization Terminology

G. Y. Onoda, Jr.
L. L. Hench

Particulate materials consist of small units of solid matter that have specific physical characteristics. A variety of small units have been described by various names, for example, primary particle, ultimate particle, particle, grain, colloid, agglomerate, conglomerate, aggregate, clump, cluster, crystallite, granule, and floc. Inconsistencies exist in the literature on the definitions and uses of these terms, partly because different disciplines (e.g., ceramics, powder metallurgy, fine particle technology) have tended to have their own preferences. In this chapter definitions are proposed that hopefully will aid in the establishment of consistent terminology in the ceramics field. These definitions are used in this book.

PRIMARY PARTICLES

A primary particle is a discrete, low-porosity unit. It may be a single crystal, a monophase polycrystal, a multiphase polycrystal, or a glass. The pores, if any, are isolated from each other and therefore the primary particle is impervious to fluids. If the primary particle is one of several that make up larger agglomerates, it is common to refer to the size of such a particle as the "ultimate" particle size.

AGGLOMERATES

"Agglomerate" is a general term describing a small mass having a network of interconnective pores. It is comprised of primary particles bonded together by surface forces and/or solid bridges (see Chapter 27). The surface forces may be electrostatic or van der Waal attractions between particles or liquid capillary forces due to the presence of liquid within the agglomerate. Solid bridges are a result of sintering, fusion, chemical reaction, or setting of a binder.

Agglomerates with solid bridges can retain their identity under a wide variety of conditions. In contrast, agglomerates held together with surface forces are much more readily disrupted by small external forces. It is often convenient to have a simple terminology that distinguishes between the two general agglomerate types. Therefore, *solid agglomerates* are defined as agglomerates with solid bridges; *weak agglomerates* are agglomerates with surface forces as bonds.

PARTICLES

A "particle" may be a single primary particle or a solid agglomerate. As such, it is a small mass that is free to move as an entity[1] when the powder is dispersed by the breaking of the surface force bonds. Most particle-size measuring techniques operate on such particles.

GRANULES

The term "granules" is frequently used in ceramics to identify agglomerates that are intentionally formed by the addition of a granulating agent to promote the formation of large agglomerates. This definition is an operational one based on a deliberate forming process. An example of a granule is the product from a spray drier.

FLOCS

Flocs are a cluster of particles that form in a liquid suspension. The floc can be dispersed by appropriate modification of the interfacial forces through alteration of the solution chemistry. The particles of a floc are held together by short-range interfacial forces or by organic flocculating agents.

COLLOIDS

Colloids are particles that, when dispersed in a fluid, are fine enough so that Brownian motion maintains the them in suspension without settling. In general, colloids are particles of submicron size.

AGGREGATE

The term "aggregate" in ceramics refers to the coarse constituent in a batch material, usually in combination with a fine constituent called the bond. Some fine-particle technologists use the term "aggregate" to represent what we define as solid agglomerates. This practice is avoided to prevent confusion within the field of ceramics.

REFERENCE

1. E. R. Stover, "A Critical Survey of Characterization of Particulate Ceramic Raw Materials," Technical Report AFML-TR-67-56, Air Force Materials Laboratory, Wright-Patterson Air Force Base, Ohio, May 1967.

6
Physical Characterization Techniques for Particles

C. Orr, Jr.

This chapter assesses and reviews the fundamental measurement techniques for particle size, surface area, shape, and pore structure of particles. The basic concepts, assumptions involved, special attributes or deficiencies, and the meaning and value of results are discussed. Some of the simplified comparison techniques that may be adapted for routine process-control purposes are outlined.

PARTICLE SIZE

Particles considered in this chapter are restricted to those having dimensions greater than colloids (~0.1 μm) and less than those that can be discerned without magnification or other aid (~1000 μm). For want of a better term, "diameter" is employed as the measure of size, although it applies strictly to circles and spheres, and very few particles are spherical. Diameter is thus subject to interpretation.

Microscopy

When measured microscopically, particle diameter is usually taken as the distance between points on opposite sides of the particle using some convention with respect to particle orientation; only rarely is an accounting of even one right-angle dimension attempted. The result of a series of such particle measurements is a statistical number-based size distribution that most likely underestimates the contribution of the smallest particles. The tendency of particles to assume their most stable position when deposited on a substrate reveals their greatest dimensions to examination, which further biases the results.

Manual measurement of size is extremely tedious. There are several sophisticated electronic commercial systems in which the taking and analysis of size data is highly automated. Whether the system is manual or automatic, great care should be exercised in preparing representative samples. Appropriate numbers of particles should be examined so that neither too many nor too few are exposed on prepared slides or photographs. Since a magnification factor is always included in microscopic measurements, this factor is subject to error which can shift a distribution either towards greater or smaller sizes.[1]

Microscopic techniques and data treatments have been described in detail in many publications.[2-7] They are not pursued further here except to note that microscopy must be regarded as very useful as a check on other methods and as a guide to method selection. Since microscopic size analysis results directly in number-based data pertaining to particle profile, a mean diameter computed on this basis can differ quite significantly from a mass mean diameter (e.g., derived from a sieve analysis).

Sieving

This is the oldest of all particle characterization methods. Sieving has often been described in the technical literature[2-5] and needs little further elaboration here. With care, a sieve analysis can be obtained down to about 1 μm in diameter[8,9]; analysis is relatively rapid, and as with microscopy, the obviousness of the method makes the results highly desirable. There are pitfalls here too, however. If analysis is carried out with a dry powder, electrostatic effects and particle agglomeration may cause problems. Particle agglomeration can still be a problem even in wet sieving, especially with the very small opening sieves. Also commercial wire-sieve openings are likely to vary several percent from the nominal size. The finer meshes, particularly, are very easily damaged by careless handling. Particles of approximately the opening size tend to be caught in the sieves, which, if it occurs over much of the powder, causes a distortion of the data. Results are highly dependent on operator technique. In general, sieving should be continued for a considerably longer time than is frequently allowed. Sieve analysis cannot yet be said to have succumbed to automation, but progress is being made.[10,11] Success here should alleviate some of the problems noted.

The diameter obtained by sieving irregular particles is that of the smallest cross section that will pass an approximately square opening. Such a diameter can differ considerably from a microscopically measured diameter, and results also differ in that they are mass based. Thus sets of data for the same powder obtained from microscopy and from sieving often appear to be very different.

Stokes' Diameter

Measurement of the sedimentation rate of particles in a quiescent liquid affords a very important means of particle sizing. The size thus determined is known as the Stokes' or equivalent spherical diameter. It is the diameter of the sphere of the same density that would exhibit the identical free-fall velocity of the particle. As such it is a measure that readily correlates with many handling and processing parameters.

Data on sedimentation rates may be obtained using initially uniform liquid suspensions by measuring the concentration of particles remaining in suspension as a function of time. Sedimentation rates can also be obtained by measuring the quantity of sediment produced as a function of time. This method is less widely used and is not described here.

Hydrometers,[12] divers,[13] light transmission,[14] and X-ray absorption[15-18] can be utilized. Hydrometers disturb the orderly sedimentation process, and use of both hydrometers and divers requires long times of analysis. Light transmission, or so-called turbidimetric analysis, is sensitive to particle refractive index unless only the light in a very small forward beam is detected,[19] and even then corrections must be applied for submicrometer diameters. The diameter finally detected is representative of particle cross section and must be multiplied by particle diameter to put the results on a mass basis. These difficulties have so far prevented the development of a wide-ranging, generally satisfactory light extinction size analyzer.

X-Ray Absorption

A narrow beam of low-power X-rays constitutes a very satisfactory means for measuring concentration since adsorption is directly dependent on interposed mass. A commercial version of an instrument, shown in Figure 6.1, utilizes low-energy X-ray detection for concentration coupled with sedimentation cell scanning to speed the analysis. It gives very satisfactory results for powders with atomic numbers greater than about 12. The concentration of particles remaining in suspension is detected as a function of time and depth within the sedimentation cell. By effectively solving Stokes' law throughout the course of an analysis, a plot of equivalent spherical diameter versus mass percent less than each diameter is developed.

A particle-size analysis with this equipment requires a dilute suspension, usually but not necessarily aqueous, of about 1.0% solids by volume. Particle- and liquid-density and liquid-viscosity data are required. Depending somewhat on density and viscosity values, particle diameters between about 150 and 0.1 μm can be measured.

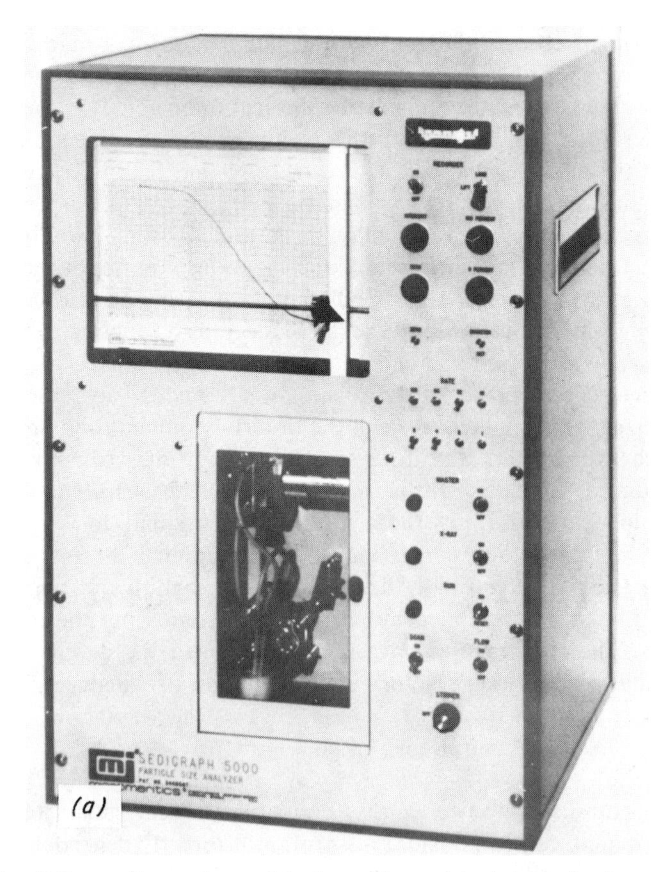

Figure 6.1. X-Ray sedimentatin particle-size analyzer: (*a*) photograph of equipment, (*b*) schematic diagram.

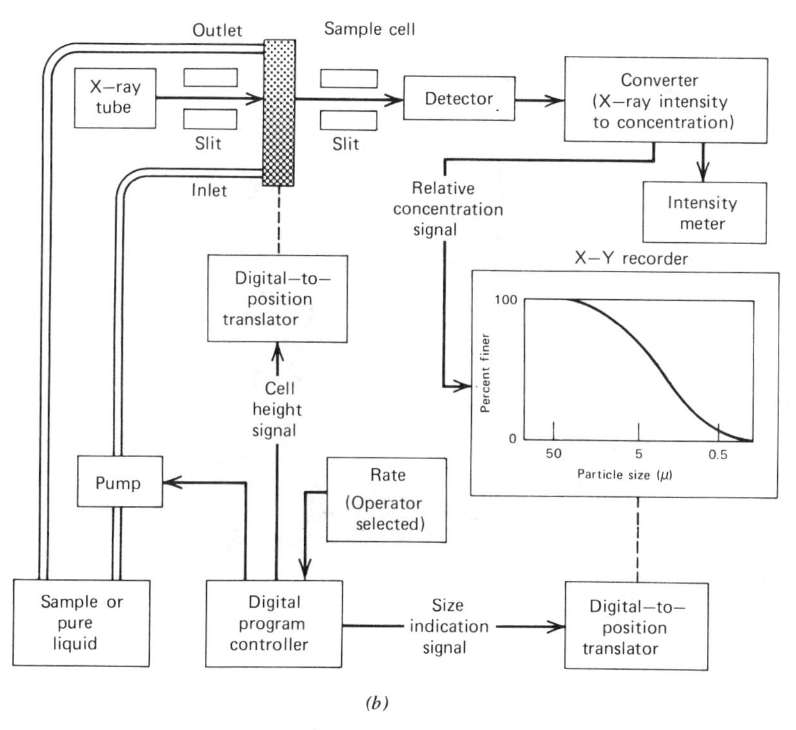

(b)

Figure 6.1 (Continued)

Electrical Resistance

The number and size of particles suspended in a weak electrolyte may be determined when the particles are made to flow through a small aperture with immersed electrodes on either side.[20-24] The electrodes carry a current that also passes through the aperture. The most familiar instrument of this type is the Coulter Counter, shown in Figure 6.2, which was originally developed for blood-cell counting but was long ago adapted for industrial use. As a particle passes through the aperture, the accompanying resistance change results in a voltage pulse of short duration proportional in magnitude to particle volume. These pulses are amplified and discriminated to give results in terms of particle number versus particle volume.

Particle volume is essentially linearly related to resistance change irrespective of particle shape, provided the particle diameter is roughly 20% of the aperture diameter. If the particles are larger, deviations from a generally spherical shape introduce an error of a few percent.[23] Particles very much smaller than the orifice are counted with decreasing sensitivity.

Figure 6.2. Coulter Counter (Coulter Electronics, Inc., Hialeah, Fla.).

Particle resistivity is seldom a problem, because, whether conducting or not, particles normally behave in an electrolyte as if they had infinite resistance.

The shape of the electronic pulse occasioned by particle passage has been the subject of much investigation. It is apparently round-topped because of particle approach and emergence influences, and not rectangular as it would be if ideality could be achieved.[22] This impedes the attainment of full amplifier response to the presence of a particle before it leaves the aperture, resulting in a slight undersizing of the coarser particles. Coincidental passage of more than one particle produces an erroneous indication of both size and number. This effect is substantially eliminated by diluting the measured suspension or correcting for coincidence by a relatively simple procedure.[25] Careful filtration and dust protection for the electrolyte are essential to a low background count. With an aqueous saline solution, a commonly employed electrolyte, it is advisable to add about 0.1% formaldehyde to prevent microorganism growth if the electrolyte is expected to be stored for a period of time.

A Coulter Counter is simple to use. A pinch of powder is added to the electrolyte and is well dispersed[26]; the aperture most likely to avoid plugging—one about $2\frac{1}{2}$ times the largest particle is chosen, and appropriate control settings are entered. The analysis is essentially automatic thereafter. The results, as noted above, represent a number-based distribution of particle volumes recorded in terms of particle diameter. The diameter spread measured with one aperture should not exceed one order of magnitude, for example, from 100 to 10 μm. By using several apertures, particle diameters from several hundred to about 0.5 μm may be encompassed.

Permeametry

Among simplified techniques for sizing particles, one of the most convenient is permeametry.[27] This technique is sometimes portrayed as indicating mean size and sometimes specific surface area. As ordinarily practiced, it probably measures neither. However, it does afford a relatively rapid and inexpensive means for establishing relative fineness and, as such, is useful in some routine control applications. One commercial permeameter requires the formation of a small compressed bed of dry powder through which dry air is forced at a fixed rate. An average particle size is then deduced from the pressure loss experienced by the air and the degree of particle packing. A still simpler version is known as the Blaine Fineness Tester[28,29]; a glassblower can fabricate one of these devices in a very short while.

SURFACE AREA

The definition of surface area for a truly plane surface is clear. Mica, for example, can be obtained molecularly flat by careful cleaving.[30] Most materials, unlike mica, do not have the centers of all atoms in the surface layer lying in one plane. Rather they have irregularities ranging from atomic scale to gross defects, including crystal imperfections, cracks, crevices, and pores. Their actual surface area can be many times the apparent geometric area.

Simple geometric considerations show that very small particles present a proportionally large fraction of the total surface area of a powder mass. The small particles are often overlooked in a size analysis but are included in a total surface area evaluation. Also, in many instances where powder-size distribution is essentially identical, surface-area measurement reveals distinct differences, because one powder has impervious particles and the other has particles containing cracks, cervices, or pores.

BET Gas Adsorption

Standard means for surface-area evaluation utilizes low-temperature absorption, or uptake on the powder surface, of a gas.[31] Most commonly nitrogen is used at the temperature of liquid nitrogen, although a number of other gases and temperatures may be employed. Adsorption reveals itself as the removal of a portion of a gas when a solid is exposed to it. At relatively low pressures the gas forms an incomplete layer of molecules attached to the solid surface. As the pressure increases a layer several molecules deep is formed. The critical factor is determining the conditions under which a layer of adsorbed gas precisely one molecule thick is formed.

The volume of gas adsorbed per unit mass of solid depends on the gas pressure, the absolute temperature, and the nature of the gas and solid. When the gas is adsorbed below its critical temperature it is convenient to express the pressure dependence in terms of the so-called relative pressure p/p_0, where p_0 is the saturation vapor pressure of the adsorbing gas that is available in various handbooks. Adsorption isotherms are obtained by measuring at constant temperature the volume of gas adsorbed per unit sample weight versus relative pressure.

Tens of thousands of adsorption isotherms have been determined using a variety of gases and solids. A typical adsorption isotherm is shown in Figure 6.3; the asymptote point *B* is approximately where a single gas layer occurs. From the volume of gas required to attain this condition as read from the ordinate, the number of gas molecules per unit weight of solid can be calculated. Then the specific surface area of the solid can be computed by including the area occupied by each gas molecule, which is believed to be 16.2 $Å^2$ for nitrogen. Areas occupied by a number of other adsorbed gas molecules have also been determined.[32]

Since locating an asymptotic point such as *B* is subject to error, much theoretical and experimental effort has been expended over the past years in devising more satisfactory mathematical treatments for adsorption data. These are not discussed here, but they have resulted in relatively simple expressions from which the single-layer condition, hence sample surface area, can be calculated directly.

Apparatus for making gas adsorption determinations of powder surface area range from the homemade glass systems of earlier days to quite sophisticated devices (e.g., the computer-controlled system shown in Figure 6.4). The preferable technique involves measuring equilibrium gas pressures as outlined above, but by a gravimetric technique in which the weight of the adsorbed gas is determined and modified gas chromatographic systems have also been employed. The gravimetric procedure has certain advantages, but the very small weight changes involved make the procedure quite tedious and lengthy. Gas chromatographic systems must employ a carrier gas along with the gas to be adsorbed which leads to proportioning and other problems.

Gas-adsorption measurement of surface area is not above questioning. Particle surfaces are not energetically uniform. Along grain boundaries and

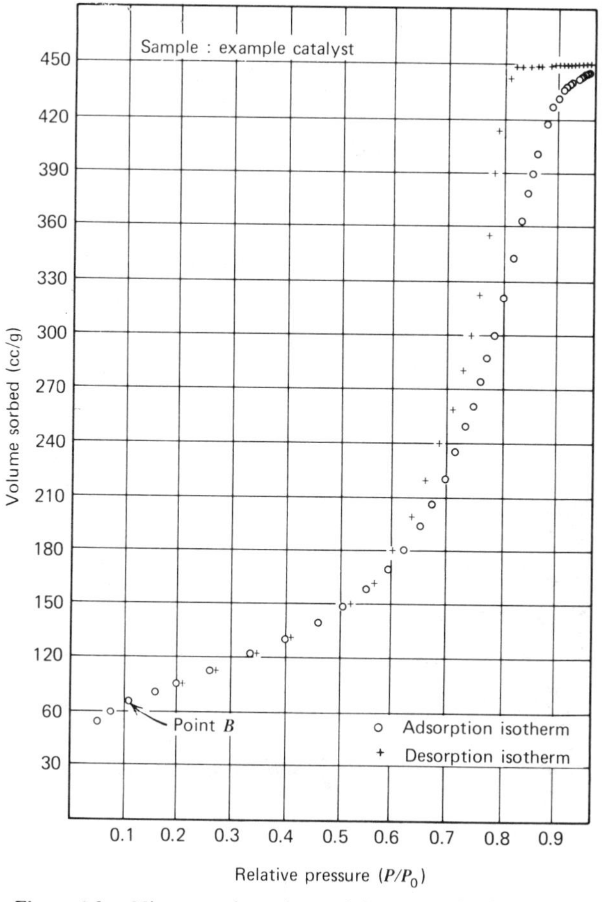

Figure 6.3. Nitrogen adsorption and desorption isotherms.

at the bottom of cracks there appear to exist enhanced adsorption sites, while less than normal energy sites are represented by protrusions. At these first regions it is entirely possibly for second, and even higher, layer adsorption to begin before single-layer coverage is completed elsewhere, and it is equally possible that some of the protrusions remain bare until most of the other surface is covered. This condition, of course, contributes to error, but one effect tends to counter the other. Overall, it appears that abnormal adsorption is small in the context of total surface area. Surface-area values have been reproduced within a few percent by different investigators separated widely in both space and time. BET results have been confirmed many times by a number of other tests utilizing a variety of procedures.[3,31] Low-temperature gas adsorption thus must be considered the standard to

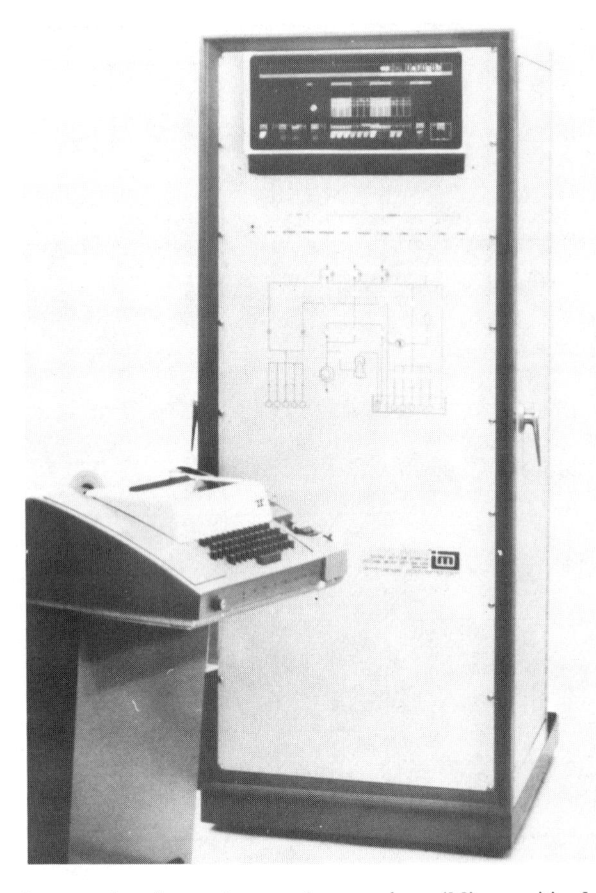

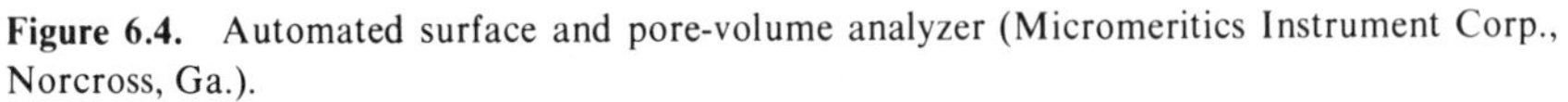

Figure 6.4. Automated surface and pore-volume analyzer (Micromeritics Instrument Corp., Norcross, Ga.).

which all other surface measurements are compared and, indeed, must be considered one of the few standards available for particle characterization.

Liquid-Phase Adsorption

The adsorption (or sorption as the phenomenon is often termed when involving liquids) of dyes, iodine, and a few other dissolved substances onto an immersed powder is often indicated as being a simple surface-measuring means. This is usually far from true. First, there is competition between the dissolved molecules and the dissolving liquid molecules for the available powder surface. Second, the area occupied by molecules in a liquid on a surface varies widely with concentration. There are also other factors to be considered, such as molecule orientation when attached to a surface. However, if there are overriding reasons that make liquid-phase sorption attractive, if relative results are sufficient, as they might be in a process control application, and if results are initially correlated with low-temperature gas adsorption, then the suggested system would be nitrophenol dissolved in water or one of several other solvents. The nitrophenol molecule attaches to either polar or nonpolar solids; its concentration in solution is readily detected colorimetrically; and it has been shown to give fairly reliable results with several solids.[34,35] Its constant-temperature sorption with increasing concentration often yields an isotherm somewhat resembling that in Figure 6.3 with one-layer coverage given by the point of inflection (point *B*). Single-molecule coverage, however, depends strongly on the nature of the solid, with values having been found to range from 15 to 50 Å^2, hence the necessity for correlation with gas adsorption. Results depend greatly on the care and uniformity with which the tests are conducted.

Low-Angle, X-Ray Scattering

X-Rays are partially deflected when they encounter an electron-density change such as they do upon passage into and out of powder particles. The extent of this density change can be related theoretically[36] to particle surface area. By measuring the intensity of the undeflected portion of an X-ray beam and energy scattered out to approximately 3°, it is possible to calculate a surface-area value.[37,38] The calculations are somewhat tedious, however, and demand a detailed knowledge of the material not readily available. Another approach to determine directly the unknown surface area is to relate the deviated beam intensity for a material of unknown surface area to that for another sample of the same material having had its surface area established by low-temperature gas adsorption.[39] This technique has the potential for yielding surface areas in a minute or two, including sample preparation, and could become very useful in process control.[40]

SHAPE

Particle shape is recognized as influencing powder flowability, bulk density, angle of repose, and packing characteristics, and may be as important as particle size in many cases. Shape is rarely measured, not because of a shortage of so-called shape-factor definitions, but because no one has yet found a completely satisfactory means for measuring shape, especially for particles with reentrant contours.

There are two-dimensional shape parameters based on microscopic determinations of the maximum particle diameter (i.e., length) and two maximum perpendicular radii, one one either side of the maximum diameter.[41] "Elongation" and "flatness" ratios are defined in terms of the length, breadth, and thickness of a particle.[42] Other considerations of shape involve particle perimeter and surface area. "Roundness" is the ratio of the perimeter of a circle of the same area as the particle to the actual perimeter of the particle, while "sphericity" is the ratio of the surface area of a sphere of the same volume as the particle to the actual surface area of the particle.[43] Such definitions are deficient for many purposes in that they do not adequately characterize the gross shape of a particle.

Other classifications categorize coarse grains as either generally spherical, intermediate in shape, or tabular, depending on the time required to slide, roll, or bounce down an inclined plane.[44] The suggestion has been offered[45] that measurement of the rate of sieve passage might be developed into a shape description, it having been found that elongated cylindrical particles pass a sieve at a slower rate than do short cylindrical ones. Finally, producing sieves with rectangular, triangular, and round instead of square openings has been suggested.

Perhaps the most fruitful approach to shape characterization is a dynamic method based on the representation of an irregular particle as an ellipisoid having equivalent radii of gyration about its principal axes.[46] Both two- and three-dimensional representations might be employed. Obtaining the two-dimensional data is straightforward; how to arrive at dimensions in three planes obviously presents problems. Treating the data, however generated, would be quite tedious were it not now possible to utilize electronic digitizing equipment and computers. The summary of the method given here considers only manual analysis in two dimensions to present the rudiments of the technique. The starting point is a photograph of the particle.

The outline of the particle is traced on graph paper (Figure 6.5). Then X and Y coordinates of small areas within the particle are tabulated. From these measures are calculated radii of gyration of the plan figure about its principal axis and the axis at right angles to the principal axis, assuming the figure actually to be a thin sheet of material of uniform thickness and density. Next is calculated an equivalent ellipse having these same radii. The ratio of the longer axis of the ellipse to the shorter is taken as the "anisometry," or unequal nature, of the particle. The area of the ellipse will be greater than that of the figure representing the particle; the ratio of the ellipse area to the figure area is taken as a measure of the "bulkiness" of the particle.

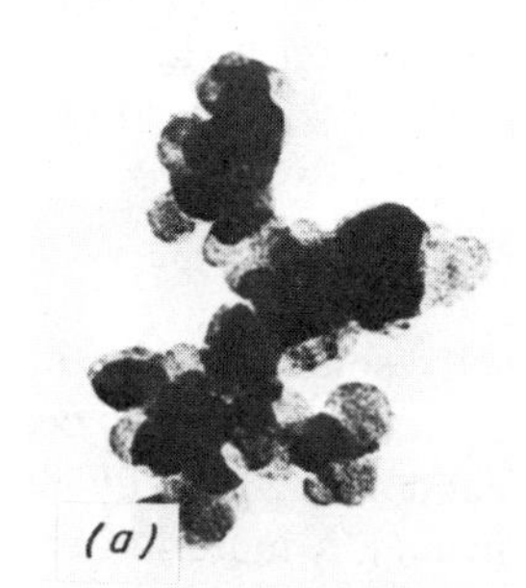

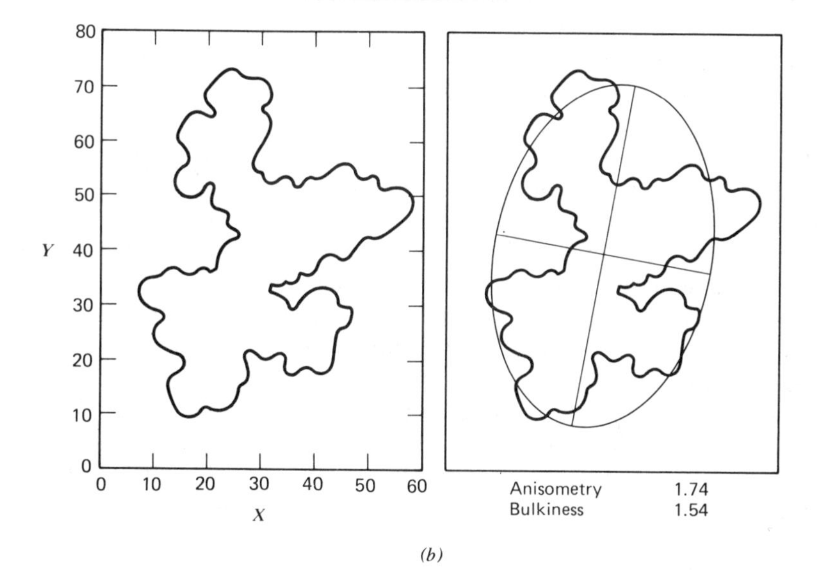

(b)

Figure 6.5. Shape factors for irregular particle.[46]

Extension of the evaluation to a figure in three dimensions requires computation of radii of gyration about three principal axes coinciding with geometric axes. The "bulkiness" of the particle is then defined as the ratio of the volume of the equivalent ellipsoid to that of the particle, and "anisometry" is defined in terms of ratios of the lengths of the three axes by reducing these to two dimensionless factors: an "elongation ratio" and a "flatness ratio." An equivalent shape other than an ellipsoid, perhaps a prism, could be used as the basic geometrical form.

While the delineation of dynamic shape factors as given above represents a considerable advance, techniques and instrumentation suitable for routine use are not currently available. Much work remains to be done to reduce particle shape to the useful parameter it should be.

PORE STRUCTURE

Quite a number of apparently solid particles have cracks, crevices, holes, and fissures within their structure, and, of course, objects formed from powders by consolidation and firing usually have internal spaces. While pores might run straight through a material, they are more likely to twist, turn, branch, and interconnect. They may start on one side and emerge on the same side, or they may never emerge. They may decrease in dimensions with depth or they may enlarge, giving so-called "ink-well" pores.

Pores are openings in rigid objects, while the term "voids" refers to the spaces abounding among the grains of an unconsolidated powder, for example. Void volume is thus easily altered; changing pore volume requires at least partial destruction of the porous object. A consolidated mass produced by compressing and sintering a powder is considered to contain only pores, even though some of the openings may be among individual grains and some in fissures within the grains themselves. Pores constitute the primary concern here. Unfortunately, measurements with loose powders may detect both pore and void volumes, so judicious examination of measurement data is necessary to distinguish the two.

Pore systems have to be described in terms of a geometric model, the right-cylinder model being most utilized. This convention is followed here. Thus pores hereafter are considered as though all were straight cylindrical openings having one of several diameters and various total lengths. There are basically two practical ways of measuring pore structure: (1) by forcing a nonwetting liquid into the pores under pressure and then analyzing the pressure–volume penetration curve and (2) by condensing a liquid within the pores, subsequently analyzing these data. Of course, particle and particle-formed structures can be evaluated by sectioning and microscopic examination. Although obvious, this is such a tedious procedure it is rarely employed.

Mercury is a nonwetting liquid for most solids, and it is used exclusively in the penetration technique. Mercury under pressure but outside a cylindrical pore experiences a force tending to push it into the cylinder expressed by the product of pressure and cylinder cross section. The opposing force tending to prevent the mercury from entering is the product of mercury surface tension, the perimeter of contact, and a term known as the contact angle. Equating these forces results in a simple expression between applied pressure and pore diameter involving as other terms the mercury surface tension and its contact angle with the solid. At normal temperature the surface tension of mercury is about 474 dynes/cm, and mercury usually exhibits a contact angle of about 130°. The pressure–pore diameter relationship becomes then a simple proportionality, some values for which are given in Figure 6.6.

Appropriate apparatus for conducting pore evaluations by mercury penetration are simple in concept but complex in practice because they are required to operate at both low and high pressures. Every sample has to be evacuated initially to free it of atmospheric gases and vapors. Sensitive volume-measuring equipment is required to indicate accurately from very small to rather sizable volume changes of mercury. Present commercial equipment requires operator attendance and involves some sample manipulation; automated instrumentation is under development, however.

Typical penetration results are shown by Figure 6.7. The data of curve *1*, for example, apply to an unconsolidated powder having a mean particle diameter of about 50 μm, and the first step in the curve is seen to correspond approximately to this diameter. This portion of the curve must then apply to void, or interparticle, spaces. The second step in the curve, applying to considerably smaller pores, must represent pores within the particles.

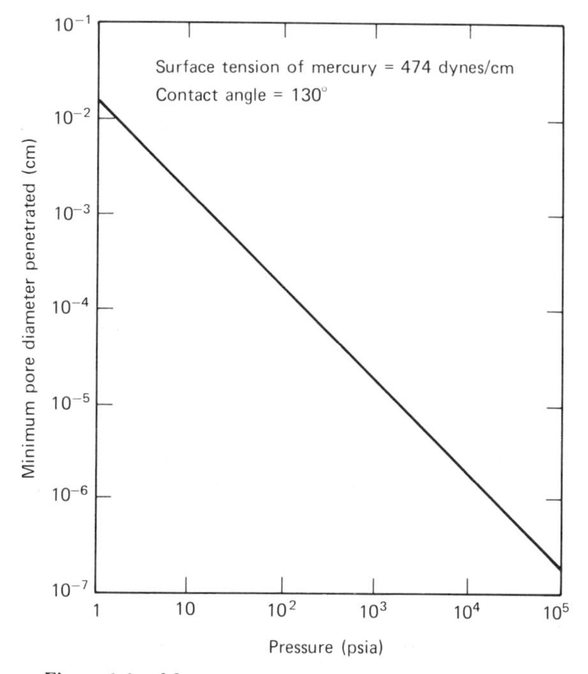

Figure 6.6. Mercury pressure–pore diameter relationship.

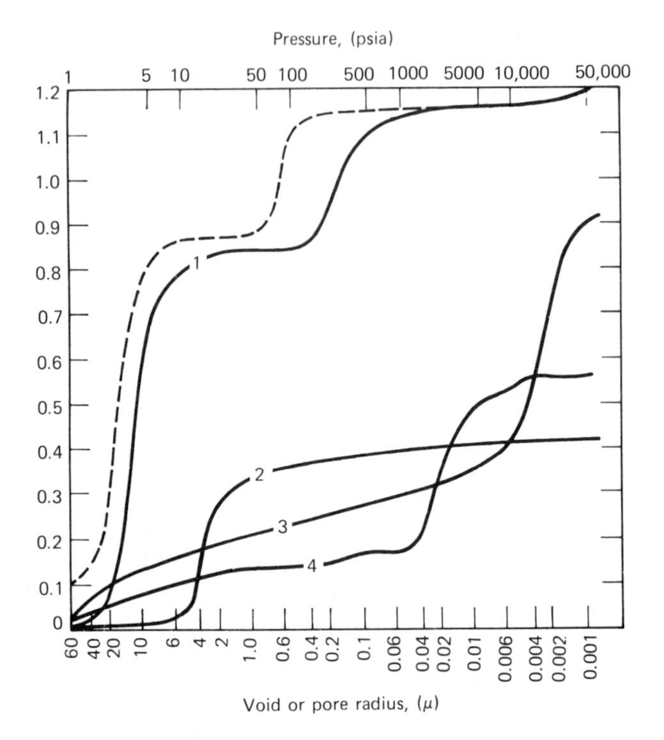

Figure 6.7. Mercury penetration results.

Many users go no further than this with a pore analysis, being content with information easily read from such a graph. In this example pore diameters are primarily between 1.2 and 0.2 μm and they represent a volume of approximately 0.30 cc/g of sample. Various ways have been suggested to fit penetration data to pore distribution functions from which the surface area associated with various pore sizes, for example, can be calculated. It is also possible to follow the expulsion of mercury from pores as the pressure is reduced to gain insight into pore shape.[47] Space precludes following these developments here.

Mercury penetration is fairly rapid, requires little sample preparation, covers a wide range of pore sizes, and gives results of considerable reliability. There are difficulties and uncertainities to be sure. With nonrigid materials, distortions may arise because of sample compression under high pressure; the mercury contact angle is difficult to evaluate accurately and may actually change with pressure; mercury surface tension may vary with radius of curvature in an unknown manner; and mercury amalgamates with some metals. This last item is often not an insurmountable problem; a thin layer of fatty acid deposited on the metal surface will usually protect it for the duration of a test.

The other method of pore evaluation, an extension of the low-temperature adsorption technique for surface-area analysis, is applicable only to pores between about 20 and 600 Å in diameter. This makes it particularly attractive for catalysts and catalyst-support analysis but eliminates it where large pores are encountered.

If the analysis that led to the results described in conjunction with Figure 6.3 is continued to high relative pressures and is then reversed, leading again toward lower pressures, a complete curve of data such as that given in Figure 6.3 will most likely be obtained. Preceding material described the first break in such a curve as signifying approximately the completion of a single layer of adsorbed gas molecules and subsequent portions to correspond to the formation of still thicker layers. Upon continuing to add gas almost to saturation, the layers within pores are thought to merge and eventually to fill the pores with condensed liquid. The curve for the reverse process, that is, the desorption curve, must then represent the evaporation of condensed liquid and the emptying of the pores. Both the filling and emptying curves contain data pertinent to the size and volume of the pores. Because the analysis of the desorption curve is perhaps easier to explain, it will be utilized here. The model retains cylindrical pores of various diameter and lengths as in the mercury penetration analysis.

It is well known that the tendency for any liquid, including liquid nitrogen, to evaporate is greater when the exposed liquid surface is flat (as it would be in a large open container) than when the liquid is confined in a small capillary tube. This means that nitrogen condensed in large pores evaporates more readily than that in small pores. The desorption curve then represents a progressive depletion of condensed liquid beginning with large pores and proceeding to smaller ones. When any pore empties of its condensed liquid, it still has on its surface an adsorbed layer of gas that may be several molecules thick. Accounting for these remnants of remaining gas complicates the calculations so that giving details will not be attempted here even though the procedure is quite straightforward. Basically it means that the calculations have to be performed in a stepwise manner.[48] Results from gas-desorption tests can be cast as a plot of volume represented by pores of decreasing diameter like the mercury penetration data or in several other ways. The surface area represented only by pore walls can also be computed.

The experimental procedure is straightforward but involved, and attempting to maintain stable conditions and carry out the several manipulations by hand inevitably leads to errors. For this reason an automated instrument under computer control, as mentioned earlier, is almost essential. One or two attempts to carry out an analysis manually usually is convincing of the necessity for instrumental control.

A simple test based on powder-flow properties[49] indicates total pore volume with acceptable accuracy for some routine control purposes, particularly with highly porous powders. The applicability of the test under use conditions would, of course, need to be established by comparison with, say, mercury intrusion before adoption. The test is based on the fact that free-flowing powders appear dry and remain free-flowing with the addition of water up to a point, after which further additions bring about a marked change. Such behavior is attributed to the first additions of water filling pore spaces within the particles and later additions filling the void spaces among the particles. To carry out a test, about 25 g of dried powder is weighed and transferred to a screw-top bottle. A small quantity of water, or perhaps another liquid, is then added from a burette and the bottle is capped and shaken vigorously. The bottle is uncapped and lumps, if any, are broken with a spatula or by more vigorous shaking. The process is repeated with water being added in small increments until the powder does not flow freely when the bottle is upended, this condition indicating the filling of the pores and the beginning of spillover into the voids. The specific pore volume is calculated by dividing the volume of water to the end point by the weight of the sample.

Sorption from solution, for reasons given previously, must be approached with caution in attempts to apply it to pore evaluaton. Nevertheless, there are apparently a few instances of reliable application.[50] Here the technique is based on the premise that the size of the dissolved molecule, usually a dye, determines the dimensions of the pore it might penetrate. There is considerable uncertainity in the size of the various dye molecules, so careful standardization against other pore measurements is essential in any attempted application of this technique.

REFERENCES

1. J. A. Davidson and H. S. Haller, "Latex Particle Size Analysis. V. Analysis of Errors in Electron Microscopy," *J. Colloid Interface Sci.*, **47**, 459–472 (1974).
2. J. M. DallaValle, *Micromeritics: The Technology of Fine Particles*, 2nd Ed., Pitman, New York, 1948.
3. C. Orr, Jr. and J. M. DallaValle, *Fine Particle Measurement*, Macmillan, New York, 1959.
4. G. Herdan, *Small Particle Statistics*, 2nd Ed., Academic, New York, 1960.
5. R. R. Irani and C. F. Callis, *Particle Size: Measurement, Interpretation, and Application*, Wiley, New York, 1963.

6. R. D. Cadle, *Particle Size: Theory and Industrial Applications*, Reinhold, New York, 1965.
7. T. T. Mercer, *Aerosol Technology in Hazard Evaluation*, Academic, New York, 1973.
8. J. D. Zwicker, "Sieve Analysis to Below Two Microns Using Micromesh Sieves," *Ceram. Bull.*, **45,** 716–719 (1966).
9. H. B. Carroll and I. B. Aksi, "Sieving and Particle Size Distribution in 10 through 1 Region," *Rev. Sci. Instrum.*, **37,** 620–623 (1966).
10. K. Schönert, W. Schwenk, and K. Steier, "A Fully Automatic Device for Screen Analyses," *Anfbereitungs—Technik*, **7,** 368–372 (1974).
11. C. Orr, Jr., Unreported development pending patent protection.
12. E. E. Bauer, "Hydrometer Computations in Soil Studies Simplified," *Eng. New-Rec.*, **118,** 662–664 (1937).
13. S. Berg, "Determination of Particle Size Distribution by Examining Gravitational and Centrifugal Sedimentation according to Pipet Method and With Divers," Symposium on Particle Size Measurement, ASTM Spec. Tech. Publ. No. 234, 1959, pp. 143–171.
14. C. C. McMahon, "Particle Size Analysis By the Method of Musgrave and Harner," *Ceram. Bull.*, **49,** 794–796 (1970).
15. J. Kalshoven, "Fast and Automatic Sedimentation Analysis," *Proc. Conf. Particle Size Anal., London*, **1967,** The Society for Analytical Chemistry, pp. 197–204.
16. J. P. Olivier, G. K. Hickin, and C. Orr, Jr., "Rapid Automatic Particle Size Analysis in the Subsieve Range," *Powder Tech.*, **4,** 257–263 (1971).
17. W. P. Hendrix and C. Orr, Jr., "Automatic Sedimentation Size Analysis Instrument," *Proc. Conf. Particle Size Analysis 1970*, The Society for Analytical Chemistry, London, 1972, pp. 133–144.
18. P. Sennett, J. P. Olivier, and G. K. Hickin, "Application of Rapid Automatic Particle Size Analysis to the Paper Industry," *Tappi*, **57,** 92–95 (1974).
19. J. R. Hodkinson, "Dust Measurement by Light Scattering and Adsorption," Ph.D. Thesis, University of London, 1962.
20. W. H. Coulter, "High Speed Automatic Blood Cell Counter and Cell Size Analyzer," *Proc. Natl. Electron. Conf.*, **12,** 1034–1042 (1956).
21. R. H. Berg, "Electronic Size Analysis of Subsieve Particles by Flowing Through a Small Liquid Resistor," ASTM Symposium on Particle Size Measurement, Spec. Tech. Publ. No. 234, 1959, pp. 245–255.
22. H. E. Kubitschek, "Electronic Measurement of Particle Size," *J. Res. Natl. Bur. Stand.*, **A13,** 128–135 (1960).
23. T. Allen, "A Critical Evaluation of the Coulter Counter," *Proc. Conf. Particle Size Anal., London*, **1967,** The Society for Analytical Chemistry, pp. 110–127.
24. S. Kinsman, "Instrumentation for Filtration Tests," *Chem. Eng. Prog.*, **70,** 48–51 (1974).
25. H. Bader, H. R. Gordon, and O. B. Brown, "Theory of Coincidence Counts and Simple Practical Methods of Coincidence Count Correction for Optical and Resistive Pulse Particle Counters," *Rev. Sci. Instrum.*, **43,** 1407–1412 (1972).
26. R. W. Lines, "Some Observations on Sampling for Particle Size Analysis with the Coulter Counter," *Powder Tech.*, **7,** 129–136 (1973).
27. F. M. Lea and R. W. Nurse, "The Specific Surface of Fine Powders," *J. Soc. Chem. Ind.* (*London*), **58,** 277–283 (1939).

28. R. L. Blaine, "A Simplified Air Permeability Fineness Apparatus," *ASTM Bull.*, **123,** 51–55 (1943).
29. N. F. Schulz, "Measurement of Surface Areas by Permeametry," *Int. J. Miner. Proc.*, **1,** 65–79 (1974).
30. F. P. Bowden, "The Nature and Topography of Solid Surfaces and the Study of Van der Waal's Forces in their Immediate Vicinity. The Surface Decomposition of Solids," *Fundamentals of Gas–Surface Interactions,* H. Saltsbur, J. N. Smith, Jr. and M. Rogers, eds., Academic, New York, 1967.
31. S. Brunauer, P. H. Emmett, and E. Teller, "The Adsorption of Gases in Multimolecular Layers," *J. Amer. Chem. Soc.*, **60,** 309–319 (1939).
32. A. L. McCellan and H. F. Harnsberger, "Cross-Sectional Areas of Molecules Adsorbed on Solid Surfaces," *J. Colloid Interface Sci.*, **23,** 577–599 (1967).
33. S. J. Gregg and K. S. W. Sing, *Adsorption, Surface Area, and Porosity,* Academic, New York, 1967.
34. C. H. Giles, T. H. MacEvan, S. N. Nakhwa, and D. Smith, "Studies in Adsorption, Part XI. A System of Classification of Solution Adsorption Isotherms and its Use in Diagnosis of Adsorption Mechanisms and in Measurements of Specific Surface Areas of Solids," *J. Chem. Soc.*, Part III, **1960,** 3973–3993.
35. C. H. Giles and S. N. Naklwa, "Studies in Adsorption XVI. The Measurement of Specific Surface Areas of Finely Divided Solids by Solution Adsorption," *J. Appl. Chem.*, **12,** 266–273 (1962).
36. E. D. Eanes and A. S. Posner, "Small-Angle X-Ray Scattering Measurements of Surface Areas," *The Solid–Gas Interface,* Vol. 2, E. A. Flood, ed., Dekker, New York, 1967, Chap. 33.
37. D. N. Winslow and S. Diamond, "The Specific Surface of Hydrated Portland Cement Paste as Measured by Low-Angle X-Ray Scattering," *J. Colloid Interface Sci.*, **45,** 425–426 (1973).
38. D. N. Winslow and S. Diamond, "Specific Surface of Hardened Portland Cement Paste as Determined by Small-Angle X-Ray Scattering," *J. Amer. Chem. Soc.*, **57,** 193–197 (1974).
39. R. A. Van Nordstrand and K. M. Hack, "Small Angle X-Ray Scattering of Silica and Alumina Gels," presented before the Division of Petroleum Chemistry, American Chemical Society, Chicago, Sept. 9–11, 1953.
40. R. W. Camp, private communication.
41. C. F. Royse, Jr., *Semiment Analysis,* Arizona State University, Tempe, 1970.
42. H. Heywood, "The Scope of Particle Size Analysis and Standardization," Symposium on Particle Size Analysis, *Trans. Inst. Chem. Eng.*, **255,** 14–24 (1947).
43. H. Wadell, "The Coefficient of Resistance as a Function of Reynolds Number for Solids of Various Shapes," *J. Franklin Inst.*, **217,** 459–490 (1934).
44. W. H. Glezen and J. C. Ludwick, "An Automated Grain-Shape Classifier," *J. Sediment. Petrol.*, **33,** 23–40 (1963).
45. N. S. Land, "A Feasibility Study of a Technique for Sorting Particles by Shape," *Mater. Res. Stand.*, **9,** 26–29, 77 (1969).
46. A. I. Medalia, "Dynamic Shape Factors of Particles," *Powder Tech.*, **4,** 117–118 (1970/1971).

47. M. Svata, "Determination of Pore Size and Shape Distribution from Porosimetric Hysteresis Curves," *Powder Tech.*, **5**, 345–349 (1971/1972).

48. E. P. Barrett, L. G. Joynder, and P. P. Halenda, "The Determination of Pore Volume and Area Distributions in Porous Substances. I. Computation from Nitrogen Isotherms," *J. Amer. Chem. Soc.*, **73**, 373–380 (1951).

49. W. B. Innes, "Total Porosity and Particle Density of Fluid Catalysts by Liquid Titration," *Anal. Chem.*, **28**, 332–334 (1956).

50. C. H. Giles, A. P. D'Silva, and A. Cameron, "Determination of Pore-Size Distributions of Powders from Solute Adsorption," *Chem. Ind.* (*London*), **1969**, 239–240.

7

Structure and Properties of Agglomerates

D. E. Niesz

R. B. Bennett

The structure and properties of solid agglomerates in a powder often dominate powder-processing parameters, microstructure, and bulk properties. The origin, characterization, elimination, and effects of various types of solid agglomerates are discussed in this chapter. All agglomerates referred to in this chapter are solid agglomerates, unless otherwise stated.

ORIGIN OF SOLID AGGLOMERATES

The most common type of agglomerate in a ceramic powder is one bonded by a diffusion bond formed during calcination. Such agglomerates are strong enough to retain their identity during green forming. However, many reactive powders contain more than one type of agglomerate; therefore, reporting properties that reflect only the average properties of the various agglomerates types is inadequate. Figure 7.1 shows two types of agglomerates that make up one alumina powder. This situation often occurs in unmilled reactive alumina because of the numerous transition phases that

Figure 7.1. Electron micrograph of alumina I agglomerates.

can be present, as discussed in Chapter 9. Here the agglomerate containing the 100 to 200 Å particles is gamma alumina and represents 10 to 20 wt % of the powder. Because of its low bulk density, it leads to a very nonuniform green density even though it is weak enough to be broken down by a kitchen blender. Quantitative determination of the relative amounts of each type of agglomerate in this powder is essential for evaluating lot-to-lot variability.

PHYSICAL CHARACTERIZATION OF POWDERS AND AGGLOMERATES

The following physical characteristics of agglomerates are important:

1. Microstructure of agglomerates.
2. Percentage of each type of agglomerate in a powder.

3. Size distribution of the ultimate particles.
4. Size distribution of the agglomerates.
5. Bulk density of the agglomerates.
6. Strength of the agglomerates.
7. Character of the agglomerate bond.

In evaluating the features listed above, care must be taken not to alter the powder character by the characterization technique. The first characterization step is microscopic evaluation of the agglomerate. The depth of field of a scanning electron microscope (SEM) is valuable, but a transmission electron microscope is required to resolve the structure of very fine powders (Chapter 8). Some effort has been made toward automating the scanning electron microscopy technique,[1] and continued effort in this direction is needed if analysis of fine powders is to become more routine. The visual observation is invaluable in determining appropriate characterization techniques and in interpreting the results of various characterizations.

The agglomerates in calcined powders normally are broken down to their ultimate particles by ball milling in a water medium followed by granulation with a spray drier. Control of the characteristics of the spray-dried powder is a key element in many ceramic powder processes. The granules in a spray-dried powder are held together by organic binders, plastic materials such as clay or talc, and by decomposed gels or hydrates precipitated from solution during drying.

The bulk density of agglomerates has a marked effect on the bulk density of the powder and the compaction ratio. Surface area and the bulk and true density of a powder are also important, but they are related more to the bulk powder than the agglomerates in the powder. If the bulk density of agglomerates in a powder vary, the agglomerates with a bulk density well below that of the major agglomerate lead to a low local density and thus to a nonuniform die fill and green compact. Techniques for measuring the bulk density of agglomerates are lacking, although mercury porosimetry[2] has been used for this purpose (Chapter 6). For an agglomerated powder, a bimodal pore-size distribution is obtained when pore diameter is plotted against pore volume. In the porosimetry technique, if the pores of the agglomerate are infiltrated by mercury without crushing the agglomerate, the bulk density of the agglomerate can be calculated.

Agglomerates in calcined powders can also be broken down by dry milling, as discussed in Chapter 10. The resulting powder is composed of weak agglomerates or granules bonded by the small amount of dry milling aid used to facilitate dry grinding.

A type of agglomerate that is not widely recognized is one bonded by decomposed gels or hydrates. These agglomerates form by precipitation of cations from solution when water slurries are dried. Some bonding may also occur between the hydrated surface layers of the powder particles. The strength of the resulting agglomerate depends on the percentage of bond phase. For high-surface-area powders, this bonding can produce strong agglomerates.

The lack of a good technique to characterize agglomerate bulk density is not a serious limitation on powder characterization, since strong, porous agglomerates cannot be tolerated in a powder to be used in powder processing. It often is enough to know that a powder does or does not contain strong, porous agglomerates. Agglomerates usually can be observed with a microscope and their strength can be measured. Powders must be pretreated by some technique, such as ball milling, to eliminate such agglomerates if an optimum powder process is desired.

The strength of agglomerates is also an important factor in determining a powder's compaction behavior and in determining the amount of milling required to eliminate strong agglomerates. It is important to characterize the agglomerate bond in powders to be used in powder processing. This is discussed in Chapter 27 as well.

AGGLOMERATE STRENGTH

The strength of agglomerates in a powder can be determined from powder-compaction data.[3] By this technique the relative strength of the agglomerates in a powder is related to a break in the compaction curve when the logarithm of pressure is plotted against percentage of theoretical density. This relation can be derived from the formula first suggested by Duckworth[4] in discussing Ryshkewitch's[5] paper on the strength of porous sintered alumina and zirconia:

$$\sigma = \sigma_0 e^{-bp} \tag{1}$$

where σ = strength of a porous body
σ_0 = strength of a nonporous body of the same material
b = an empirical constant
p = porosity expressed as a fraction

If the strength is replaced by the pressing pressure, then

$$P = P_0 e^{-bp} \tag{2}$$

Taking the logarithm of both sides gives

$$\ln P = \ln P_0 - bp \tag{3}$$

This equation is a straight line when the logarithm of pressure is plotted against the volume fraction of porosity. The logarithm of pressure is also linearly related to porosity expressed as a percentage and to density expressed as a fraction or as a percentage of theoretical density.

Figure 7.2 shows the compaction data for several materials plotted as the logarithm of pressure versus percentage of theoretical density. The breaks in the curves for aluminum and copper indicate the pressure at which plastic deformation of the powder particles at their contact points begins to control the compaction behavior.

Alumina I is an unmilled, calcined alumina containing two distinct types of agglomerates. An electron micrograph of this powder is shown in Figure 7.1. The break in the curve at 300 psi for this powder indicates the pressure at which crushing of the contact points of the porous agglomerates begins to control the compaction behavior of the powder.[3] Thus the strength of the agglomerates composed of 100 to 200 Å ultimate particles is indicated by this break point. The minimum amount of this agglomerate that can be detected in this powder is about 5 wt %.

There is a slight break in the curve for alumina II[3] at 12 psi. This powder is a dry-milled reactive alumina, and the break indicates the relative strength of the agglomerates bonded by the milling aid. The curve is straight between the 12 psi break and 100,000 psi. This type of curve indicates the absence of strong, porous agglomerates, such as those formed during calcination. During calcination the bond area at particle contact points is increased by diffusion processes. The sensitivity to change in agglomerate strength with calcination temperature was evaluated by calcining alumina II at progressively higher temperatures. The break points occurred at 60, 150, 875, and 2000 psi for calcination temperatures of 1800, 2000, 2200, and 2400°F, respectively.[3]

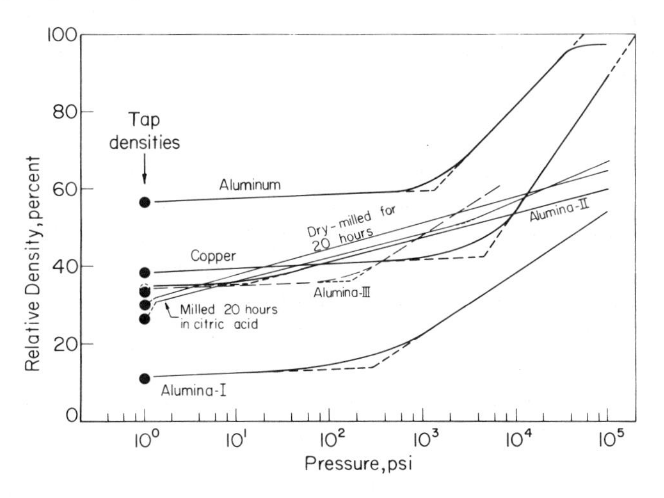

Figure 7.2. Compaction data for several powders.

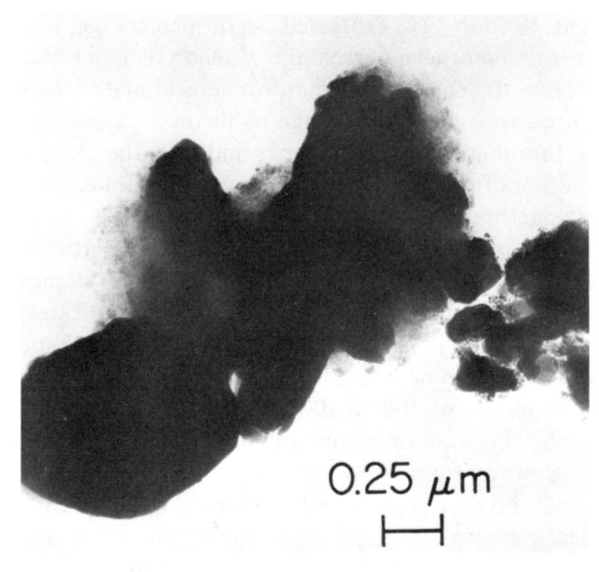

Figure 7.3. Electron micrograph of alumina I powder after water milling for 20 hours.

The curve for alumina I after milling in water for 20 hours shows a break point at approximately 3000 psi.[6] This indicates the presence of strong agglomerates. Figure 7.3 shows an electron micrograph of this powder. A translucent, weblike material that can be seen around the edges of the agglomerates has been identified as aluminum monohydrate by thermogravimetric analysis. This hydrate acts as a cement that bonds the agglomerate particles together. The mechanism by which this bond forms is discussed later.

Alumina III is a wet-milled and spray-dried 96% commercial alumina body. The break point at 200 psi represents the strength of the spray-dried granules. This technique is quite sensitive to variations in the strength of spray-dried granules, and considerable variation can be detected in nominally identical lots of powder.

The curve for alumina I after dry milling for 20 hours does not exhibit any breaks between 10 and 100,000 psi.[6] The only granules in this powder are those held together by the small amount of dry-milling aid. These are so weak that their presence is undetectable by this technique.

There is an upper limit of agglomerate strength that can be evaluated by this technique, and the presence of such agglomerates must be determined microscopically. The coarse-particle agglomerate in alumina I is an example of a agglomerate whose strength cannot be determined by this technique at pressures below 100,000 psi.

EFFECT OF AGGLOMERATES ON MICROSTRUCTURAL DEVELOPMENT

As an example of the significance of agglomerate structure and properties on microstructural development, the microstructural development during sintering is illustrated for three powders.[6] All three powders were derived from alumina I. Electron micrographs of the three powders are shown in Figures 7.4 through 7.6 The as-received powder contains two types of agglomerates, as discussed earlier. In the dry-milled powder, the agglomerates are essentially all broken down into their primary particles. The 100 to 200 Å particles are dispersed on the surfaces of the larger particles. The wet-milled powder is composed of agglomerates bonded by aluminum monohydrate, as discussed earlier.

The differences in the microstructures developed from these three powders after isostatic pressing and sintering can best be illustrated by examining the fracture surfaces of compacts with densities between 90 and 95% of the theoretical value. The compact prepared from the as-received powder is shown in Figure 7.7. The very nonuniform microstructure consists of dense, coarse-grained regions and porous, fine-grained regions. On a polished surface the porous regions are isolated. The porous areas are remnants of the porous agglomerates in the powder that had 100 to 200 Å

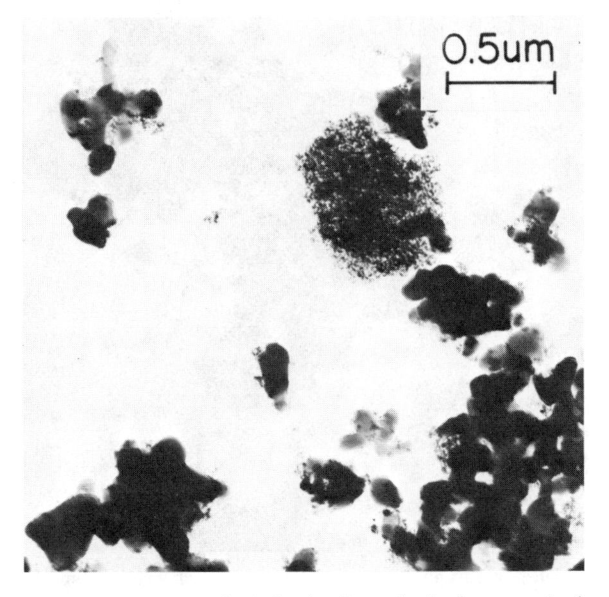

Figure 7.4. Electron micrograph of alumina I powder in the as-received condition.

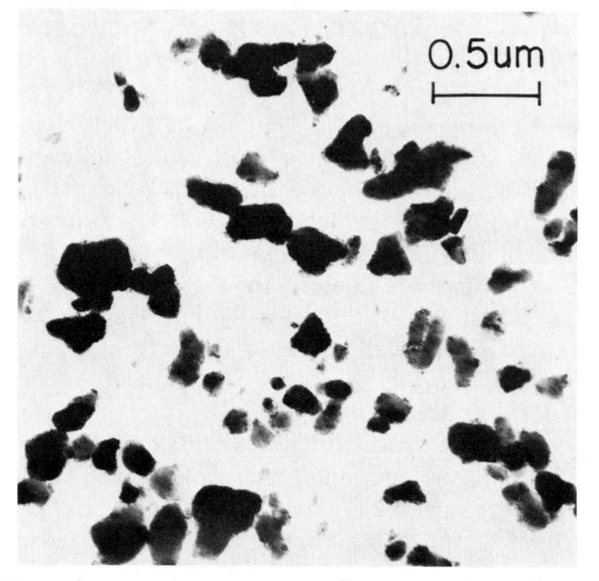

Figure 7.5. Electron micrograph of alumina I powder after dry-ball milling for 20 hours with napththenic acid as a dry-grinding aid.

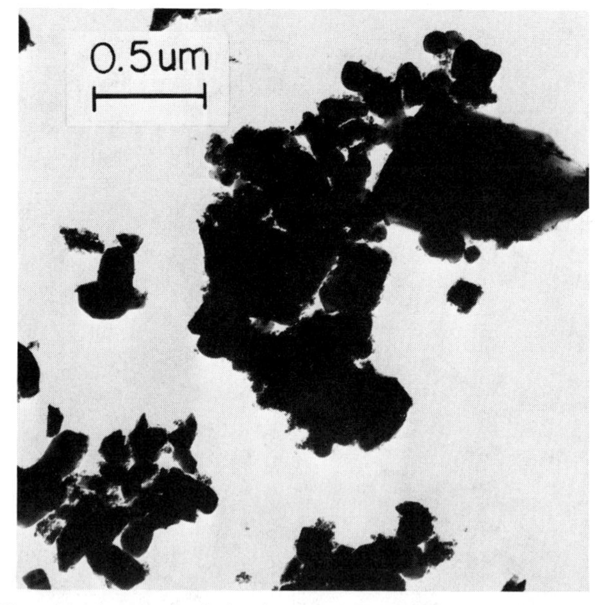

Figure 7.6. Electron micrograph of alumina I powder after water milling for 20 hours with citric acid.

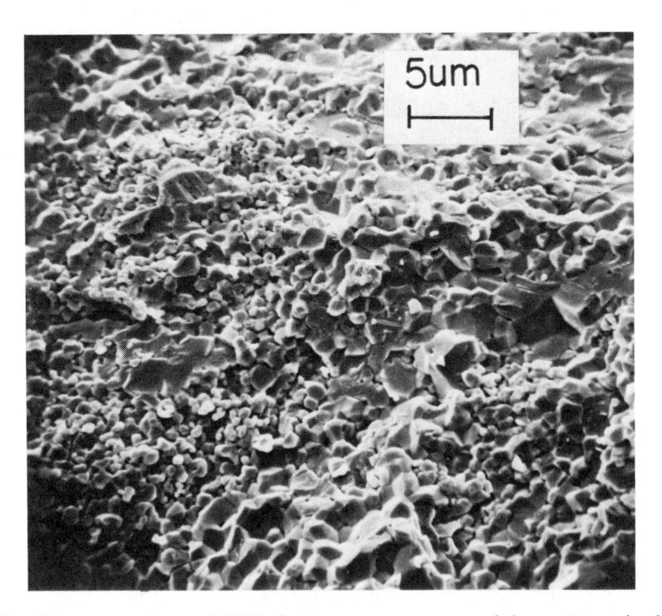

Figure 7.7. Fracture surface of 95% dense compact prepared from as-received alumina I powder.

primary particles. As sintering proceeds, the denser areas reach a limiting density and pockets of porosity remain. These pockets of porosity limit the attainable density, act as light-scattering centers, and act as critical flaws for fracture initiation.

By contrast, the fracture surface of the compact prepared from the dry-milled powder is uniform (Figure 7.8). This type of microstructure is required if a compact is to be sintered to theoretical density to achieve translucency and optimum mechanical strength. The microstructure of a theoretically dense compact prepared from this dry-milled powder is shown in Figure 7.9. Final microstructures of this quality cannot be obtained from powders that result in nonuniform microstructures, such as that shown in Figure 7.7.

The fracture surface of the compact prepared from the wet-milled powder appears to be quite uniform. However, closer examination reveals numerous separations in the microstructure (Figure 7.10). These separations appear to be the result of disruption of the particle contacts due to shrinkage and phase inversions in the hydrated phase that served as the bond phase in the powder agglomerates. Except for small samples, the disruption leads to bulk rupture of the compact during sintering.

Figure 7.8. Fracture surface of 96% dense compact prepared from dry-milled alumina I powder.

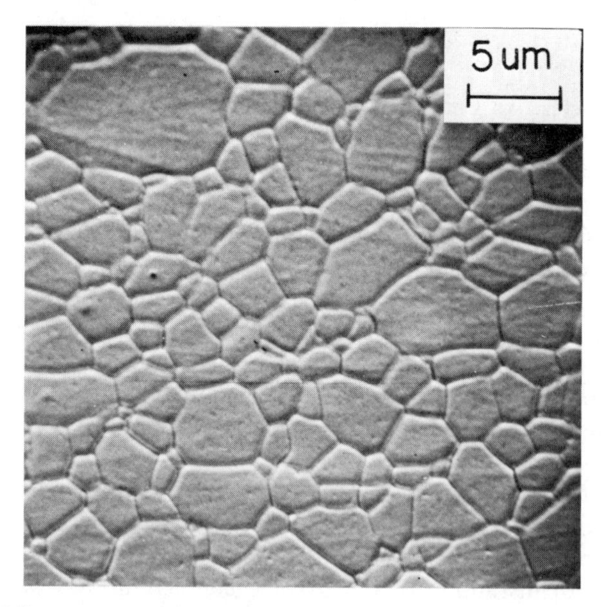

Figure 7.9. Microstructure of theoretically dense compact prepared from dry-milled alumina I powder.

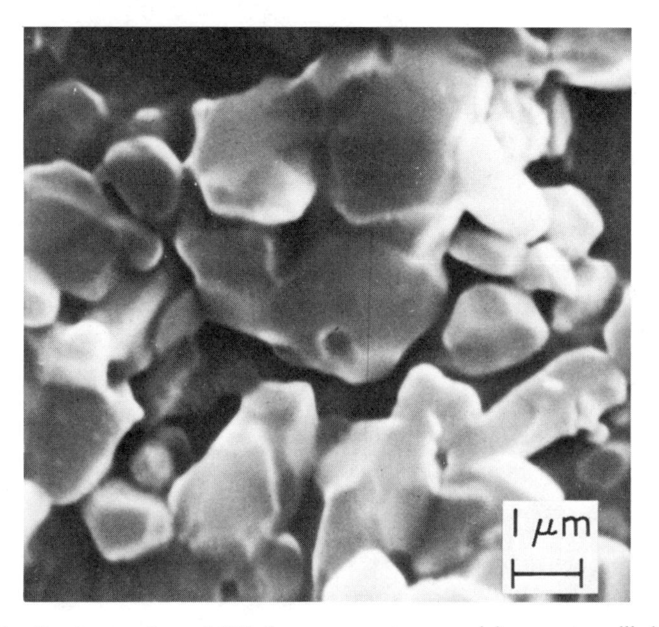

Figure 7.10. Fracture surface of 92% dense compact prepared from water-milled alumina I powder.

MECHANISM OF AGGLOMERATE FORMATION DURING WET MILLING

The mechanism by which the agglomerates form during wet milling is shown in Figure 7.11.[6] The left view depicts a water-solvated alumina particle, with hydrogen ions preferentially absorbed on the surface in a double-layer effect. The probable surface reaction is modeled in the right view, which depicts a fresh fracture surface exposed by milling attrition.

Exchange of hydrogen ions for aluminum ions permits the reaction to penetrate below the outer surface of the particle. The ion exchange normally produces an aluminum monohydrate layer on the particle[7] and releases aluminum ions into the water medium. Normally this reaction is quite limited, since the monohydrate acts as a reaction barrier, but transition aluminas can be hydrated by aging in water.[7] Thus the gamma alumina content of alumina I would be expected to have a strong influence on this reaction. For alumina I the extent of this reaction in 20 hours is normally slow. However, attrition during milling continuously removes the monohydrate reaction barrier, and the extent of the reaction during 20 hours of milling is quite significant for this powder. During drying, the aluminum-ion concentration in the liquid phase increases, and an aluminum

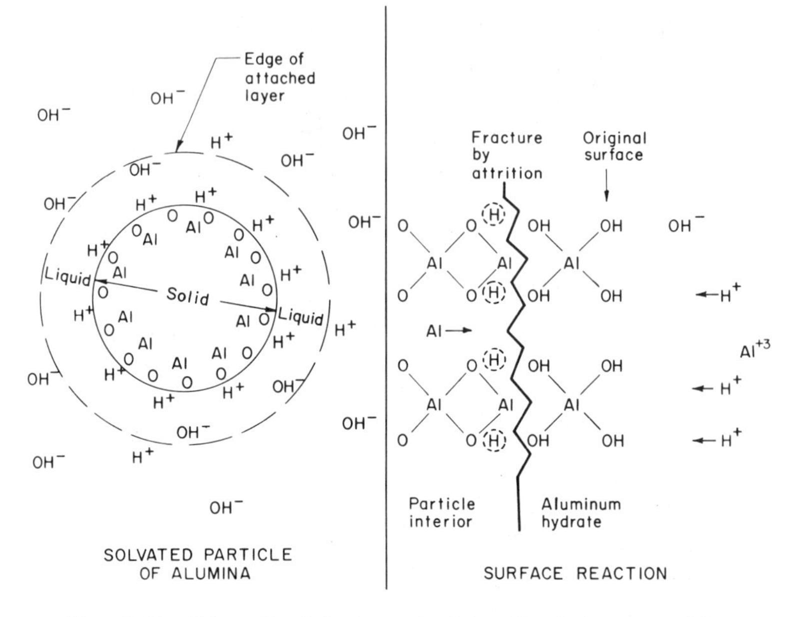

Figure 7.11. Schematic of alumina–water interaction during water milling.

hydroxide gel precipitates. With continued drying, the gel partially dehydrates, the extent depending on the drying conditions for the powder. Normally the gel ends up as a monohydrate with a surface of approximately 500 m^2/g and a capillary pore structure. It serves as a bond for the agglomerates in the dried powder.

During heating to the sintering temperature, the remainder of the structural water is evolved, after which the bond goes through several intermediate phases. More importantly, however, it begins to shrink because of elimination of the capillary porosity by sintering at temperatures well below that at which the body begins to shrink. The sintering shrinkage and the volume changes associated with the phase changes lead to disruption of the particle contacts and a weakening or rupture of the compact. The mechanism described is believed to be the cause for poor sinterability of high-surface-area powders milled in water.

For alumina powders composed of 100% alpha alumina, the alumina–water reaction is of little significance, since they normally have surface areas below 20 m^2/g.[8] If the alumina is to be milled in water, particular emphasis should be placed on eliminating all transition phases, such as gamma, since they normally have surface areas above 100 m^2/g. The high surface area of the hydrated reaction products also suggests that surface-area data are inadequate to unambiguously characterize the effects of wet milling a reactive alumina powder, since a small amount of aluminum monohydrate can override the surface area of the bulk of the powder.

REFERENCES

1. J. Lebiedzik, K. G. Burke, S. Troutman, G. G. Johnson, Jr., and E. W. White, "New Methods for Quantitative Characterization of Multiphase Particulate Materials Including Thickness Measurement," *Scanning Electron Microscopy,* Illinois Institute of Technology, Chicago, Illinois, 1973, pp. 121–128.
2. M. J. Orr, et al., "Applications of Mercury Penetration to Materials Analysis," Micrometrics Instrument Corporation, Norcross, Georgia, undated.
3. D. E. Niesz, R. B. Bennett, and M. J. Snyder, "Strength Characterization of Powder Aggregates," *Bull. Amer. Ceram. Soc.,* **51** (9), 677–680 (1972).
4. W. Duckworth, "Discussion of Ryshkewitch paper," *J. Amer. Ceram. Soc.,* **36** (2), 68 (1953).
5. E. Ryshkewitch, "Compression Strength of Porous Sintered Alumina and Zirconia—9th Communication to Ceramography," *J. Amer. Ceram. Soc.,* **36** (2), 65–68 (1953).
6. R. B. Bennett and D. E. Niesz, "Effect of Surface-Chemical Reactions During Wet Milling of Alumina," paper presented at the 74th Annual Meeting of the American Ceramic Society, Washington, D.C., May 9, 1972.
7. J. W. Newsome, H. W. Heiser, A. J. Russell, and H. C. Stumpf, "Alumina Properties," Alcoa Research Laboratories Technical Paper No. 10, 2nd revision (1960).
8. R. B. Bennett and D. E. Niesz, "Relation of Calcination to Character of Reactive Alumina," paper presented at the 76th Annual Meeting of the American Ceramic Society, Chicago, Illinois, April 29, 1974.

8

Characterization of Agglomerates with Transmission Electron Microscopy

L. L. Hench
E. J. Jenkins

In the previous chapter the importance of ceramic processing of agglomerates in powders was described. It was indicated that transmission electron microscopy is required to resolve the structure of fine powders.

This chapter describes methods for obtaining detailed information on localized, microstructural features by transmission electron microscopy. Means for characterizing solid agglomerates, weak agglomerates, and primary particles are discussed and limitations are noted.

TRANSMISSION ELECTRON MICROSCOPY (TEM) METHOD

In using transmission electron microscopy to characterize powders, severe approximations and assumptions are inherent in the analysis as it is usually applied. Dry dusting of powders onto TEM grids preserves some of the structural relations. However, the observations may not be representative of the relations in bulk powder because the time and expense involved in the method necessitates a small sampling number. Single-plane TEM imaging of agglomerates is usually insufficient to obtain representative data because agglomerates are three-dimensional structural entities.

As an example of the limitation of single-plane imaging, consider Figure 8.1. Both pictures are the same agglomerate of Linde A (α-Al_2O_3), but at two different angles of tilt (0 and 45°) in a Phillips Model 301 Scanning Transmission Electron Microscope (STEM) at 80 kV electron accelerating

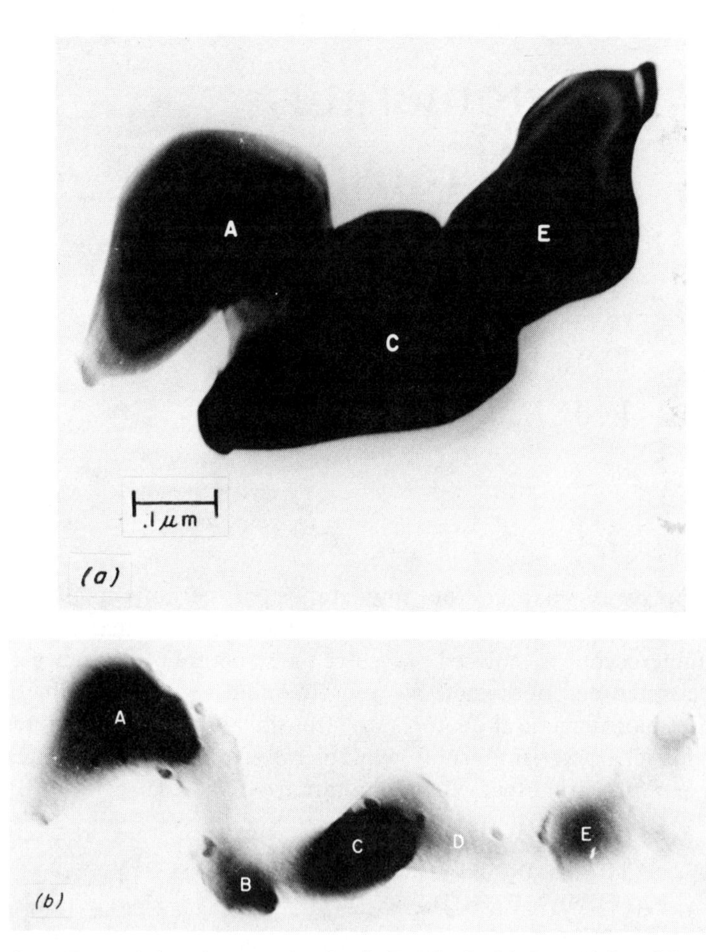

Figure 8.1. Transmission electron micrograph (TEM) of a Linde A Al_2O_3 solid agglomerate at (*a*) 0° tilt and (*b*) 45° tilt to the EM beam.

voltage. The agglomerate was one of several obtained by dry dusting Linde A onto a Cu EM grid.

The 0° tilt angle in Figure 8.1 suggests that the agglomerate consists of three particles (*A*, *C*, and *E*). The size of the agglomerate from this view can be characterized by maximum and minimum axes lengths of 0.9 and 0.5 μm. Changing the tilt angle to 45° reveals that the agglomerate is very much different than expected from the 0° tilt view alone. At least five particles are visible. Particles *B* and *D* were obscured by the other particles in the 0° view. Maximum and minimum axes of 1.3 and 0.3 μm, respectively, are observed from the 45° view.

Additional information, such as whether the particles are attached by weak or solid bonds, cannot be easily judged from one view direction alone. A weak agglomerate bond has a point contact character that can be seen only if viewed at the proper angle. Tilting provides different views to aid in the interpretation. For the agglomerate shown in Figure 8.1 the particles are clearly seen to be sintered together. As a larger number of particles per agglomerate are considered, it becomes even more difficult to distinguish the intraagglomerate features required for total characterization.

Tilting larger agglomerates in the TEM often provides less information because of particle overlap in the path of the EM beam, as shown by Figure 8.2. A major change in the agglomerate structure is apparent after tilting, but clear identification of even the number of particles present in the agglomerate is not possible. By comparing views at different angles, particles *A*, *B*, and *C*, *D* are found to exist as separate particles. Very careful rotation and tilting of the EM stage simultaneously might eventually reveal all the particle–particle interfaces in this agglomerate.

Since the particles in Figure 8.2 are sintered together, the only correct particle-size distribution for this powder clump would require a laborious analysis as outlined above. Particle-size analysis by any sedimentation-type technique will yield only a distribution of solid agglomerate sizes, not a distribution of particle sizes. Also, considering the wide range of axial ratios of agglomerates observed in a powder, the meaning of agglomerate-size distribution achieved by a sedimentation-type technique is unclear.

TEM ANALYSIS OF Al_2O_3 POWDER

Figures 8.1 and 8.2 show two small agglomerates representative of Linde A α-Al_2O_3 taken as a random sample from a commercial bag of powder. The agglomerates were shown to be solid agglomerates by use of the tilting stage. A larger agglomerate of Linde A is shown in Figure 8.3. This is also a solid agglomerate and is representative of the Linde A agglomerate struc-

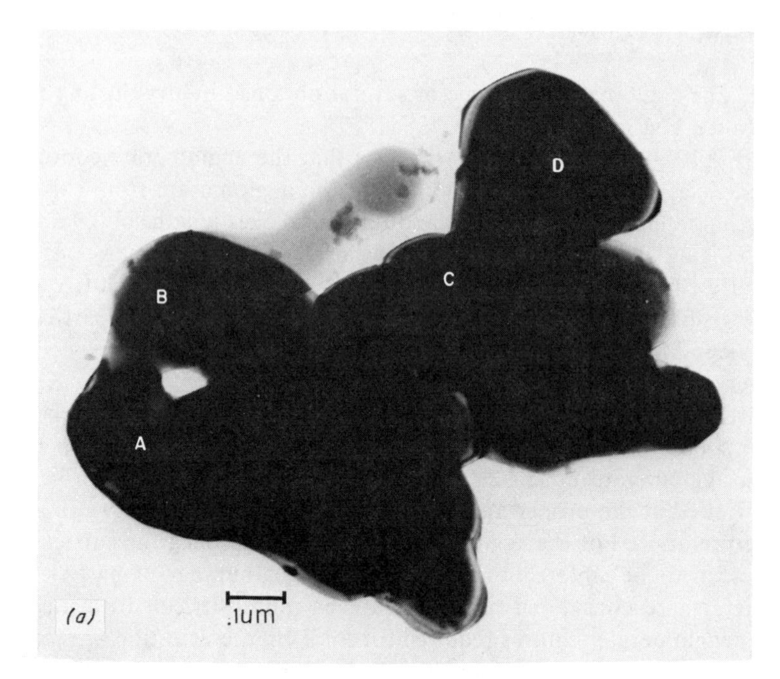

Figure 8.2. TEM of a larger Linde A Al_2O_3 agglomerate at (*a*) 0° tilt and (*b*) 45° tilt.

tures. Although tilting does "open-up" some apparent particle–particle contacts, there is always at least one sinter bond holding a particle to the large agglomerate. The agglomerates have a large external and internal surface area because of the low-density packing of the particles in the agglomerate. The agglomerate shown in Figure 8.3 is especially dramatic with respect to this stringy-like feature and high surface area characteristic of the Linde A agglomerates.

The agglomerates also show attachment, probably by weak bonds, of ultrasmall (0.1 μm) debris particles. Quantitative characterization of features of this type of multiagglomerate system is a formidable challenge.

Altering Al_2O_3 production to change the size of primary particles has a strong influence on agglomerate features as well. Larger-particle-size aluminas such as A-17 have fewer particles per agglomerate, as shown in Figure 8.4. The particles also tend to be weak bonded as shown through TEM tilting (Figure 8.5). Primary particles contain substructural features, probably as remnants of their processing history.

Tabular alumina T-61 exhibits very little agglomeration tendency and remains mainly as large isolated particles (Figure 8.6).

EFFECT OF A BINDER ON POWDER STRUCTURE

Addition of 0.2 wt % of acrylic ester resin binder (acryloid B-7, Rohm and Haas) to A-17 alumina results in binder-bonded granules, as seen by comparing Figure 8.7 (A-17 with binder) with Figure 8.4 (A-17 without binder). However, complete dispersion of the binder in the powder is not achieved. In a TEM tilting stage analysis of a granule from the A-17 binder sample

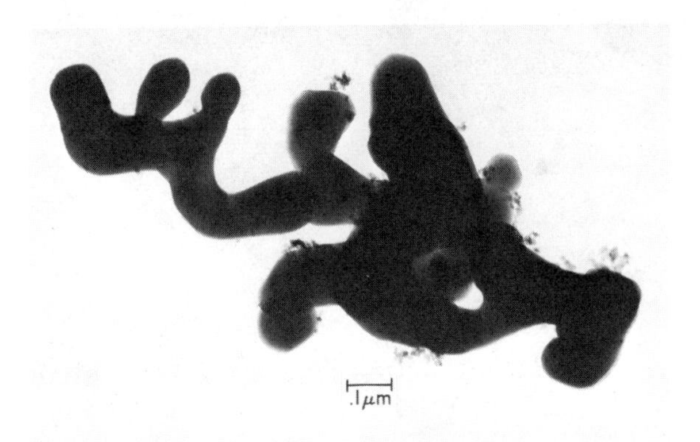

Figure 8.3. TEM of a large Linde A Al_2O_3 agglomerate.

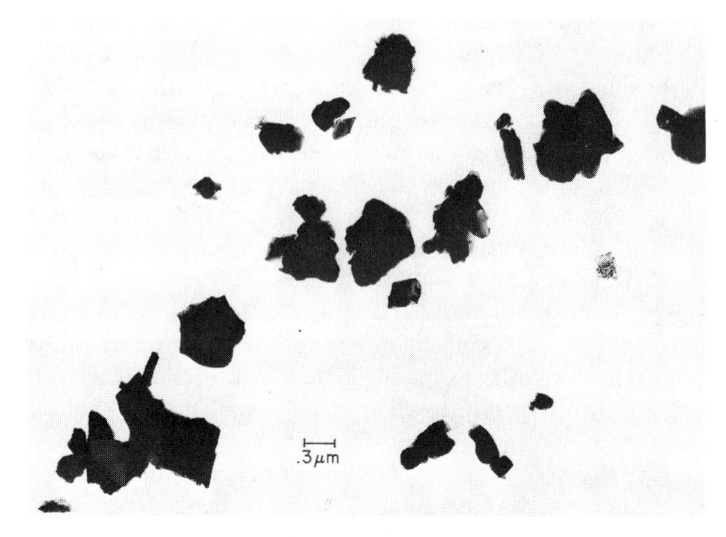

Figure 8.4. TEM of A-17 Al_2O_3 agglomerates.

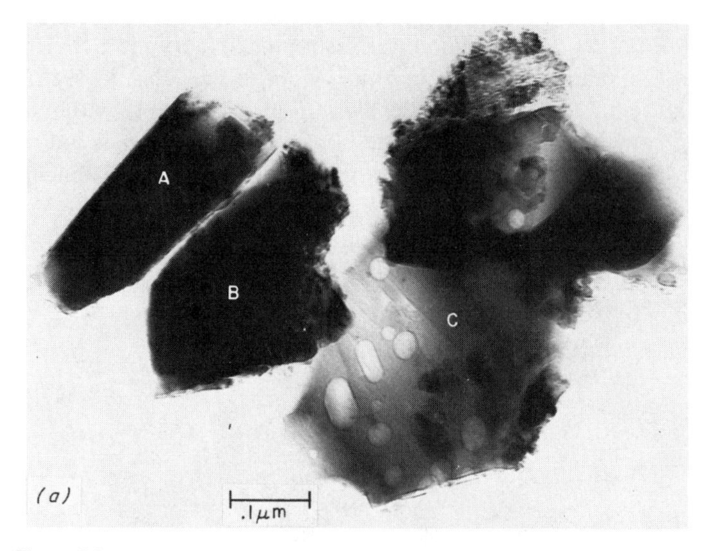

Figure 8.5. TEM of an A-17 Al_2O_3 agglomerate at (*a*) 0° tilt and (*b*) 16° tilt.

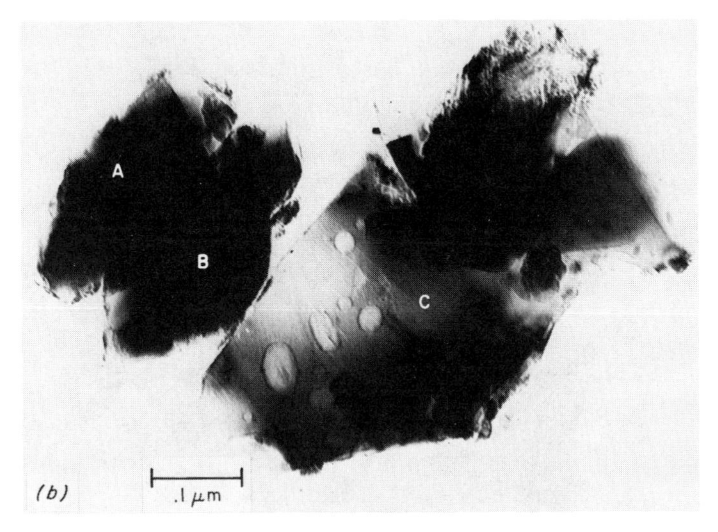

Figure 8.5 (Continued)

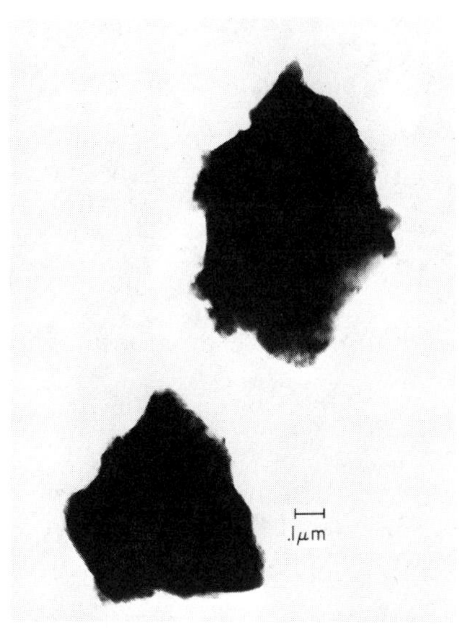

Figure 8.6. TEM of T-61 Al_2O_3 single particles.

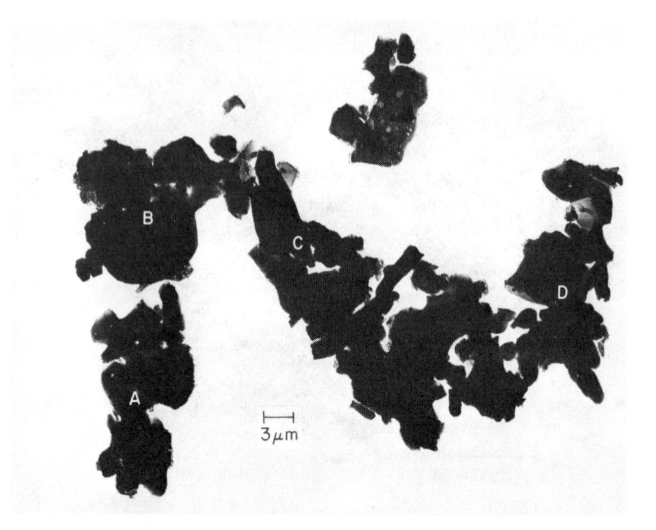

Figure 8.7. TEM of a granule of A-17 Al_2O_3 with binder. A through D are individual agglomerates.

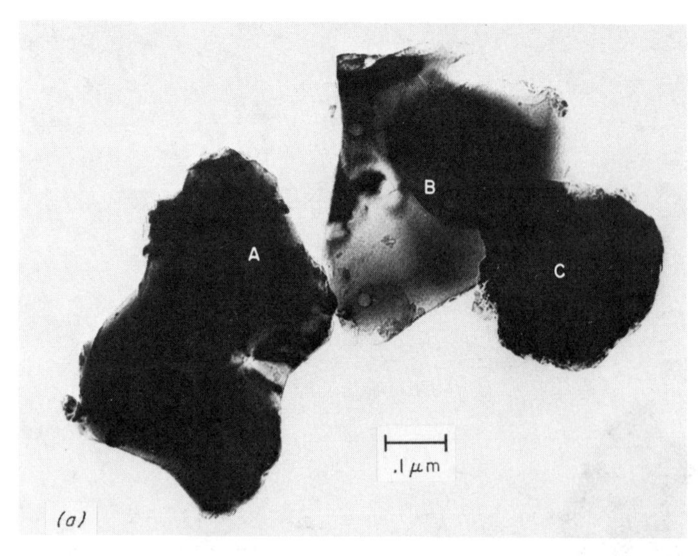

Figure 8.8. TEM of an agglomerate from an A-17 binder granule at (*a*) 0° tilt and (*b*) 45° tilt.

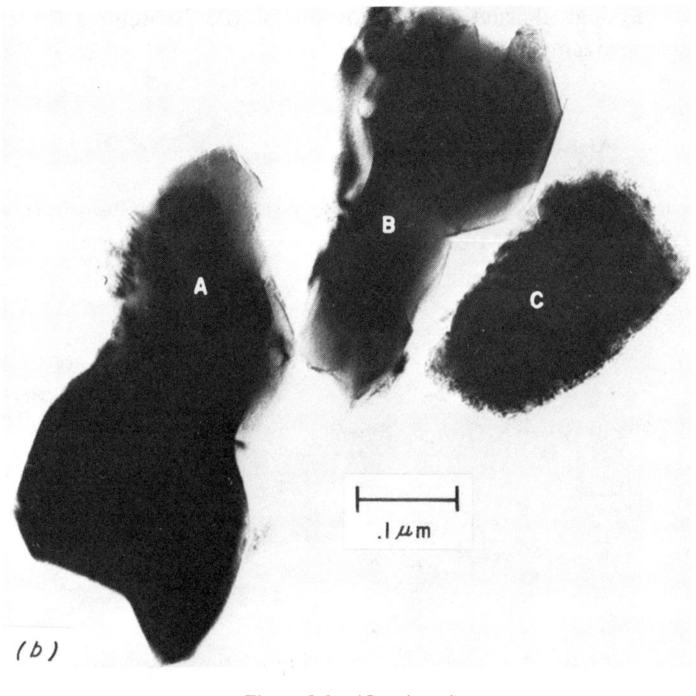

Figure 8.8 (Continued)

(Figure 8.8), particles *A*, *B*, and *C* all separate with tilting with no evidence of the binder being present.

SUMMARY

Transmission electron microscopy of ceramic powders can establish whether strong or weak agglomerates are present by tilting small powder agglomerates in the EM beam. Binder distribution in granules can also be analyzed. Particle- and void-size distributions in the agglomerates can be measured with fair accuracy by exposing the agglomerates to various beam orientations. The size distribution and axial ratios of agglomerates can also be determined with fair accuracy with the same technique. However, determination of the characterization features of large agglomerates and a cluster of agglomerates is only marginal because of particle overlap in the EM beam. Because submicron ceramic powders contain all the above complexities, physical characterization of the powder structures leaves much room for improvement at this time.

ACKNOWLEDGMENT

The authors gratefully acknowledge the partial financial support of NIH grant NIGMS 1 PO1 GM 21056-01.

9

Bayer-Processed Aluminas

W. M. Flock

A wide variety of alumina powders are available for ceramic processing applications. The endless generation of new-product code numbers can be overwhelming to process and material engineers. It is hoped, at least for Bayer aluminas, that the characterization data presented in this chapter will impart some insight into the various differences among aluminas.

To define the material parameters to be measured for as-received aluminas, a brief review of the Bayer process is necessary and thus is presented first.

THE BAYER PROCESS

The Bayer process begins with bauxite ore, a naturally occurring weathered residue consisting of hydrated aluminum oxide, kaolin clays, hydrated iron oxides, and titanium dioxide. The bauxite is digested with caustic soda to dissolve aluminum into solution. Other undissolved matter is filtered off, and the filtrant solution is precipitated to yield aluminum trihydrate. The trihydrate is calcined to produce alumina.

PURITY OF ALUMINA TRIHYDRATES

The potential chemical contaminants in the final alpha alumina are restricted to silica from the kaolin clay, iron, titanium, and sodium that is added as caustic soda during the digestion process. The amount of chemical contamination is largely determined in the precipitation step by the rate and manner of the trihydrate precipitation, that is, the amount of adsorbed or included contamination. Since these impurities are largely adsorbed, chemical purity dependence on this process step can be reduced by subsequent trihydrate washing. Another technique, which can be used to increase the chemical purity, is to precipitate the trihydrate from specially refined solutions, that is, those very low in silica, iron, and titanium.

The precipitation rate can also indirectly affect later calcining conditions and the resultant phase purity. Since trihydrate agglomerates are converted to alpha alumina by rotary calcining, the uniformity of phase conversion is directly affected by the agglomerate shape. Rapid precipitation produced by the addition of many nucleating sites (i.e., excessive seed crystals) can result in trihydrate agglomerate growth by accretion. This coalescence of rapidly growing agglomerates produces grapelike irregular clusters that are difficult to mix during rotary calcining. Slow precipitation, from a few nucleating sites, produces nearly spherical agglomerates, which mix more rapidly.

Trihydrate Calcination

During calcining, the alumina trihydrate, which consists of approximately 40% water, is converted to the final ceramic-grade alumina by the addition of heat and mineralizers. The latter additions are patented constituents (boric acid, chlorine and/or fluorine, and silica). These additions are required to reduce the sodium content (the major chemical impurity in Bayer-process aluminas) and to control the grain size. It should be noted that the various mineralizers produce differences in particle shape. Under controlled conditions, these slight variations in shape are not sufficient to exclude a mutual substitution of one supplier's alumina for another. It is, however, a factor that adds complexity to material characterization.

From a ceramic users point of view, two critical crystal structure changes occur during calcining that have a profound effect on later characterization and final ceramic character. These structural changes are the mode of trihydrate decomposition and the decomposition path. The decomposition path is through one or more of five well-defined intermediate crystalline phases (gamma, chi, delta, theta and kappa[1]). Trihydrate decomposition occurs by way of a topotactic transformation.

Topotactic phase transformations have been described by Taylor[2] and Nicol.[3] A topotactic transformation is one in which an earlier structure, in this case the close-packed oxygen ions of the OH radical, is maintained during dehydration. The hydrogen ions are lost and the aluminum ions rearrange without destroying the close-packed oxygen array. The crystallographic orientation of the trihydrate structure is therefore maintained through these intermediate phases to the final alpha-phase agglomerates. These agglomerates are therefore defined as pseudomorphic structures.

It is this pseudomorphic trihydrate structure that forms the basis for the classification of Bayer aluminas presented in this chapter. The pseudomorphic structure preserves the Bayer-process history and therefore provides insights into the precipitation rate and calcining conditions. By studying the pseudomorphic agglomerates, estimates can be made concerning chemical and phase purity, particle shape, and size. Such information is useful in the design of more quantitative material characterization tests. The relations between individual alpha crystallites and the Bayer agglomerates are described in this chapter in the section on Surface Area and Its Significance.

The trihydrate decomposition path is of significant importance to ceramic processing since it encompasses a series of lower-molecular-density, high-surface-area phases. These phases are a potential source of contamination in as-received aluminas. The presence of such phases can result in misleading characterization data, and in some cases, adversely affect ceramic processing.

Commercial Aluminas

Considerable progress has been made in the characterization of Bayer aluminas. The three major alumina suppliers have made major strides in the development of tests for the routine characterization of aluminas. Table 9.1 is a tabulation of data supplied by Alcan, Alcoa, and Reynolds Aluminum Companies. The data cannot be directly compared in all cases, since several test methods are not standardized. This problem will soon be rectified, since a joint effort is being made by the alumina suppliers and the Alumina Ceramic Manufactures Association (ACMA) to standardize these tests. The significant point of this table is that a great wealth of characterization information is available and that these data are routinely generated by the alumina suppliers. However, these data are not sufficient to make decisions regarding possible material substitutions. Further classification required for this purpose is presented in the remainder of this chapter.

CHARACTERIZATION OF BAYER-PROCESSED ALUMINA

The commercially available Bayer aluminas (Table 9.1) are characterized with respect to the expanded process model depicted in Figure 9.1. The major characterization tool is the petrographic microscope, and additional

Table 9.1. Ceramic grade Bayer aluminas

Type[a]	% Na_2O	Average Particle Size (μm)	Compaction	Fired Density	Surface Area (m^2/g)
RC-20	0.40	2.7	2.31[b]	3.37[c]	0.87
RC-24	0.23	2.9	2.35[b]	3.50[c]	0.53
RC-25	0.31	2.6	2.33[b]	3.52[c]	0.95
RC-122	0.03	2.6	2.35[b]	3.46[c]	0.35
C-70	0.50	2.5	2.28	3.18[c]	0.66
C-71	0.75	2.2	2.24	3.33[c]	0.71
C-72	0.03	1.61	2.27	3.58[c]	1.28
C-73	0.02	3.1	2.28	2.99[c]	0.34
C-75	0.01	2.8	2.25	2.97[c]	0.49
A-2	0.46	3.25	2.21	3.22[c]	—
A-5	0.35	4.7	2.30	3.07[c]	—
A-12	0.24	—	—	—	0.5
A-14	0.06	—	—	—	0.6
A-10	0.06	—	—	—	0.2
A-3	0.36	0.63	1.79	3.64	9.0
A-15	0.07	—	—	—	—
A-16	0.06	0.6	—	—	6.5
XA-139	0.008	0.42	2.00	3.90[d]	6.5
RC-23	0.30	0.6	1.91[b]	3.86[c]	7.5
RC-152	0.05	1.5	2.30[b]	3.46[c]	2.6
RC-172	0.04	0.6	2.21[e]	3.94[d]	4.0
ERC-HP	0.008	0.55	2.15[e]	3.95[d]	7.4

[a] C = Alcan, A = Alcoa, RC = Reynolds.
[b] 20 g pellet pressed at 4000 psi, 4 hour grind.
[c] Fired densities obtained at 1620°C for 1 hr soak.
[d] Fired densities obtained at 1510°C for 2 hr soak.
[e] 10 g pellet pressed at 5000 psi, 4 hour grind.

characterization data are obtained by particle-size analysis, surface-area measurements, and compactional and functional tests such as shrinkage and fired density. A summary of these data is given in Table 9.1. A recent Alcan publication[4] presents an excellent summary of the test methods used by both the suppliers and many ceramic users. Similar information can be obtained from Alcoa and Reynold's brochures.

The characterization objectives are to provide information from which ceramic processing can be predicted and/or adjusted to meet the desired ceramic character. The concept used is that the material character is a

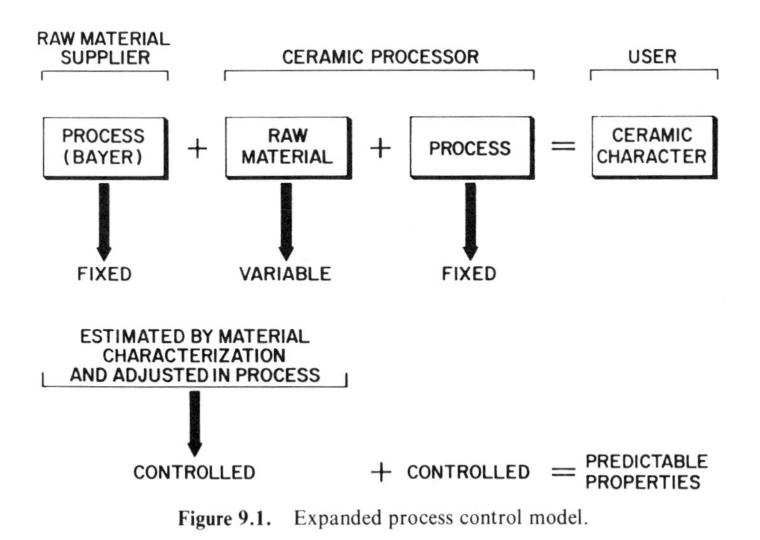

Figure 9.1. Expanded process control model.

function of the shape, size, orientation, packing, and composition of its constituents. The principal characterization critera employed to estimate these parameters are refractive index and agglomerate structure (see Figure 9.2).

Definition of Terms

Before discussing the characterization criteria, several specific terms are defined:

Agglomerate. A genetic term used to describe mesh-size particles formed during the precipitation of alumina trihydrate.

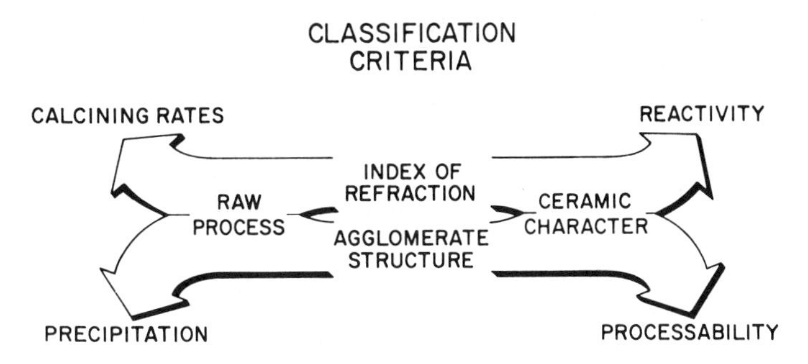

Figure 9.2. Genetic classification of Bayer agglomerates.

Pseudomorphic agglomerate. An agglomerate that has been subjected to partial or complete dehydration, but which retains the original trihydrate morphology.

Nonalpha phase. Any transition metastable aluminum oxide phase, such as gamma, kappa, chi, theta, or delta.

Alpha phase. The high-temperature, stable phase of aluminum oxide (Corundum mineralogical designation).

Reactive agglomerate. Alpha-phase agglomerates with high surface area and sufficient free energy to sinter to essentially theoretical density without the addition of glass-flux modifiers.

Beta alumina. The 12:1 sodium aluminate phase.

RELATION BETWEEN ALPHA CRYSTALLITES AND AGGLOMERATE STRUCTURE

The pseudomorphic trihydrate structure, which is described earlier, permits the characterization of alpha crystallites to be carried out as if they were orders of magnitude larger than their actual size of 0.2 to 10 μm. The occurrence of crystallites in 200 mm diameter agglomerates produces an effective magnification of 20,000×. This inherent magnification places the material into the ideal working range of the petrographic microscope. The additional magnifications of 50 to 200× obtained with the microscope produce a total useful magnification of one million to four million times.

The ability to use the petrographic microscope is a major asset, since it provides a link among the atomistic (angstrom), microstructural (micrometer), and phenomenological levels. It enables the simultaneous integration of texture parameters (size, shape, orientation, and packing) and chemical and mineralogical phase analysis. Such analyses for fine-grained material are normally possible only through the combined use of the electron probe and scanning microscopes. The petrographic microscope has, however, for the last 100 years provided the same information for coarse-grained material (i.e., 50 μ grains or larger) that these instruments provide today for fine-grained materials.

Throughout the remainder of this chapter, when agglomerate properties such as surface area, chemical purity and phase are discussed, it is implied that these are the properties of the individual primary particles.

Surface Area and Its Significance

The reactivity of an alumina is not simply related to its BET surface area. At one time it was generally believed that high surface area should be correlated with high sintering reactivity. With alumina agglomerates, however, nonalpha phases (transition aluminas) may be present that have very high surface area but contribute little to reactivity. The transition aluminas have surface areas above 100 m^2/g. Reactive aluminas (alpha phase) have surface areas of 4 to 7 m^2/g, while nonreactive alpha aluminas have much smaller surface areas. The surface area contribution of the alpha aluminas is defined as "useful," while nonalpha-phase contributions are defined as "nonuseful."

The surface areas given in Table 9.1 are for as-received powders. Milled powders (deagglomerated) in general have approximately 2½ times the surface area of the powder before milling, provided no contamination has occurred. Since this increase is consistent and predictable, the surface areas stated in this chapter are for as-received aluminas.

Since the transition aluminas have surface areas equal to several hundred times those of the nonreactive alpha phase, the presence of 1 or 2% of these phases has a significant effect on the measured surface area. The A-2 and RC-20 alumina types typically have 1 or 2% nonalpha phase and yield surface areas of 0.7 to 0.9 m^2/g. A-12 and RC-24 types are typically free of nonalpha phase and have surface areas of 0.4 to 0.7 m^2/g. However, the RC-24 type is thermally more reactive than the A-2 or RC-20 types. Because of the nonalpha phases the surface area of the A-2 or RC-20 aluminas does not provide a meaningful index for describing thermal reactivity. Thus, in nonreactive aluminas, surface-area contributions from transition-phase agglomerates are a characterization nuisance and have little effect on final ceramic character. They are a serious nuisance, since the true physical surface area is used to predicate such processing parameters as binder concentration, drying time, and mill packing. The tape-casting process is very sensitive to surface area, since it affects such properties as drying time, belt release, and tape fixability.

Surface-area contributions from nonalpha-phase agglomerates in reactive aluminas such as A-16 or RC-172 are much more serious because they have a direct influence on ceramic processability. Since these materials have surface areas 10 to 15 times as high as nonreactives, a greater percentage of nonalpha-phase material is required to change the surface area. However, since reactive aluminas are calcined at a lower temperature, the possible occurrence of nonalpha phases is much higher. In addition to destroying the correlation between surface area and thermal reactivity, nonalpha material seriously effects processability. The reactive surfaces readily "cake" or reagglomerate during milling and cause incomplete size reduction. Furthermore, they adsorb large quantities of binder and affect, for example, viscosity and drying rates. These reactions are very detrimental in the tape casting of super-smooth alumina substrates. Finally, transition phases are believed, by the writer, to produce spotty porosity in 99+% dense bodies. Since transition agglomerates are difficult to mill, and since they are of lower phase density, their sintering to the higher-density alpha phase results in a large volume reduction and produces localized high shrinkage.

For the above reasons, it is important to identify the surface-area source particularly for reactive aluminas. The presence of transition agglomerates in Bayer aluminas is a serious problem. Therefore, identification of these agglomerates with the petrographic microscope is an essential characterization process.

Characterization Criteria

Refractive Index. Refractive index differences are the basis of the optical-microscopy technique used to differentiate the transition phases and beta alumina from the stable alpha phase. The specimens are prepared as loose-grain mounts with the powder immersed in 1.720 refractive index oil. Analyses are carried out at low magnification (50 to 200×).

The selection of 1.720 index oil is of importance, since alpha (corundum) alumina has indices of refraction of 1.760 and 1.768 and the common phase impurities, beta alumina and the anhydrous transition phases, have refractive indices of 1.720 or less. Phase identification is based on this difference in refractive index (1.760 to 1.720). The optical principle employed is that transparent solids become practically invisible when placed in liquids with similar refraction index. Their boundaries form a continuous medium and light passes with little or no reflection or refraction. Secondly, as the differences in refractivity (refractive index) between the liquid and the solid become greater, the amount of the boundary light (refraction and reflection) increases. This increase in refractivity is observed as "relief"; the solid becomes plainly visible from the liquid as the difference in refractive index increases. By using the Becke method or central illumination method of refractive index measurement, a solid can be determined to have indices greater or less than the liquid. By careful measurement of these refractive indices, precise statements can be made concerning the chemical composition and internal structure of the solid phase.

In the present analysis the liquid refractive index was fixed at 1.720 and the difference in refractivity (relief) was used to identify the phases. Alpha-phase material presents distinct boundaries and blue colors, while nonalpha-phase agglomerates have indistinct boundaries and light brown colors. The color is a function of the amount of light dispersion.

Alpha-phase agglomerates are identified by their refractive index. These agglomerates have a refractive index slightly lower than the stable alpha phase and therefore exhibit a distinctive pale blue color.

It can be argued that another source for the distinctive pale blue color of the reactive agglomerates is the finer crystallite size. It is possible that a portion of the color dispersion is due to the greater number of crystal boundaries in reactive agglomerates. Whichever mechanism is active, a distinct color difference between reactive and nonreactive agglomerates exists and the author has found this to be the only definite method for clearly differentiating reactive materials.

Of the two mechanisms the lower-refractive-index mechanism is favored, since it also accounts for the high surface area of these agglomerates. A relation between refractive index and surface area can be formed on the basis that refractive index is a direct measure of electron density, and if cation or anion defects exist, the electron density would decrease, which in turn would slightly reduce the refractive index.

Birefringence. In addition to refractive indices, birefringence and optic axes orientation are other optical techniques used to characterize these aluminas. Because of the small differences in refractive index between the crystallographic axes (1.760 versus 1.768) of alpha alumina, its birefringence is increased to a workable range by inserting a gypsum plate into the optical path of the polarizing microscope with crossed Nicols. The first-order red of the gypsum plate provides a background color in the micrographs. For additional information on optical properties and use of the petrographic microscope, the reader is referred to any text on optical mineralogy[5] or to a paper by Allen.[6]

It is the interaction of light with variations in the crystal electron density that produces birefringent colors that aid phase identification. Alpha-phase material exhibits low first-order grays and whites, while the other more anisotropic transition phases exhibit high first-order colors.

Agglomerate Structure and Shape. These are parameters that estimate chemical purity and, indirectly, phase purity. Agglomerate structure is a direct estimate of precipitation rate. In the Bayer process, as previously described, precipitation rate is a function of pH, temperature, and the number of nucleation sites, that is, seed crystals. If the rate is low because of any of the above conditions, agglomerate growth is slow and radiant growth occurs from the central nuclei. If precipitation is rapid, the radiant growth is replaced by coalescing agglomerates, which result in irregular, grapelike agglomerates with internal structures lacking areas of optical continuity.

The major significance of these two structures designated as random (grapelike with no initial crystal orientation) and oriented (spherical with distinct internal crystal orientation) is that they directly affect the chemical purity of the final material. Since precipitation occurs from a sodium aluminate solution containing finely suspensed iron, silica, and titinia particles, inclusion due to physical adsorption in the nonoriented agglomerates is the contamination source.

An indirect effect of agglomerate shape is phase uniformity. Since the trihydrate conversion is accomplished in a rotary kiln, uniform agglomerate mixing is critical. Nonspherical, irregularly shaped particles are more difficult to mix and can result in incomplete or excessive phase conversion, that is, grain growth. The author has never observed a uniformly calcined product that did not have controlled trihydrate as its source material.

CLASSIFICATION OF BAYER ALUMINAS

The aluminas given in Table 9.1 are genetically classified in this section. The classification purpose is to relate raw-material process history to ceramic processability and final character. Secondly, the aluminas are classified to reduce the spread in properties so that quantitative characterization can be carried out. The final purpose is to aid in the selection of alumina types, that is, material substitution.

The classification criteria are shown in Figure 9.2. The interaction among these criteria, the previous raw material processing, and their ceramic behavior are discussed in a previous section. In this section eight alumina types are classified according to reactivity, agglomerate structure, and

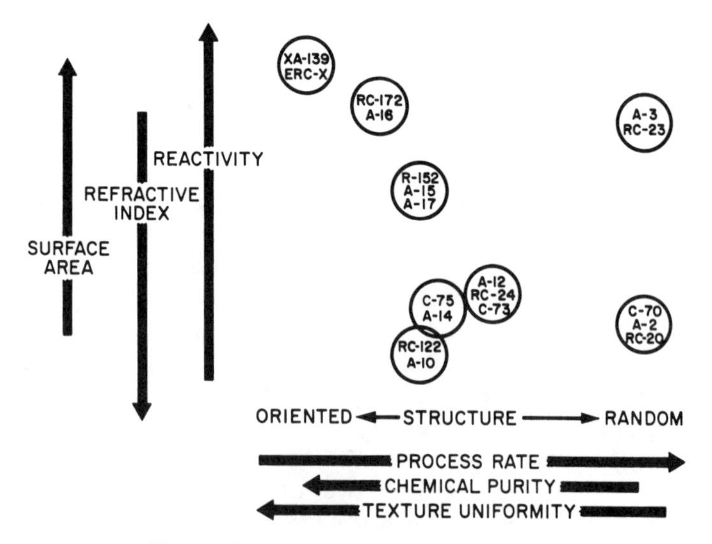

Figure 9.3. Classification of Bayer aluminas.

phase content, as shown in Figure 9.3. Surface areas, soda contents, and the range of average particle sizes are also stated. The variation in reactivity among types is indicated by an arrow. The direction of the arrow indicates increasing reactivity and the arrow length reflects the spread in reactivity. Several typical alumina types are placed on the reactivity arrow at their appropriate positions. A tight cluster of types indicates little or no difference in reactivity.

For discussion purposes, the eight alumina types are grouped into the following four larger classes:

1. Nonreactive; high purity.
2. Nonreactive; low purity.
3. Reactive; high purity.
4. Reactive; low purity.

The nonreactive alumina are discussed first.

Nonreactive Aluminas

Oriented—High-Purity Types. One of the first commercial Bayer aluminas marketed in the early forties was Alcoa A-10. Prior to A-10, tabular alumina was the only source of high-purity aluminum oxide. Tabular is a Bayer product that is further processed by sintering and is then reduced to a fine grain size by steel-ball milling. With respect to thermal reactivity, tabular alumina is completely inert, having an unmeasurable low BET surface area. A-10 does not require extensive ball milling, but like tabular, A-10 is a very nonreactive material, having surface areas of 0.15 to 0.25 m^2/g and an average particle size greater than 5 μm. There is no direct commercial substitute for A-10 alumina. Alcan C-6 was an early attempt and Kaiser KC-8 is a product with some similar properties.

Based on historical usage, A-10 has been produced in a greater volume for high-alumina bodies (refractory applications not included) than any other material. It has an excellent record of uniformity but is being replaced with lower-cost, more thermally reactive materials, such as A-12, C-73, and RC-24.

A-10 is characterized petrographically by a uniform deep blue alpha conglomerate. The majority of these agglomerates have oriented structures indicating slow precipitation. The chemical and phase purity confirm that A-10 is a highly controlled product. Nonalpha-phase agglomerates are never observed. The grain size is coarse, but the uniform size distribution results in excellent green densities and low-fired shrinkages but, conversely, produces a low surface area and low thermal reactivity, that is, a high firing temperature. Although A-10 stands in a class alone, it has been combined with the alumina family of C-75, A-14, RC-122, and B-4. Like A-10 these aluminas are characterized by very uniform, nonreactive alpha agglomerates and high chemical purity. The agglomerates of B-4, RC-122, and A-14 are primarily the oriented type, while C-75 is closer to the random agglomerate. Because of similar calcining conditions, particle size, and distribution, A-14 and C-75 are very similar and result in materials capable of mutual substitution in most ceramic processes. RC-122 has an improved size distribution and yields higher compactions and lower-fired shrinkages than either A-14 or C-75. RC-122 and B-4 are both closer to A-10 but are more reactive. The surface areas of the four materials are a true reflection of the physical surface, since these aluminas are free of surface-active nonalpha-phase agglomerates. The surface area varies from 0.35 to 0.55 m^2/g and the average particle size is 2.6 to 2.8 μm.

Like A-10 these materials have been the standard for high-quality electrical grade ceramics and, like A-10, they are being replaced with either a lower-cost nonreactive alumina or a higher-cost transition reactive alumina.

The second family in the nonreactive, high-purity group is transitional between the high-purity and low-purity aluminas. These aluminas have intermediate soda levels but a highly uniform phase and particle size. The three members of this class are A-12, RC-24, and C-73.

This class is unique in many respects. All three aluminas can be substituted for one another. Their higher soda contents make them lower in cost than the previous materials, but their closely controlled agglomerate and phase composition make them well suited for close-tolerance technical ceramics. The increase in soda does lower the TE value several hundred degrees but has no effect on the dielectric strength of electrical ceramics made from these aluminas.

Their excellent particle size distribution yields compactions much higher than A-14 and C-75 but lower than A-10, hence shrinkage is moderate to low.

The surface area is real (physical only) and varies from 0.4 to 0.7 m^2/g. Compared to the A-14 types, the thermal reactivity of this group is higher. The agglomerates are both random and oriented, reflecting lower chemical purity. RC-24 is predominantly an agglomerated material.

Commercially these materials are increasing in usage owing to their lower cost, higher compactions, thermal activity, and uniformity. They are used in practically all types of high-alumina ceramics.

Random—Low-Purity Types. The final nonreactive family is the A-2, RC-20, and C-70 type. Other members of this group are RC-25, A-5, and C-71. All these aluminas have random agglomerates (rapid precipitation) and high soda contents and contain nonalpha-phase agglomerates. The surface areas are therefore high, 0.6 to 1.2 m^2/g, with 0.9 to 1.0 m^2/g being typical. Based on particle size, the true physical surface area should be 0.5 m^2/g or less. The powder compaction of these aluminas is fair to moderate.

Their low phase and chemical purity exclude their use for close tolerance electrical grade ceramics. These aluminas, however, are widely used in lower-cost ceramics where the size control and electrical properties are not critical. Typical uses are for abrasion-resistant material, grinding media, and low-temperature refractories. A significant recent use has been as high-modulus fillers in whiteware products. The substitution of one material for another in these less-critical ceramic applications is usually possible.

Reactive Aluminas

There is no question that future ceramic bodies will be formulated with thermally reactive aluminas. Reactive aluminas are the newest Bayer products, having been introduced in 1966. Several product types are still in experimental pilot production (XA-139 and Reynolds ERC-HP). The production experience in the manufacture of these aluminas is much less than that in the case of nonreactive types, and the controls required for their production are much more extensive. Based on these two conditions, variations in as-received materials can be expected. Reactive alumina agglomerates are readily identified by their refractive index and light blue color in singularly polarized light (plain light).

Four families of reactive aluminas are recognized and distinguished based on phase purity, chemical purity, and the number of reactive agglomerates.

The highest purity, most-reactive aluminas require the tightest process control. Precipitation rates are greatly reduced and calcining is carried out using proprietary mineralizers to control grain growth. Because of these conditions and lower volume usage, reactive aluminas cost 3 to 10 times as much as the nonreactive aluminas. The benefits to the ceramic users are increased thermal reactivity and greatly improved ceramic character. Because of lower product maturity, process complexities, and the failure of the user to adequately characterize reactive materials, the expected ceramic benefits are not always achieved.

In fairness to the alumina suppliers, it should be noted that the alumina material improvements have not been matched by ceramic-process improvements and hence the problem of large-volume ceramic fabrication with reactive materials still faces the industry.

Transition Reactive. Because of the ceramic users inability to handle the high-surface-area reactive aluminas, the original product had to be scaled down to the users capability level. The high surface area and fine particle size resulted in poor milling, and inadequate knowledge of binder-powder interactions resulted in poor green densities, laminations in parts, and low green strengths. These process defects produced high shrinkage (greater than 22%) and cracked and warped parts. RC-152 and A-15 were adjusted to provide a compromise between high-surface-area reactive aluminas and the low-surface-area nonreactives.

Light-microscopy analyses clearly show that these materials are mixtures of reactive and nonreactive agglomerates that have been precipitated under similar conditions but that have been subjected to different calcining temperatures. Such materials can be produced by the physical mixing of reactive and nonreactive agglomerates or by calcining at the critical conversion point and producing both reactive and nonreactive materials. From light-microscopy analyses, it is not possible to determine which procedure was used, nor is it critical to know. The important fact is that these materials are mixtures of two distinct agglomerate populations—reactive and nonreactive.

Both agglomerates have oriented structures and very good chemical purity. They are free of nonalpha-phase materials and the surface areas of 1.5 to 2.5 m^2/g represent the true physical surface. The average particle size is 1.5 μm which is significantly finer than the previous nonreactives but much coarser than full reactive aluminas.

Commercially these aluminas are very important in that they are used for practically all 99.5% alumina bodies. The dual-agglomerate population produces a wide particle-size distribution, which in turn produces excellent compactions and low-fired shrinkage. However, since they are mixtures of reactive and nonreactive materials, wide variations in reactivity are possible, making surface area an important characterization tool.

Reactive—High Purity. With improved ceramic process experience and capability, the fully reactive aluminas were again placed on the market. These materials are A-16 and RC-172. They contain essentially all reactive agglomerates and, like the transition types, contain oriented agglomerates. Since they are calcined at a reduced temperature, some lots contain a percentage of nonalpha-phase material. Typical material surface areas are 4.0 to 6.5 m^2/g. Higher values due to the nonalpha phase have also been measured. The average particle size is extremely fine, being less than 0.6 μm, and the soda level is very low (around 0.05%).

The major commercial use for these materials has been for super-smooth substrates (as-fired surface finishes of less than 5.0 μ in.). The need for super-smooth substrates produced to very close specifications has greatly improved the state of the art of reactive materials. Another important application has been for high-density, 99.5% grinding media. The 99+

material improves grinding efficiencies, greatly reduces ball wear, and improves final ceramic purity.

Reactive Mixed Phase—Low Purity. Two very interesting reactive aluminas are A-3 and RC-23. These materials are truly reactive, having surface areas greater than 7.5 m^2/g, but are not produced with precise controls required for the other reactive materials. They are therefore produced at much lower costs and have excellent potential use in less-critical ceramic applications.

The problems in working with these materials are numerous. The abundance of nonalpha phase, particularly the beta phase, caused by the high soda content, makes milling extremely difficult. Mill problems, plus fine particle size, result in very poor compactions, low green strengths, and shrinkages greater than 25%. Standard material preparation procedures and forming by tablet pressing are impossible. However, additions of these materials to improve reactivity of other aluminas are possible.

As an interesting aside, the high reactivity of these materials points out that chemical purity itself is not a prerequisite for reactivity. Reactivity is a function of the degree of phase conversion, that is, defect structure.

Reactive Ultrahigh Purity. The finest example of controlled Bayer aluminas is found in the last family, which consists of two experimental production materials, XA-139 and ERC-HP. These aluminas have the highest purity (< 0.01% Na_2O) and finest grain size (0.5 μm) and are the most reactive Bayer materials produced. XA-139 is particularly interesting in that the precipitation rate is so slow that single or twinned trihydrate crystals are developed instead of the typical trihydrate agglomerate.

This controlled precipitation is reflected in the materials high chemical and phase purity. The high surface area (6.5 to 7.5 m^2/g) is a true reflection of the reactive alpha phase, since nonalpha-phase material has not been observed.

These aluminas are excellent base material for sintering studies. Their commercial usage is not known but they could be used to provide additional surface-finish improvement (super-smooth type) or for very-high-strength materials. Their typical properties are given in Table 9.1.

Classification Summary

The Bayer aluminas given in Table 9.1 have been classified into eight families based on agglomerate structure and refractive index. These criteria, their relations to other material characteristics, and ceramic reactivity are shown in Figure 9.3. The relations of the eight classes to one another and with respect to their properties are also illustrated.

REFERENCES

1. H. P. Rooksby, "Oxides and Hydroxides of Aluminum and Iron," *The X-Ray identification and Crystal Structures of Clay Minerals,* G. Brown, Ed., Mineralogical Society, London, 1961, pp. 354–392.
2. H. F. W. Taylor, "Crystallographic Aspects of High Temperature Transformations of Clay Minerals," *Proc. Natl. Conf., Clay and Clay Minerals, 12th,* **1964,** Pergamon, London, pp. 9–10.
3. A. W. Nicol, "Topotactic Transformation of Muscovite under Mild Hydrothermal Conditions," *Proc. Natl. Conf. Clay and Clay Minerals, 12th,* **1964,** Pergamon, London, pp. 11–19.
4. "Characterizing Aluminas," Alcan Company, 1973.
5. E. H. Kraus, W. F. Hunt, and L. S. Ramsdell, *Mineralogy,* McGraw-Hill, New York, 1951.
6. A. W. Allen, "Optical Microscopy in Ceramic Engineering," *Ceramic Microstructures,* R. M. Fulrath and J. A. Pask, eds., Wiley, 1966, pp. 71–158.

10

Grinding of Aluminas

M. Berg

Bayer-processed alumina, as is discussed in Chapter 9, consists of solid agglomerates rather than individual primary particles. Breaking down the agglomerates by grinding results in many advantages, including denser compacts, lower firing shrinkage, lower firing temperatures, lower tendency for lamination and warpage, and, most important of all, better-fired microstructures. Grinding does not break down the individual primary particles but instead fractures the agglomerates to form a combination of individual or several individual primary particles.

HISTORY OF ALUMINA-GRINDING TECHNOLOGY

Alumina grinding started in the early 1930s with two inventions. One invention was used at RCA for making a ceramic coating for vacuum tube filaments. The other related to work by Schwartzwalder on alumina at AC Spark Plug. Schwartzwalder's work was basically a continuation of the work by Seamen in Germany. These studies involved ball milling a combination of fused and calcined aluminas in steel mills, acid bleaching the iron that was trapped on the alumina during grinding, and then decanting and washing repeatedly. This work led to the first alumina spark-plug insulators in the United States. At RCA, because the process was too costly, mullite was eventually used for filaments.

A spark-plug insulator fits in a steel shell and serves as both a structural member and an insulator at high temperatures. In the early 1930s, spark-plug insulators were high alumina mullite compositions. Alumina insulators became important in the early 1940s. With the addition of organic resins to the aluminas, the finest aircraft spark plugs became available for bombers in World War II. The insulators were typically 94 to 96% alumina.

Two schools developed in the approach for grinding aluminas, wet grinding and dry grinding. Wet grinding was done in a ball mill, which was followed by filter pressing and drying, or spray drying. In our case we used a continuous drum filter, pugged the material, and made it into an insulator.

The dry-grinding approach was carried out by Schwartzwalder in the early 1940s. This approach was adapted at AC Spark Plug and continues to be the method utilized. Suitable liners and balls for mills had to be developed that would not contaminate the alumina. They were usually matched very closely to the composition of the final batch. Special grinding aids were developed to prevent the rebonding of alumina in the mill.

Injection molding was the forming process for spark plugs at AC during the 1940s. This process was eventually phased out in favor of isostatic pressing because it was more costly to make the ribs of the insulator than to form the whole insulator.

In the 1950s isostatic pressing of insulators began. After the grinding operation, organic materials (binders) were added to a water–alumina slip and the slip was spray dried. The spray-dried granules were fed to the pressing operation.

During the 1960s Sommer received four patents in the area of dry grinding. In one contribution he changed the ball-mill loading ratio of alumina to balls. The normal ratio at that time was 3:1 or 4:1. Sommer showed how the ratio could be increased to between 10:1 and 20:1. With suitable grinding aids, he enhanced the milling and obtained more compactable powder.

PRESENT STATUS

Most grinding of aluminas today takes place dry in ball mills. Where wet grinding is being used, the vibratory mill is common. The fluid energy mill is being examined to grind various materials, such as gamma alumina for use in catalysts.

In milling an alumina, probably some of the most important items to consider are the alumina characteristics obtained from the incoming material. The bulk density will provides information concerning powder handling in the plant and on loading a mill. The flowability will indicates how the material will flow in your mill and will give insight on the particle distribution. The grindability of the material is determined at AC by a test that is a spinoff of Sommer's so-called super grinding. This involves a very high loading (10 to 1) in a small gallon mill, and milling for 4 hr. The milled powder is tested for compaction behavior, surface area, and size distribution. With this knowledge large mills can be tailored by predicting to some degree how the powder will compact and how much grinding time is required. The surface area of the material yields information not only on the particle, but also on whether other materials are present (such as a gamma alumina in an alpha). Chemical composition is important to determine the sintering behavior of the final product. Recognizing the important characteristics of the incoming alumina, it is possible to work with the various alumina companies so as to control this product and to use material from three or four alumina suppliers.

Grinding aids are used to promote flowability in the ball mill and to prevent caking. In theory the grinding aid tends to be a polar molecule that attaches itself to the active bond site that has been broken and prevents that grain from rebonding. Grinding aids include esters, organic salts such as sodium liquisulfonates, stearic acid, oleic acid, and monostearates. Even ethylene glycol, added as a binder, functions as a grinding aid as well. Also in grinding, lubricants are added so that we actually have a combination of binders, lubricants, and grinding aids.

11

Theories of Grinding

P. Somasundaran

The previous chapters have dealt with agglomerates and the breakdown of hard agglomerates by grinding. The grinding of coarse-grain material to produce fine powders is also of concern in many areas of ceramics. The science of fine grinding in ceramics appears to have lagged considerably behind the technological advances. Considering the large energy consumption and capital equipment costs in grinding, and the importance of the size and size distribution of the ground powder, it would be helpful to understand more about the mechanisms of grinding and the kinetics involved. These factors have received much attention in the field of mineral processing. This chapter reviews concepts in grinding and discusses the theoretical progress in fine grinding and effects of grinding aids.

STRENGTH OF INDIVIDUAL PARTICLES

The fracture of a particle involves the propagation of cracks that are present or initiated in the particle. The stress σ required for fracture is given by the Griffith relationship[1]

$$\sigma = \sqrt{\frac{2E\gamma}{L}}$$

where E = Young's modulus
γ = fracture energy
L = crack length

For brittle materials, γ is between 10^3 and 10^4 erg/cm^2. With plastic deformation, γ is much greater than 10^4 erg/cm^2.

When a particle is repeatedly fractured, each new particle (fragment) tends to be stronger. The larger cracks existing in the original particle propagate first, leaving the finer cracks in the new particles. The probability of finding a flaw of a given minimum fracture stress decreases. As fragmentation continues, eventually the fracture stress required may increase to the extent that some plastic deformation is possible. With plastic deformation occurring, the particle cannot be ground further; consequently, a limit of fineness in grinding exists. This limit is reported to be 1 μm for quartz and 3 to 5 μm for limestone.

The possibility that new flaws are created during the fragmentation process has not been considered by past investigators. However, it can expected that movement and merger of dislocations and other defects can produce additional flaws and can retard the tendency of increasing fracture strength with decreasing particle size. Another overlooked factor is that the amount of elastic energy that must be stored to propagate a crack is limited to the volume of the particle, and very small particles may not have sufficient stored energy.

The effect of rate of stress application is reported to be a gradual increase in the fracture stress initially when the rate is increased from that of slow compression tests to that of fast compression tests. Fracture stress further increases with low-velocity impact but then decreases when the velocity of the impact is further increased.[2] The initial increase is attributed to larger plastic deformation before failure when a higher rate of compression is used.

Increase in temperature of the fracture environment can be expected to cause an increase in plastic deformation. In this regard it is interesting to note that even under low-temperature conditions, the temperature near a propagating crack-tip can be very high owing to release of large quantities of energy in the form of heat. The magnitude of this energy is 10 to 10^5 times that of the surface energy requirement for fracture.[3] The temperature at the crack tip under such conditions can even be above the melting point.[3] After the propagation of the fracture, such a region, probably of the order of 100 A to 1 μ, cools down rapidly, freezing amorphous or other high-energy structures at the fracture surface.

Newly forming, high-energy surfaces can react with the surrounding environment if the possible rate of penetration of the environment is equal to or larger than the speed of crack propagation. Figure 11.1 shows the effect of water vapor on the specific crack-extension energy of glass to be significant, but only when the crack velocity is less than about 0.1 to 10^{-4} cm/second. Below such crack velocity, cracking is assisted by water penetration

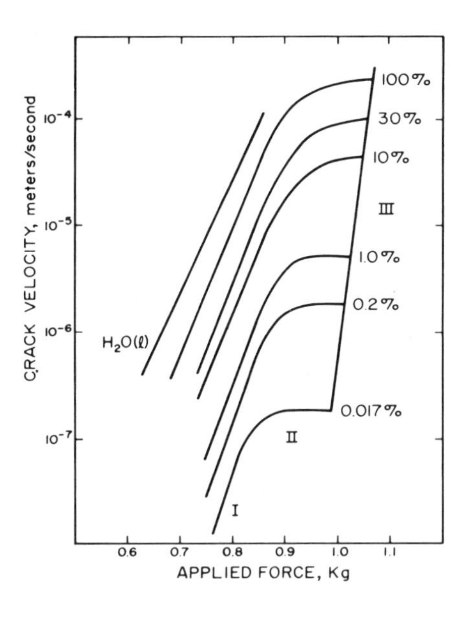

Figure 11.1. Crack velocity v in soda-lime glass as a function of the applied force, with relative humidity shown for each line. [S. M. Wiederhorn, *J. Amer. Ceram. Soc.* **50**, 407 (1967).]

followed by "hydrolytic stress corrosion," whereas above this velocity water penetration is apparently not fast enough for it to be in contact with the advancing crack tip.

GRINDING MECHANISMS

During grinding several particles are simultaneosly and repeatedly subjected to stress application in the grinding zone. With each stress application, several fractures may occur in each particle. Theories of crack initiation and propagation discussed above are therefore compounded by the interaction of flaws in a particle, secondary breakage, interaction of particles with each other and with the surface of the container, secondary interactions between particles and the grinding media, and physical and chemical interactions between particles and the grinding environment. In addition, the type of transport of the material through the grinding zone and size classification of it, if any, in the mill will also affect the nature of the product obtained. The present understanding of the above interactions is limited and in several cases is intuitive or speculative.

The distribution of cracks and the interaction among them during propagation will essentially determine the size distribution of the particles obtained during fracture of a single particle. Interaction of a propagating crack with a dormant crack can accelerate, decelerate, or terminate the propagation and can possibly activate the dormant one during the process, depending possibly on the relative orientation between them. Shock waves generated in the specimen as a result of stress application or crack propagation can also activate dormant cracks. Shock waves that diverge from a peripheric point of contact in a sphere or disc can be focused at a point a distance of one-third the diameter from that point on the symmetrical axis to cause tensile fracture.[5]

Fracture fragments possessing sufficient kinetic energy can undergo further fragmentation by impacting on the walls of the container. Slow compression tests on single spheres have shown that the particles, when allowed to undergo secondary breakage, produced 3.6 times as much surface area for a given specific energy input than those subjected to single fracture when they were embedded in gelatin.[6] Primary fragments rich with flaws have been found to degrade much more readily than a new sphere comparatively free of surface flaws.[7,8] During grinding a significant amount of secondary breaking can occur as a result of the impact of primary fragments with the grinding media and container wall, as well as with other particles.

Interaction among particles can be significant, depending on the grinding mechanism, relative hardness and size of particles, and the extent of size classification during grinding. A number of interaction phenomena can take place. Particles can nip each other and thereby remove chips, releasing a certain amount of elastic stress. They can also reduce their rugosity and even change their shape by abrasion, which in turn will alter the effects of subsequent comminution events. Most importantly, the presence of fines in the mill can reduce the grinding efficiency considerably because of the cushioning effect produced by a bed of fine particles.[8] Energy is wasted in deformation and flow of the bed of fines. The above effect is more prominent under dry-grinding conditions than under wet-grinding conditions and all the more so for ball milling. In a rod mill coarse particles are considered to be preferentially ground, and the fines are protected by the wedging of the rods by the coarse particles.[8]

In addition to these physical interactions, chemical reactions have also been shown to occur, as in the case of prolonged grinding of a mixture of massicott and sulfur.[9]

While impact-type comminution will produce fragments with a more normal size distribution, nipping by the grinding media or the other particles will produce a distribution with more coarse particles. Also, abrasion will produce a distribution with more ultrafine particles. Intense point loading produces fine fragments of the intensely stressed region and much larger fragments of the remaining material.[2] The combination of the above types of fragmentation yields a product characteristic of the ground material and the grinding media,[10] as each type is active to a different extent with different machines and materials.

SIZE DISTRIBUTION

The size distribution of powders is most conveniently represented by comparing y versus x, where x is the size of the particle and y is the cumulative weight percent of all particles finer than size x.

A plot of log y verus log x often yields a straight line for the range $10 < y < 40$. However, for y values below 10 and above around 40, the linearity in the plot no longer applies. To describe the distribution over a broader range mathematically, several equations have been proposed.[11-21] Some are largely empirical fits to data, while others are based on models involving distribution of Griffith flaws,[12-16] probability of survival under constant stress,[11] and statistical consideration of random division of particles.[18-20] Most equations are of the form

$$y = 1 - f(x/a)$$

where $f(x/a)$ is a function that involves exponents or power relationships and constants that determine the average size and breadth of the distribution. The various equations fit specific situations and no universal equation has yet been widely accepted.

RATE OF GRINDING

The size reduction as a function of time has been considered mainly in terms of the energy consumption. If the rate of energy consumption is constant with time, then the time and total energy consumption are directly proportional to each other. Major relationships between particle size and time or energy have been recently discussed by Agar and Somasundaran.[22] Attempts have been made to relate energy consumption to surface area,[23,24] volume, or weight of the particles,[25] size and size distribution parameters,[3,26] and fracture stresses.[27] The more recent treatments satisfy specific experimental observations that were considered by the investigator, but none appear widely acceptable for all conditions.

Other approaches have involved the monitoring of the rate of disappearance of material coarser than a certain size[28,29] or the rate of production of fine material from a narrowly sized feed.[30,31] In some cases the forms of the equations can be shown to be equivalent to certain size versus time relationships mentioned previously.

Changes in the entire size distribution of ground material with time have been studied[32-48] extensively. One of the simpler equations presented is one obtained by curve fitting a three-parameter equation[49]:

$$y(x, t) = 1 - \left[1 - \left(\frac{x}{x_0}\right)^s\right]^{\frac{(t)p}{t_0}}$$

where t is time and s, p, and t_0 are constants. This equation fits several sets of data in the literature, particularly those for ball milling.

ULTRAFINE GRINDING

As grinding proceeds into an ultrafine region, it becomes more and more difficult to obtain further reduction in size because a grind limit is approached. A practical grind limit exists for most systems. This is most importantly determined by the tendency of the product particles to re-aggregate and establish a physical equlibrium between aggregation and fragmentation. In addition, the probability of a particle becoming involved in a comminution event, as well as that of its fracture when it is involved in an event, decreases with decreasing particle size. Depletion of flaws during the grinding as well as a decrease in volumetric capacity to store elastic stress energy, increases the required stress for initiating fracture. Difficulty in obtaining particles below a limiting size has also been attributed to "excessive clearance between impacting surfaces," diminished utilization of energy due to transmission of forces "through a long chain of particles few of which suffer sufficient strain to shatter," semifluid nature of the final product, and protection of smaller particles by the larger ones.[50]

Additional flaws can probably be generated in the fine particles by applying thermal shocks. It is of interest to note that laser techniques, even though still uneconomical, have been studied for reducing the drilling strength of rocks.[51]

Aggregation can be retarded using a number of techniques. It can be minimized by removing the finest particles continuously using closed-circuit grinding. It is also considered advisable to successively reduce the size of the grinding medium (balls, pebbles, etc.) as the grinding proceeds into fine and ultrafine regions, since the ratio of the size of the grinding medium to the size of the particles should be kept within certain limits for maximum grinding efficiency. Cooling the machines by improving the ventilation or by external or internal watersprays minimizes agglomeration due to rising temperature.[52] Addition of dispersive chemical agents is found to be beneficial for ultrafine grinding.[53] Both grinding aids and grinding liquids have significant influence on ultrafine grinding.[54] Inorganic salts with multivalent ions or complex anions were found to be the most effective grinding aids. Adsorption of multivalent ions on the particles to increase the electrical repulsion between them is possibly the major reason for their influence. Formation of brittle or corrosive surface films has also been considered as a reason for their effect.

An "attritor ball mill" in which balls are rotated with a stirring arm was found to be significantly more efficient than use of a standard ball mill for fine grinding.[54] Milling time was 5 to 10 times less with the attritor ball mill for the same output. This grinding method is stated to be good also for alloy powders and refractory compounds. Vibration grinding,[55-57] centrifugal grinding,[58] and fluid energy milling[59] have been considered recently for efficient fine grinding. Grinding time with a two-chamber (1.5 in. wide diameter) experimental vibratory ball mill was found to be less than 3 hours for producing graphite fines with a specific surface area of 400 m^2/g.[55] This is to be compared with 26 hours of grinding with a 15 in. ball mill for the same surface-area development. Theories of rate of vibrational milling have been presented along with experimental analysis of the theories.[60]

Relations described earlier become usually inapplicable in the fine-size region. The following equation, which takes into account the possible existence of a grind limit, was proposed[61]:

$$S = S_m\,[1 - \exp(-KE)] \tag{20}$$

where F is again the energy input, S is the specific surface area reaching a limiting value S_m, and K is a constant. Harris[62] has a more general equation in which E is raised to a constant. The data fit however, is, better with another relation of Harris', developed on the basis of a logistic growth function model for fine grinding:

$$\frac{\Phi_m}{\Phi} = 1 + \left(\frac{h}{t}\right)h'$$

t is grinding time and h and h' are two constants. Φ is a measure of fineness reaching a maximum value of Φ_m. It is assumed to be proportional to specific surface area or to the inverse of size modulus. None of these relations, however, represent cases where the Φ_m exists at $0 < t < \infty$.

It is important to note that surface area and related surface activity are the parameters of importance in a system using very fine particles. It is therefore most meaningful to represent the fine-grinding system by surface area versus time or even cumulative surface area versus size models instead of weight versus time or cumulative weight versus size models.

MECHANOCHEMICAL EFFECTS

Both physical and chemical characteristics have been found to undergo significant changes during powder preparation by prolonged grinding.[9] There is sufficient evidence in the literature[53,63] that, in addition to desired and expected changes in physical properties such as specific surface area, changes occur also in shape, sintering activity, chemical reactivity, and so forth. The exact nature of alterations is dependent on, among other things, the conditions of grinding and the method of grinding used. Using gravimetric, thermogravimetric X-ray diffraction and electron microprobe analyses, the change in various properties of samples of quartz, calcite, and massicot ground in a pebble mill for several hundred hours was studied.[9] It was found that the density of the particles decreased in the case of quartz as a function of grinding time (or particle size) apparently owing to the creation of deep amorphous layers on the particles. This effect however, was, absent when the grinding was done in water, possibly because of continuous dissolution of amorphous material in water. More than half of the quartz particles with diameters of 200 μ was converted to amorphous quartz during prolonged ball milling. Such effects have been also reported in several other cases.[64-66] During grinding of hematite, a distinct change in color was observed owing to the production of ferrous oxide at the surface.[65] Work with clay minerals has provided additional evidence for the effects of prolonged grinding. Bloch,[67] for example, found that prolonged grinding of montmorillonite caused disruption of its crystalline structure and release of some alumina and magnesia. Reactivity of materials is in general found to be enhanced by comminution, as in the case of kaolinite and molybdenum sulfide, the latter becoming particularly reactive towards oxygen when ball milled in air. Researchers on catalytic activity of metallic oxides, ionic crystals, metal sheets, and wires have observed a favorable increase in such activity due to grinding. Ceramic powders activated by prolonged grinding are reported to sinter more readily. Snow and Luckie[63] and Naeser and Fielder[68] have recently reviewed these effects. Increase in catalytic and sintering activities has been proposed to occur as a result of an increase in the dislocation density of the materials. Even though no simultaneous work on both reactivity and dislocation density has been carried out by any of the above workers, it is not unreasonable to expect a change in sample reactivity owing to an increase in dislocation density that naturally occurs during most mechanical treatments.

Polymorphic transitions have been reported to occur during grinding[9,69-73] possibly owing to temperature and pressure changes that take place locally as a result of the grinding process. Thus, during grinding or massicot and calcite, such transitions altered the structure of these materials to those of litharge and aragonite, respectively.[9]

In addition to the above physical and structural changes, even solid-state reactions have been reported to occur during grinding. Figure 11.2 shows the changes in the massicot–sulfur system when it is ground for several hours.[9] Massicot converts itself slowly to litharge up to about 20 hours; a chemical reaction then follows in which the oxide in both forms and sulfur are converted to galena. Such solid-state reactions have been reported to be most prominent during the grinding of carbonates. For example, zinc carbonate[65] and cadmium carbonate[74] with relatively low decomposition temperatures give carbon dioxide by mere grinding at room temperature. In the case of carbonates such as magnesium with higher decomposition temperatures, prolonged dry grinding lowered their decomposition temperatures significantly. Another important example of chemical decomposition during grinding is that of $Na_5P_3O_{10} \cdot 6H_2O$ to form ortho- and pyrophosphates.[75] Several hydrated salts, such as $FeSO_4 \cdot 7H_2O$ and $BaCl_2 \cdot 2H_2O$, have been found to decompose during grinding.[76] These reactions are most prominent when the mill atmosphere is dry. An interesting complete chemical reaction that has been discovered to occur as a result of grinding is that between black lead sulfide and white cadmium sulfate to form white lead sulfate and yellow cadmium sulfide, the progress of the reaction being indicated by the gradual change in color.

A point to note is that contamination of the samples with the grinding medium can also occur during prolonged grinding. Mullite contamination

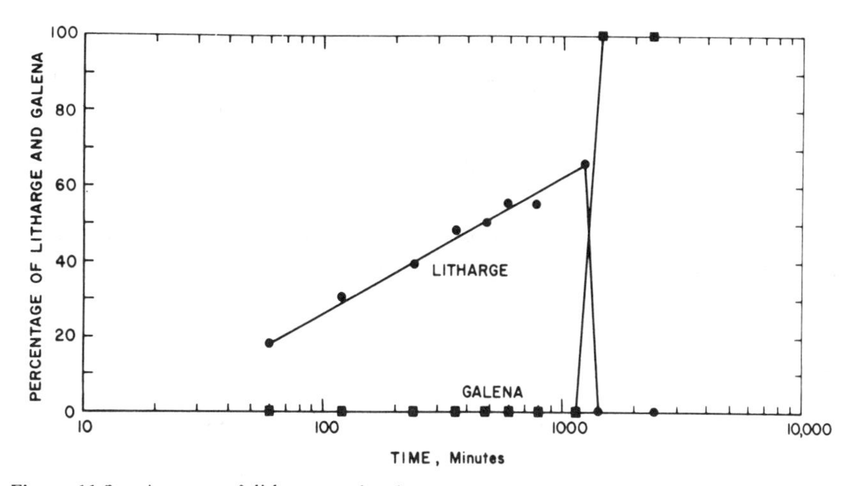

Figure 11.2. Amount of litharge and galena formed during grinding in a massicot–sulfur system.[9]

was found in a sample ground with a mortar and pestle.[77] The implications of these effects on powder preparation for various purposes, including sample preparation by grinding for chemical analysis, must be recognized.

GRINDING AIDS

Use of grinding aids and the mechamisms by which they act are discussed in a recent review by Somasundaran and Lin.[53] Grinding aids in vapor form, such as ethylene glycol, proplylene glycol, butylene glycol, are used commercially in Germany and Yugoslavia for improving the efficiency of cement grinding.[78-80] Amine acetates and diethylene glycol are used in Japan as grinding aids in plant-size mills.[81,82] Any such attempt to improve the grinding efficiency is highly desirable, since the current grinding operations are notorious for their efficiencies, which are of the order of 1%. Parts of our review[53] relevant to grinding aids are given below.

WATER (MOISTURE)

Grinding in water in usually more efficient than dry grinding.[83-85] This effect of water has been ascribed by Lin and Mitzmager[86] to a reversible reaction between unsatisfied surface bonds and water molecules. Water in the form of vapor also should be expected to produce such hydrolytic corrosive effects. Even though there is no grinding work in the literature reported as a function of humidity, some evidence exists that this factor does affect the process. For example, the grinding rate of soda lime glass is higher in humid air than in a vacuum.[87] It must be noted that the increased efficiency of wet grinding can also be due to physical reasons. Cushioning effects due to the presence of a bed of fines will be less during wet grinding than during dry grinding, since the fine particles tend to remain suspended in the water in the former case. This would of course cause an increase in the efficiency of the grinding. In addition, effects of viscosity and specific gravity of the medium can also be significant.[88,89]

Organic Liquids

Grinding in organic liquids is reported to be more efficient than in water. A 12-fold higher production of surface area for grinding in organic liquids, such as isoamyl alcohol, than for that in water was shown.[90] Higher grinding rates were obtained in carbon tetrachloride and methylcyclohexane than in nitrogen.[86] An interesting observation was that the grinding efficiency was lower in the two organic liquids than in water but became the same when small amounts of water were present in the organic liquids in dissolved form.

Surface Active Agents

Surfactants have been widely reported as effective grinding aids. The effect of adding a flotation agent called Flotigam P on wet-ball milling of quartzite and limestone[91] is shown in Figure 11.3 as an example. It can be seen that as much as 100% increase in specific surface area was obtained by additions of up to 0.3%. Additions in amounts higher than 0.03% caused a decrease in specific surface area. Sodium oleate in large concentrations has also been reported to produce a net decrease in specific surface area. The effect of Armac T on the grinding of quartz in a ball mill[92] is shown in Figure 11.4. The effect is detrimental under all concentrations studied. It is not yet known whether these detrimental effects are due to experimental artifacts introduced by the aggregation of fines or are the result of change in interfacial properties due to adsorption of surfactant adsorption on particles. Flocs in the mill during grinding could consume some of the impact energy for deflocculation. In addition, hydrophobization of particles by the adsorbed surfactants can result in the attachment of air bubbles to them and consequent levitation. The grinding efficiency can be expected to be lower if the particles remain levitated. The beneficial effect of these grinding aids has been considered to be due to the reduction in surface energy upon

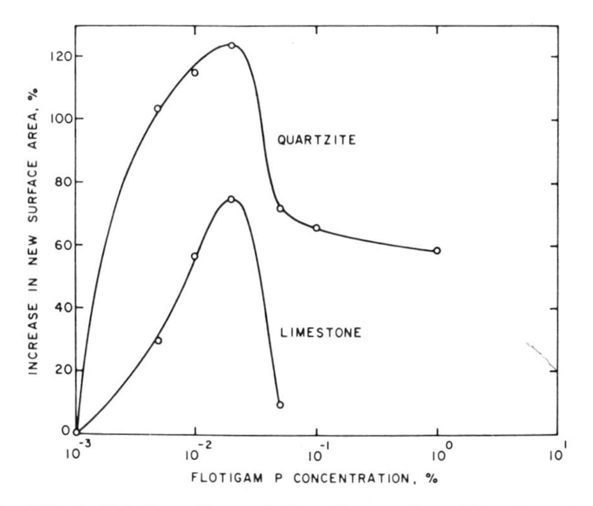

Figure 11.3. Effect of Flotigam P on grinding of quartzite and limestone in a rod mill.[53]

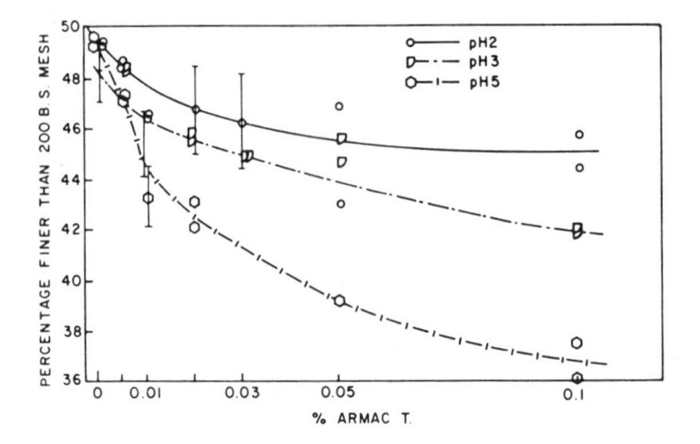

Figure 11.4. Effect of Armac T on the comminution of quartz in a ball mill at different pH values.[92]

their adsorption, making it easier to produce new surfaces under such conditions. This reasoning is in line with the explanation offered by Rehbinder and coworkers for the observed increase in drilling rate on the addition of these agents. On the other hand, it is also possible that it is an indirect result of several other phenomena that could occur in the system, such as the interaction of the surfactant molecules adsorbed on the surface and the resultant effects on various interfacial properties. In some cases it could also be the result of the ability of the reagents to enhance the dispersion of the particles and thus indirectly to facilitate fragmentation.

Examples of grinding aids used in the past include polysiloxane in the grinding of ultraporcelain and talc; silicones in the drop-weight crushing of limestone and quartz; glycols, amines, organosilicones, organic acetates, carbon blacks, and wool grease in the grinding of cement; silicones in the ball milling of quartz; acetones in nitromethane benzene, carbon tetrachloride and hexane in vibratory milling of ground glass, marble and quartz; and wool grease in the milling of gypsum, limestone, and quartz. Some of these reagents are reported to act by preventing ball coating but not aggregation of particles.[93] This effect of additives is very beneficial, since it is known that ball coating impairs the grinding efficiency.

Inorganic Electrolytes

Use of inorganic electrolytes during drilling was noted in the forties by Rehbinder et al.[94] to increase the efficiency significantly. A number of workers have attempted to establish corresponding effects during grinding.[95-100] Even though all the reported results are not in agreement, grinding is in general found to be more efficient in the presence of inorganic electrolytes. In the ceramic industry, grinding of metallic and refractory-type materials is found to be more efficient when multivalent electrolytes are used as additives.[101-109] Effect of $AlCl_3$ and $CuSO_4$ on wet grinding is shown in Figure 11.5. Both reagents aid the grinding, the extent of the effect being determined both by the valency of the active ion of the salt and the manner of grinding. In some cases use of salts above certain levels is reported to cause poor grinding.[96,99] In addition to any effect that these electrolytes might have on the hardness of the materials, the influence on the flocculation or dispersion of particles is also possibly a major reason for their overall observed effect.[53] This reasoning is supported by work that showed that the addition of dispersing agents always improved the comminution of solids.[95]

Physical Nature of the Environment

Properties such as viscosity and density of the grinding environment can be expected to have an effect on the hydrodynamic behavior of particles, as well as of the grinding medium, and therefore on the grinding performance. Grinding is normally dependent on the viscosity of the medium up to about 20,000 mill revolutions and is then independent.[110] Similar results were found for the viscosity of the environment.[89] The effect of density of the suspending fluid by grinding quartz and pyrite in air, water, and tetrabromoethane has been studied.[88] Grinding in water was more efficient than in air. The results obtained in tetrabromoethane however, were, inferior,

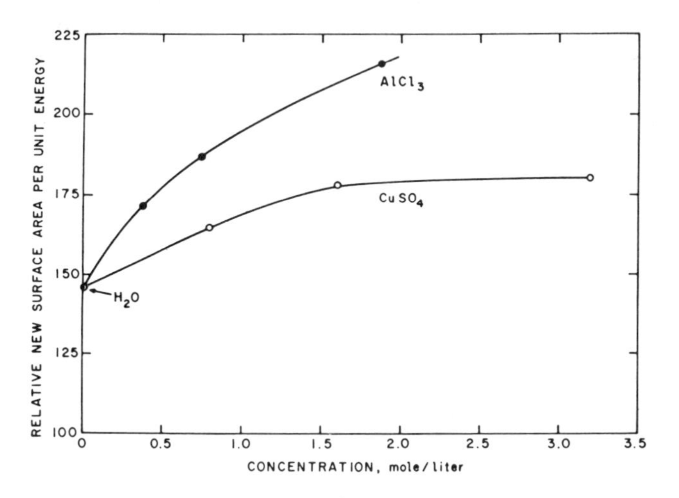

Figure 11.5. New surface area produced per unit energy applied during mill grinding in aqueous solutions as a function of $CuSO_4$ and $AlCl_3$ concentrations.[53]

possibly because quartz particles, which are lighter than tetrabromoethane, remained partially afloat and thus out of the path of the impacting grinding media.

Mechanisms

The main two mechanisms proposed to explain the effects of comminution aids are Rehbinder's[111] mechanism based on adsorption-induced surface energy changes and Westwood's[112,113] mechanisms based on adsorption-induced mobility of near-surface dislocations. Rehbinder's mechanism is based on the concept that a fruitful comminution event involves the production of new surfaces and that to accomplish this an amount of energy proportional to the free energy of the surface should be spent. Addition of chemical agents to reduce the effective surface energy of the solid particles should, on the basis of the above concept, enhance the grinding process. According to Westwood, the Rehbinder effect is more likely due to changes in the electronic states near the surface and point and line defects caused by the adsorption of the additives on the solid. Such changes are known to influence the specific interactions between dislocations and point defects that control the dislocation mobility and hence the hardness. Hardness of materials has been recently proposed by Westwood et al. to be controlled by the zeta potential of the material in solutions.[113] Since addition of surface active agents does affect the zeta potential significantly, it can be then expected to affect the grinding performance.

On the basis of the results obtained during abrasion and grinding tests on cement clinker in the presence of organic liquid vapor, it was found that the effectiveness of the vapor grinding aids in industrial ball milling is due to reduction of adhesive forces leading to prevention of aggregation of powder and of coating of balls and liners.[78] Influence of adsorption of vapors on plastic deformation itself is considered to be significant only for cases where stresses are applied for long intervals and for grinding where stresses are applied rapidly.

REFERENCES

1. A. A. Griffith, "The Phenomena of Rupture and Flow in Solids," *Philos. Trans. Roy. Soc. London, Ser. A* **221,** 163–198 (1920–1921).
2. C. C. Harris, "On the Role of Energy in Comminution," *Trans. Inst. Mining Metall. (London)*, **75,** C37–C56 (1966).
3. H. Rumpf, "Physical Aspects of Comminution and New Formulation of a Law of Comminution," *Powder Tech.*, **7,** 145–149 (1973).

4. H. Schonert, H. Umhauer, and W. Klemm, *Proc. Int. Conf. Fracture 2nd Brighton,* **1969,** p. 474.
5. H. H. Gildemeister, and K. Schonert, "Berechnung Zur Wellenausbereitung in Kugeln Und Bruchphanomente in Kreisscheiben Bei Prallbeanspruchung," 3rd European Symposium on Comminution, paper 1–8, Canner, October 1971; *Dechema Monogr.,* 69 (1972).
6. B. H. Bergstrom, C. L. Sollenberger, and W. Jr. Mitchell, "Energy Aspects of Single Particle Crushing," *Trans. AIME,* **220,** 367–372 (1961).
7. G. L. Fairs, "A Method of Predicting the Performance of Commercial Mills in the Fine Grinding of Brittle Materials," *Trans. Inst. Mining Metall.* (*London*) **63,** 211 (1953).
8. P. Somasundaran and D. W. Fuerstenau, "Preferential Energy Consumption in Tumbling Mills," *Trans. AIME,* **229,** 132–134 (1963).
9. I. J. Lin and P. Somasundaran, "Alterations in Properties of Samples During Their Preparation by Grinding," *Powder Tech.,* **6,** 171–179 (1972).
10. D. D. Crabtree, R. S. Kinasevich, A. L. Mular, T. P. Meloy, and D. W. Fuerstenau, "Mechanisms of Size Reduction in Comminution Systems 1. Impact, Abrasion, and Chipping Grinding," *Trans. AIME,* **229,** 201–206 (1964).
11. H. Heywood, "Principles of Crushing and Grinding," *Chemical Engineering Practice,* H. W. Cremer, and T. Davies, eds., Butterworths, London, Vol. 3, (1956), pp. 1–23.
12. J. J. Gilvarry and B. H. Bergstrom, "Fracture of Brittle Solids: Pt. 2—Distribution Functions for Fragment Size in Single Fracture, (Experimental)," *J. Appl. Phys.,* **32,** 391–399 (1961).
13. J. J. Gilvarry and B. H. Bergstrom, "Fracture of Brittle Solids: Pt. 3—Experimental Results of the Distribution of Fragment Size in Single Fracture," *J. Appl. Phys.* **33,** 3211–3213 (1962).
14. J. J. Gilvarry, "Fracture of Brittle Solids Two Dimensional Distribution Function for Fragment Size in Single Fracture: Pt. 4 (Theoretical), Pt. 5 (Experimental)" *J. Appl. Phys.* **33,** 3214–3217 and 3218–3224 (1962).
15. J. J. Gilvarry and B. H. Bergstrom, "Fracture and Comminution of Brittle Solids (Theory and Experiment)," *Trans. AIME,* **220,** 380–390 (1961).
16. J. J. Gilvarry, and B. H. Bergstrom, "Fracture and Comminution of Brittle Solids: Further Experimental Results," *J. Appl. Phys.,* **223,** 419 (1962).
17. E. Evans, "A Contribution to the Theory of the Size Distribution of Broken Coal," *Proc. Conf. Sci. Use Coal, London,* Institute of Fuel, **1958,** paper 14.
18. A. M. Gaudin and T. P. Meloy, "Model and a Comminution Distribution Equation for Repeated Fracture," *Trans. AIME,* **223,** 43–50 (1962).
19. T. P. Meloy, "A Three Dimensional Derivation of the Gaudin Size Distribution Equation," *Trans. AIME,* **226,** 447–448 (1963).
20. R. R. Klimpel and L. G. Austin, "The Statistical Theory of Primary Breakage Distributions for Brittle Materials," *Trans. AIME,* **232,** 88–94 (1965).
21. C. C. Harris, "The Application of Size Distribution Equations to Multi-Event Comminution Process," *Trans. AIME,* **241,** 343–358 (1968).
22. G. E. Agar and P. Somasundaran, "Rationalization of Energy—Particle Size Relationships in Comminution," *XthInt. Miner. Proc. Cong. 10th, London,* **1973,** Institute of Min. and Metallergy, paper 16.
23. P. R. Von Rittinger, *Lehrbuch der Aufbereitungskunde, Ernstand Korn, Berlin, 1867.*

24. F. C. Bond, "The Third Theory of Comminution," *Trans. AIME,* **193,** 484–494 (1952).

25. F. Kick, *Das Gesetz Der Proportionalem Widerstand und Seine Anwendung,* Arthur Felix, Leipizg, 1885.

26. R. J. Charles, "Energy—Size Reduction Relationships in Comminution," *Trans AIME,* **208,** 80–88 (1957).

27. J. A. Holmes, "A Contribution to the Study of Comminution—A Modified Form of Kick's Law," *Trans. Inst. Chem. Eng.,* **35,** 125–141 (1957).

28. N. Arbiter and C. C. Harris, "Particle Size Distribution/Time Relationships in Comminution," *Br. Chem. Eng.,* **10**(4), 240–247 (1965).

29. F. W. Bowdish, "Theoretical and Experimental Studies of the Kinetics of Grinding in a Ball Mill," *Trans. AIME,* **217,** 194–202 (1960).

30. N. Arbiter and U. N. Bhrany, "Correlation of Product Size, Capacity and Power in Tumbling Mills," *Trans. AIME,* **217,** 245–252 (1960).

31. D. W. Fuerstenau and P. Somasundaran, "Comminution Kinetics," *Proc. Int. Miner. Process. Congr. Cannes,* **1963,** Pergamon, pp. 25–33.

32. G. F. Hutting, "Zur Kinetik Der Zermahlungsvorgange," *Z. Elektrochem.* **57,** 534–539 (1953).

33. O. Theimer, "Uber Die Kinetic Von Zermahlungsvorg-Ange," Kolloid Z., **132,** 134–141 (1953).

34. L. Bass, "Zur Theorie Der Mahlvorgange," *Z. Angew, Math. Phys.,* **5,** 283–292 (1956).

35. B. Epstein, "Logarithmico—Normal Distribution in Breakage of Solids," *Ind. Eng. Chem.,* **40,** 2289–2291 (1948).

36. R. P. Gardner, and L. G. Austin, "The Use of a Radioactive Tracer Technique and a Computer in the Study of the Batch Grinding of Coal," *J. Inst. Fuel,* **35,** 173–177 (1962).

37. R. P. Gardner, and L. G. Austin, "A Chemical Engineering Treatment of Batch Grinding," *Symposium Zerkleinern,* H. Rumpf and D. Behrens, eds., Verlag Chemie, Duseldorf, W. Germany., 1962, pp. 217–248.

38. C. C. Harris, "Batch Grinding Kinetics," *Trans. AIME,* **241,** 359–364 (1968).

39. C. C. Harris, "Size Reduction—Time Relationships of Batch Grinding," *Trans. AIME,* **241,** 449–454 (1968).

40. P. C. Kapur, "A Similarity Solution to an Integro-Differential Equation Describing Batch Grinding," *Chem. Eng. Sci.,* **25,** 899–901 (1970).

41. D. F. Kelsall and K. J. Reid, "The Derivation of a Mathematical Model for Breakage in a Small, Continuous, Wet, Ball Mill," *Application of Mathematical Models in Chemical Engineering Research, Design, and Production,* Institute of Chemical Engineers, London, 1965, pp. 14–20.

42. T. S. Mika, L. M. Berlioz, and D. W. Fuerstenau, "An Approach to the Kinetics of Dry Batch Ball Milling," *Zerkleinern,* 2nd Symposium on Comminution, H. Rumpf, and W. Pietsch, eds., Verlag Chemie, Weinheim, 1967, pp. 205–240.

43. J. A. Herbst, G. A. Grandy, and T. S. Mika, "On the Development and Use of Lumped Parameter Models for Continuous Open and Closed Circuit and Grinding Systems," *Trans. Inst. Mining Metall.* **80,** C193–C198 (1971).

44. J. A. Herbst, G. A. Grandy, and D. W. Fuerstenau, "Population Balance Models for the Design of Continuous Grinding Mills," 10th International Mineral Processing Conference, London, Institution of Mining and Metallurgy, 1973, paper 19.

45. P. C. Kapur and P. K. Agrawal, "Approximate Solutions to the Discretized Batch Grinding Equation," *Chem. Eng. Sci.*, **25,** 1111–1113 (1970).

46. J. A. Herbst and D. W. Fuerstenau, "The Zero Order Production of Fine Sizes in Comminution and its Implications in Simulation," *Trans. AIME,* **241,** 538–549 (1968).

47. J. A. Herbst and D. W. Fuerstenau, "Influence of Mill Speed and Ball Loading on the Parameters of the Batch Grinding Equation," *Trans. AIME,* **252,** 169–176 (1972).

48. G. D. Gumtz and D. W. Fuerstenau, "Simulation of Locked-Cycle Grinding," *Trans. AIME,* **247,** 330–335 (1970).

49. C. C. Harris, "Relationships for the x y t Comminution Surface," *Trans. Inst. Mining Metall.* **79,** C157–C158 (1970).

50. A. L. Mular, "Comminution in Tumbling Mills—A Review," *Can. Metall. Q.*, **4** (1), 31–73 (1965).

51. W. N. Lucke, "Novel Methods of Rock Fracture," *Mining Congr. J.*, 64–69 (Aug. 1973).

52. B. Becke, *Principles of Comminution,* Akademiai Kiado, Budapest, 1964.

53. P. Somasundaran and I. J. Lin, "Effect of the Nature of Environment of Comminution Processes," *Industrial and Engineering Chemistry and Proc. Des. and Dev.*, **11,** 321–331 (1972).

54. M. Quatinetz, R. J. Schafer, and C. R. Smeal, "The Production of Submicron Metal Powders by Ball Milling with Grinding Aids," *Ultrafine Particles,* Ed. W. E. Kuhn, Wiley, New York, 1963, pp. 271–296.

55. E. A. Smith, "Some Special Functions of the Vibrational Ball Mill," *Chem Ind.* (*London*), **1967** (34), 1436–1442.

56. H. L., Podmore and E. S. G. Beasley, "Vibration Grinding in Close Packed Media Systems," *Chem. Ind.* (*London*) **1967** (34), 1443–1450.

57. A. Kirk, "Practical Review of Vibration Milling," *Chem. Ind.* (*London*) **1967** (3), 1378–1382.

58. R. Planiol, "Vacuum-Operated Centrifugal Grinder is Proposed for Powder Savings," *Eng. Min. J.*, 140–142 (Sept. 1968).

59. B. Dobson and E. Rothwell, "Particle Size Reduction in a Fluid Energy Mill," *Powder Tech.*, **3,** 213–217 (1969/1970).

60. H. E. Rose, "Some Observations on the Application of Vibration Mills," *Chem. Ind.* (*London*) **1967** (33), 1383–1389.

61. Cited by A. L. Mular, "Comminution in Tumbling Mills—A review," *Can. Metall. Q.*, **4,** (1), 31–73 (1965).

62. C. C. Harris, "On the Limit of Comminution," *Trans. AIME,* **238,** 17–30 (1967).

63. R. H. Snow and P. T., Luckie, "Annual Review of Size Reduction—1973," *Powder Tech,* **10,** 129–142 (1974); Snow, R. H., "Annual Review of Size Reduction—1972," *Powder Tech.*, **7,** 69–83 (1973); R. H. Snow, "Annual Review of Size Reduction," *Powder Tech.*, **5,** 351–364 (1971/1972); see *Ind. Eng. Chem.* for previous annual reviews.

64. R. C. Ray, "The Effect of Long Grinding on Quartz," *Proc. Roy Soc. London, Ser. A,* **A102,** 640–642 (1923).

65. Y. J. Burton, "Change in the State of Solids Due to Milling Processes," *Trans. Inst. Chem. Eng.*, **44,** 37–41 (1966).

66. C. Legrand and J. Nicolas, "Contribution of X-ray Diffraction and the Electron Microscope to the Study of Ground Kaolins," *Bull. Soc. Fr. Ceram.*, **44**, 61–69 (1959).
67. J. M. Bloch, "Effect of Grinding on the Crystal Structures and Properties of Montmorillonite," *Bull. Soc. Chimique Fr.*, 774–781 (1950).
68. G. Naesear and A. Fielder, "Mechanical Activation of Solids and Its Technical Significance," *Verfahrenstechnik*, **6** (9), 299–305 (1972).
69. J. H. Burns and M. A. Bredig, "Transformation of Calcite to Aragonite by Grinding," *J. Chem. Phys.*, **25**, 1281 (1956), D. O. Northwood and D. Lewis, *Can. Miner.*, **10**, 216–224 (1970).
70. R. B. Gammage and D. R. Glasson, *J. Appl. Chem.*, **13**, 1466 (1963).
71. A. Schleede and H. Gantzckow, *Z. Physik*, **15**, 184 (1923).
72. G. L. Clark and R. Rowan, "Polymorphic Transitions by Grinding, Distortion and Catalytic Activity in PbO, *J. Amer. Chem. Soc.*, **63**, 1302–1305 (1941).
73. F. Dachille and R. Roy, *Proc. Int. Congr. on Reactivity of Solids, 4th*, Elsevier, Amsterdam, 1960; M. Senna and H. Kuno, *J. Amer. Ceram. Soc.*, **54**, 259–262 (1971).
74. G. Naeser and W. Scholz, *Ber. Deut. Keram. Ges.*, **39**, 106 (1962).
75. D. Ocepek, "Mechanical and Mechano-Chemical Reactions in Crushing Processes," *Rudarsko-Met. Zbornik*, (1), 5–16 (1969); E. A. Prodan, et al., "Decomposition of Sodium Triphosphate Hyxahydrate During Dry Grinding," *Dokl. Akad. Nauk Belorussk.*, **14**, (6) 526–529 (1970) [*Chem. Abstr.*, **73**, 72664z (1971)].
76. I. J. Lin and A. Metzer, "Changes in the State of Solids Due to Comminution," Technion Facility Civil Engineering Internal, Publication No. 140, 1970.
77. J. C. Jamieson and J. R. Goldsmith, "Some Reaction Produced in Carbonates by Grinding," *Amer. Mineral.*, **45**, 818–827 (1960).
78. F. W. Locher and H. M. V. Seebach, "Influence of Adsorption on Industrial Grinding," *Industrial and Engineering Chemistry Process Des. Dev.*, **11**, 190–197 (1972).
79. V. Korac, "Use of Grinding Auxiliary Agents in Cement Production," *Tehnika* (*Belgrade*), **27**, 649–653 (1972), *Chem. Abstr.* **77**, 922609.
80. K. Popovic, "Effects of Grinding Aids on Portland Cement Clinker," *Cement* (*Zagreb*), **15**, 14–17 (1971); *Chem. Abstr.* **76**, 144429p.
81. T. Furukawa, A. Anan, and K. Yamasaki, "Use of Grinding Agents in Closed Circuit Grinding System for Cement Clinker," *Semento Gijutsu Nempo*, **25**, 69–75 (1971); *Chem Abstr.* **77**, 14340g.
82. T. Furukawa, A. Anan, and K. Yamasaki, "Use of Grinding Agents in Closed Circuit Grinding System for Cement Clinker," *Semento Gijutsu Nempo*, **25**, 69–75 (1971); *Chem. Abstr.* **77**, 14341g.
83. F. C. Bond, *Mining Congr. J.*, 38–40 (January 1975).
84. W. H. Goghill, and F. D. Devaney, "Ball Mill Grinding," U.S. Bureau of Mines, Tech. Paper No. 581 (1937.
85. H. E. Rose, R. M. E. Sullivan, *Ball, Tube and Rod Mills*, Chemical Publishing Co., New York, 1958, p. 30.
86. I. J. Lin, and A. Mitzmager, "The Influence of the Environment on the Communition on Quartz," *Trans. AIME*, **241**, 412–418 (1968).
87. F. W. Locher, W. Eichartz, and H. M. von Seebach, S. Sprung, "Environmental Effects in Grinding," Extended Abstracts, 163rd National Meeting of the American Chemical Society, Boston, April 10–14, 1972, p. 81.

88. P. C. Kapur, A. L. Mular, and D. W. Fuerstenau, "The Role of Fluids in Comminution," *Can. J. Chem. Eng.*, **43,** 119–124 (1965).

89. W. A. Hockings, M. E. Volin, and A. L. Mular, Effect of Suspending Fluid Viscosity on Batch Mill Grinding," *Trans. AIME,* **232,** 59–62 (1965).

90. S. Z., Kiesskalt, *Ver. Deut. Ing.*, **91,** 313–315 (1949).

91. E. Von Szantho, "Der Einflux Von Oberflachenaktiven Stoffen Beider Feinzerkleinerung," *Erzbergbau Metallhuettenwes.* **2,** 353–360 (1949).

92. L. A. Gilbert, and T. H. Hughes, "Some Experiments in Additive Grinding," *Symposium Zerkleinern I,* Verlag Chem., Duseldorf, Germany 1962, pp. 170–193.

93. L. Opoczky, *Epitoanzag,* **19,** 121–125 (1967).

94. P. A. Rehbinder, L. A. Schreiner, and K. F. Zhigach, "Hardness Reducers in Drilling," Moscow Academy of Science, 1944; *Trans Counc. Sci. Ind. Res.*, Melbourne, Australia, 1948, p. 163.

95. G. Beyer, *Rudy* (*Prague*), **12** (7–8), 296–298 (1964).

96. J. H. Brown, "The Effects of Chemical Agents in Comminution,"M.I.T. Progress Report N.Y.O. 7172, 1955.

97. V. I. Byalkovskii and I. A. Kudinov, *Keram. Sb.*, 8–13 (1940); *Khim. Ref. Z.*, **4** (5), 114 (1941); *Ceram. Abstr.*, **22,** 121 (1943).

98. A. Z. Frangiskos, and H. G. Smith, "The Effect of Some Surface Active Reagents on the Comminution of Limestone and Quartz," *Progress in Mineral Dressing, Trans. Int. Miner. Dressing Conar, Stockholm, Sweden,* **1957,** 67–84.

99. R. Mallikarjunan, and K. M. Pai, and P. Halasyamani, "The Effect of Some Surface Active Agents on the Comminution of Quartz and Calcite," *Trans. Indian Inst. Metals,* **18,** 79–82 (1965).

100. M. H. Stanzyk, I. L. Feld, U.S. Bureau of Mines Report 7168, 1968, p. 28.

101. E. R. Dawley, *Pit Quarry,* **36,** 57 (1943).

102. V. H. Dodson, and F. G. Serafin, U.S. Patent 3,443,976 (May 13, 1969).

103. R. A. Knight, and C. A. Calow, AWRE Report 039-068, United Kingdom, April 1968.

104. M. Quatinetz, R. J. Schafer, and C. Smeal, *Trans. AIME,* **221,** 1105–1110 (1961).

105. R. J. Schafer, and M. Quatinetz, U.S. Patent 3,090,567 (May 21, 1963).

106. M. J. Sinnott, "The Influence of Surfaces on the Properties of Materials," in *Properties of Crystallizing Solids,* ASTM Technical Publication, No. 283, Philadelphia, 1961, pp. 28–39.

107. F. G. Serafin, U.S. Patient 3,443,975 (May 13, 1969).

108. F. G. Serafin, U.S. Patient 3,459,570 (August 5, 1969).

109. T. Tanaka, *Zement Kalk-Gibstein,* **15,** 28 (1962).

110. H. E. Schweyer, *Ind. Eng. Chem.*, **34,** 1060–1064 (1942).

111. P. A. Rehbinder, "On the Effect of Surface Changes on Cohesion, Hardness, and Other Properties of Crystals," *Proc. Phys. Congr. State Press, 6th Moscow,* **1928,** 29.

112. A. R. C. Westwood, "Environment Sensitive Mechanical Behavior, Status and Problems," in *Environment-Sensitive Mechanical Behaviors,* A. R. C. Westwood and N. S. Stoloff, eds., Gordon and Breach, New York, 1966, pp. 1–65.

113. N. H. MacMillan and A. R. C. Westwood, "Surface Charge-Dependent Mechanical Behavior of Non-Metals," Office of Naval Research Project NR-032-524, RIAS, Martin Marietta Corp., Baltimore, September 1973.

12

Reactive Powders from Solution

D. W. Johnson, Jr.
P. K. Gallagher

A variety of techniques for preparing ceramic powders involve a liquid solution as an initial step. The aqueous or organic-based solutions contain dissolved salts with appropriate cations or anions. The solvent is removed, leaving a residue that is the final powder or that is thermally converted to the desired form.

The most critical step in solution techniques is the removal of the solvent. The solutions, if well prepared, are homogeneous on at atomic scale. The degree of homogeniety of the residue, however, depends on the mechanism of the solvent-removal technique.

Although generally straightforward, solution preparation requires careful consideration of the solubility of the critical cations and anions. Problems can arise, for example, in preparing barium ferrite; ferrous sulfate and barium nitrate salts cannot be used because the highly insoluble barium sulfate phase would precipitate out.

This chapter describes a number of solution techniques for preparing powders. The techniques involve solvent vaporization, solution combustion, or precipitation–filtration.

SOLVENT VAPORIZATION

The solute is separated from solution by removing the solvent as a vapor phase, either by evaporation or sublimation.

Direct Evaporation

Solvent can be removed using as simple a technique as heating in a beaker on a hot plate. Direct evaporation may be suitable for single-component solutions. With multicomponent solutions, large-scale segregation of the components often results. An exception to this problem is when a solid solution of multicomponent salts is formed. Ferrites have been prepared by using ammonium sulfate (schoenite type) salts that form solid solutions when precipitated.[1] Another exception involves a technique in which a filter paper is soaked, dried, and ashed to give presumably equilibrated phases in the subsolidus region of the $NaAlO_2$–Al_2O_3 system.[2] The filter paper aids in preventing large-scale segregation of the components.

Spray Drying

This common technique prevents segregation during evaporation by breaking the liquid into very small droplets. This promotes rapid evaporation to minimize segregation and assures that any segregation is confined to the small droplets, since no mass transport from droplet to droplet takes place. Spray driers usually consist of an atomizer, a drying chamber, and a powder collector. The atomizer is either a pneumatic nozzle or a centrifugal disc. Droplet diameters are on the order of 10 to 20 μm. The droplets are intercepted in the drying chamber by a heated air stream that quickly evaporates the solvent. The dried powder is collected in a cyclone separator.

Ferrite compositions and magnesium aluminate have been successfully produced from sulfate solutions using spray drying.[3] Decomposing the sulfates at 800 to 1000°C gave agglomerates with crystallites about 0.2 μm in diameter. These could be pressed and sintered to practically the theoretical density. Nickel–zinc ferrites have also been prepared by spray drying sulfates.[4] Sintered nickel–zinc ferrites had lower coercive force, higher initial permeability, and more-controlled stoichiometry and grain size when made from spray-dried powder rather than from conventionally ball-milled powders.

By increasing the temperature of the gases in the spray drier the salts can be dried and decomposed directly in the spray drier in a single step. This technique is termed "spray roasting" and has been used for the preparation of ferrite powders using nitrates or chlorides.[5,6]

Fluid Bed Drying

This technique is similar to spray roasting except that the solution is trapped in a fluid bed for drying and decomposition. A pneumatic nozzle injects solution droplets into a heated fluid bed of the solid product. By varying the temperature, fluidizing gas flow rate and solution injection rate, the process could be made continuous, with a portion of the calcined product being removed from the bed periodically. It has been used to prepare and calcine uranyl nitrate and aluminum nitrate.[7]

Advantages of fluid bed drying include: (1) the absence of moving parts, (2) large capacity per unit volume of equipment, and (3) the formation of granulated powders of relatively high bulk density and large particle size. It appears well suited to powders for which the conditions for stable fluidization are known, but it may not be convenient to change compositions on a small laboratory scale. It is possible to obtain a controlled discharge of solid particles that have reached a specific size.[8]

Codecomposition

In this technique a solution (typically nitrates) is atomized and blown against a hot platinum (or other) surface to evaporate the solvent and decompose the salt in a single step.[9] It appears to be most suitable for the preparation of laboratory batches.

Emulsion Drying

A solution can be separated into small droplets prior to drying as an emulsion of an aqueous solution and an immisible liquid such as kerosene. This is stirred while being dried in a vacuum dryer to give a rather stable suspension of solid salt particles in kerosene. The particles are deflocculated, precipitated, filtered, washed, and decomposed. The behavior of the powders is comparable with that of spray-dried powders.[10] The technique allows efficient transfer of heat to the emulsified droplets, but the homogeneity is probably dependent on the degree of emulsification that can be stabilized.

Gelation–Evaporation

A variety of techniques have been reported in which gravitational settling of precipitated phases or diffusion of components in an amorphous phase is hindered during solvent evaporation by the formation of a gelatinous or glassy matrix. Many processes termed "sol–gel techniques" are included under this heading. Sol–gel usually refers to the preparation of spheres of nuclear reactor fuels. The process generally infers the preparation of a sol (an aqueous colloidal suspension of the desired oxides in hydrated form) from the nitrates or other salts by controlled precipitation or digestion of precipitates.[11,12] The sol is converted to a gel by partial dehydration or by adding ammonia in various forms. In some cases the dehydration is achieved by evaporation and in other cases by dispersion of the sol as droplets in a column containing a long-chain alcohol and a surfactant. The gel is completely dried and can be fired to a very dense ceramic with no further processing.

The sol–gel technique is particularly useful for preparing dense microspheres at low sintering temperatures.[13] In general the technique provides highly reactive agglomerates of very fine particles, but it is difficult to reform the agglomerates into other sinterable shapes.

Other gel processes do not involve the preparation of a sol. Silicates have been prepared where ethyl orthosilicate is dissolved in absolute ethyl alcohol and is added to a solution of the nitrate of other metals.[14] The solution is hydrolized with water, and silica precipitates to a gel. This is then evaporated slowly to a dry powder and is calcined to give reactive oxides.

$MgAl_2O_4$ has been prepared by dissolving aluminum hydroxychloride in water and blending in a slurry of magnesium hydroxide.[15] This mixture becomes a gel, which is dried, ball milled, and calcined.

A number of evaporative methods involve the immobilization of ions in a glassy or amorphous matrix other than those commonly considered to be gels. In the simplest case some salts can be dissolved in their waters of hydration by heating and then quenching to a glass without phase separation and can be carefully decomposed to the oxides. This technique has been used to prepare $MgAl_2O_4$ to $Y_3Al_5O_{12}$.[16]

Another technique involves the processing of a glassy matrix using organic polyfunctional acids possessing at least one hydroxy and one carboxylic function, such as citric, malic, tartaric, glycolic, and lactic acid.[17] A solution is formed with the metallic salts and the organic acid. The solution is rapidly dehydrated in a rotating evaporator at a pressure of a few torrs. The solution is discharged from the evaporator before the viscosity is too high for removal and is further dehydrated in a vacuum oven. This forms a transparent glass that has mixing on an atomic scale and can be pyrolized to the oxides. Ferrites have been prepared by a similar technique using glasses made from a solution of maleic anhydride.[18] $Y_3Fe_5O_{12}$ has been prepared using citric acid as the glass former.[19]

Freeze-Drying

Freeze-drying or cryochemical preparation of ceramic oxides was introduced by Schnettler et al.[20] A solution is separated into small droplets by spraying it through a hydraulic nozzle at a pressure of a few psi. At this point the technique differs from the other volatilization methods in that segregation is prevented by a rapid freezing step. Solution droplets are sprayed into a bath of immisible liquid, such as hexane chilled by Dry Ice–acetone, or by spraying directly into liquid nitrogen.[21] The rapid freezing step is very important because the degree of ice–salt segregation is minimized. The hexane is not as cold as the liquid nitrogen but has better thermal contact with the droplets, since a gaseous layer of nitrogen around the droplets in liquid nitrogen impedes heat transfer. A continuous method of freezing involves introducing droplets of solution into the bottom of a chilled bath of refrigerant with a density higher than that of the frozen solution.[22] The frozen product is skimmed from the top of the refrigerant. These freezing methods generally produce spherical frozen beads of solution with diameters in the 0.01 to 5 mm range.

The drying step[20] involves the sublimation of water from the frozen solution without melting. The frozen sample is introduced into a vacuum chamber that is evacuated to a pressure of about 1 torr or less, at which the water sublines rather than melts. Heat can be applied to aid sublimation. To keep the partial pressure of water in the system as low as possible, a refrigerated condensing coil is usually introduced to collect the sublimed water. The drying can be done in a commercial freeze-drier or in a small glass apparatus for laboratory use consisting of a chamber with a sample and a liquid nitrogen trap all evacuated by a mechanical pump.

The freeze-drying technique gives spheres of aggregated crystallites that are replicas of the frozen beads of solution and that have a very low bulk density.[23] The crystallite size can be varied by changing the calcination step or the concentration of the solutions.[20,23]

Reactive powders of Al_2O_3, prepared by freeze-drying, were easily sintered to 99.9% density.[24] A modified nickel ferrite composition was sintered to a high density using freeze-dried powders.[25] Drying, spray drying, and precipitation for the preparation of $LiFe_5O_8$ were compared and it was found that freeze drying allowed the most flexibility in terms of sintering temperature and grain size.[26] The sintering temperatures needed were much lower than those needed for conventionally prepared powders, and a large degree of control over grain size was available.

Lithiated NiO for catalytic studies has been prepared using freeze drying.[27] The very uniform mixing allowed the Li_2O to diffuse into the NiO lattice at about 400°C as compared to 950 to 1000°C for conventionally prepared mixtures. With such low-temperature preparation, surface areas up to 60 m^2/g were preserved compared with less than 1 m^2/g for conventionally prepared NiO.

Some low-freezing-point solutions have a tendency to form glasses when cooled rather than to freeze into ice and salt. When these are heated in a freeze-drier they tend to flow and in some cases the homogeneity of the solution is lost. The addition of ammonium hydroxide to concentrated solutions of ferric sulfate has been shown to result in a raising of the freezing point and the promotion of salt–ice phase separation.[28] A variation of freeze-drying has been introduced for preparing actinide metal oxides from nitrate solutions, avoiding the problem of glass formation during freezing.[29] The nitrate solution is atomized, chilled to near Dry Ice temperatures, and contacted with NH_4OH to convert the nitrates to insoluble hydroxide at low temperatures. This is freeze-dried and decomposed to the oxides.

SOLUTION COMBUSTION

Solution combustion encompasses those preparation techniques in which solutions are actually burned to form solid particulates. It also includes to some degree those techniques that volatilize liquids for hydrolysis, decomposition, or oxidation from the gaseous state.

Alcohol Solutions

Ferrites have been produced by dissolving nitrates in alcohol and burning the solution in an atomizing burner.[30] The stoichiometric ratios of nitrates are dissolved in alcohol, the solution is atomized in oxygen, the dispersion is burned, and the powder is collected using a cyclone chamber. This technique is reproducible and more convenient and controllable than conventional mixing or coprecipitation.[30] The same technique has been used for ferrites except for the substitution of a water-spray tower for powder collection.[31] The phases present in the ferrite were dependent on the oxygen pressure in the burner; the higher oxygen pressures produced single-phase ferrites.

Barium and titanium alcoholates mixed in an organic solvent have been burned in air or oxygen to give $BaTiO_3$.[32]

Organometallics

While some organometallics, such as metal alkyls, can burn in air or oxygen to give oxide particulates, it is an expensive process and is not widely used. Some organometallics are not readily combustable but are mixed with alcohols for burning,[33] as is duscussed in the previous sections. However, some organometallics can be decomposed in a hot gas stream to give the oxides, as in the case of some transition metal alkoxides.[34] For zirconium tetratertiary butoxide, the material is vaporized, mixed with an inert carrier gas, and fed into a decomposition chamber where it is contacted with another stream of hot (325 to 500°C) inert gas. The oxide, recovered by an electrostatic precipitator, has an average particle size of less than 100 Å and a purity greater than 99.95%.

Chlorides

The vapor-phase decomposition or hydrolysis of metal chlorides in a flame is a fairly common technique. It is used commercially to make high-surface-area SiO_2 powders.[35] Silicon tetrachloride is reacted at high temperatures with hydrogen and oxygen (a highly exothermic reaction) to give silica and hydrochloric acid. The silica can be collected by a cyclone separator and calcined to remove traces of residual hydrochloric acid. This produces a very fine powder with a surface area in the range 175 to 200 m^2/g. Properties of silica and alumina made from chlorides have been reported.[36]

A burner designed to prevent clogging of the orifice by the oxide has been described.[37] The burner has four concentric tubes admitting from the center out chloride vapor in oxygen, nitrogen, hydrogen, and oxygen. The nitrogen layer delays the hydrolysis until the gas is a few centimeters from the tip. Mixed oxides or doped oxides have been prepared in the flame reactor by using mixtures of chlorides.

PRECIPITATION–FILTRATION

Precipitation with subsequent removal of the solids by filtration is one of the most widely used techniques for the preparation of ceramic powders from solutions. Usually the equipment needed is very simple and can be found in most chemical laboratories. However, the precipitation step involves a large number of variables that can affect the process. Some of the more important ones are discussed briefly here.

The pH of an aqueous solution is important for several reasons. The influence of pH on the precipitation of hydroxides is obvious in that the OH^- concentration appears in the solubility product. For example, in the precipitation of $Al(OH_3)$ the equilibria is:

$$Al^{3+} + 3OH^- \leftrightarrows Al(OH)_3$$

If the pH is too low, there are insufficient hydroxide ions in solution. At high pH the soluble complex $Al(OH)_4^-$ is formed and again precipitation is not accomplished. The precipitation of $Al(OH)_3$ is best done at a solution pH in the range of 4 to 9.

pH also plays a more subtle role in the precipitation of salts other than hydroxide. For instance, ammonium oxalate and oxalic acid are commonly used precipitating agents and the relative concentrations of $H_2C_2O_4$, $HC_2O_4^-$, and $C_2O_4^{2-}$ are strongly affected by pH. It can also affect the oxidation state of the ions of interest. At high pH, iron(II) will react rapidly with atmospheric oxygen because the iron(III) that is formed is removed from solution as the hydroxide.

Solution concentrations, temperature, and atmosphere also affect the equilibrium of a precipitation. However, equilibrium is seldom achieved during precipitation, and a number of nonequilibrium factors can also affect the precipitation process.

The order of mixing of the solution and the precipitating agent is important. Generally the solution of cations is added slowly with stirring to a solution containing the precipitating agent. This allows an excess of precipitating agent, and the solubility products of all the cations are likely to be exceeded simultaneously. Adding the precipitating agent to the cation solution will often cause the cations to be precipitated stepwise with the resulting inhomogeneities.

The rate of mixing is a variable that can affect the particle size of the precipitate by way of the degree of supersaturation and thus the number of nuclei formed. Finely divided precipitates are formed by rapidly mixing cold concentrated solutions.

Stirring rate can also affect the degree of supersaturation and the particle size. Stirring also prevents the formation of large agglomerates, which can hinder later steps of powder processing. The application of ultrasonic fields during precipitation can also reduce the agglomerate sizes.[38]

The effect of other ions on the surface chemistry of precipitates can be important and is treated more completely elsewhere.[39] An important example involves the electrical double layer that surrounds many fine precipitates in suspension. The heavy concentration of electrolyte surrounding the particles after the precipitation often destroys this layer and allows the particles to flocculate and settle. When the precipitate is then washed free of this excess electrolyte the particles may deflocculate and form a stable colloid that is difficult to filter. This can be avoided by the addition of a small amount of an ammonium salt which can subsequently be removed by calcination.

Impurities in a precipitate can be incorporated though a precipitation from the solution either as a separate phase or in solid solution. Impurities can be adsorbed on the surface of a precipitate and subsequently incorporated by further growth or whole pockets of solvent can become occluded. In general, conditions that favor large particle size, slow growth, and equilibrium will produce the purest precipitates.

Precipitates can be conveniently broken into three categories: (1) single-compound precipitates, which decompose directly into the desired stoichiometry; (2) solid-solution precipitates, for which the desired stoichiometry can be incorporated into a single-phase solid solution precipitate; and (3) mixtures or polyphase precipitates.

Single Compounds

When possible this is the ideal form of a precipitate. The cation ratio of the desired product is incorporated directly into the precipitate as a compound. There is no need for accurate assays, since the stoichiometry is assured and the homogeneity of the precipitate is on the atomic scale. Some examples are $BaTiO_3$ from $BaTiO(C_2O_4)_2 \cdot 4H_2O$;[40] $BaSnO_3$ from $BaSn(C_2O_4)_2 \cdot 0.5H_2O$;[41] and $LaFeO_3$ from $LaFe(CN)_6 \cdot 5H_2O$.[42]

One advantage of the homogeneity is the low-temperature reactivity. When $BaSnO_3$ is prepared by the reaction of milled $BaCO_3$ and SnO_2, the reaction temperature is near 1000°C and Ba_2SnO_4 as a reaction intermediate is detected.[43] Using $BaSn(C_2O_4)_2 \cdot O.5H_2O$ as a precursor, $BaSnO_3$ can be formed at about 700°C without any detectable intermediates.[41,44] Also, the low reaction temperature widens the range of particle sizes that can be made.

With barium titanate, the precipitation process has been extended to include the direct doping with lanthanum during the precipitation process.[45] The availability of the rare earth, transition metal cyanides has proven useful in the preparation of highly reactive catalysts, such as $LaCoO_3$, $PrCoO_3$, and $LaMnO_3$.[46,47]

The principal disadvantage of the technique is the rarity of its applicability. Most practical compositions have no precursor of the same composition and other types of precipitation must be used.

Solid Solutions

A single-phase solid-solution precipitate assures homogeneity on an atomic scale but does not assure the proper cation ratio. The ratio must be set by very careful control of the precipitation process.

During the precipitation of a solid solution, the cation ratio in the solution is not necessarily found in the precipitate. For instance a barium titanate with partial substitution of strontium and lead for barium can be prepared using the standard oxalate process, which assures a 1:1 ratio of divalent metal to titanium.[5] Generally it is necessary to establish by careful chemical analyses the divalent metal ion ratio necessary in the solution to give the desired product. Again, the pH of the solution is important and must be controlled, since the cation ratios in the precipitate change with pH.[48,49] Generally the higher the pH, the more closely the cation ratios of the precipitate follow those of the solution.

Common examples of solid-solution precipitates are the divalent metal oxalates, which can be used for the preparation of ferrites.[1,49-51] Sulfates also frequently form solid solutions that can yield oxide ferrites or sulfides.[1]

Mixtures

In contrast to the first two types this precipitation technique does not give a single phase. The homogeneity of the original solution is partly lost, since segregation occurs during precipitation. However, the degree of mixing normally achieved is still far superior to that found by conventional mixing techniques. This type of precipitation is the most common, since systems in which the single compound or solid-solution precipitation will work are comparatively few. The cation ratio is usually easily controlled, because precipitation takes place under conditions where the solubility product of all components is exceeded and the desired cations are precipitated nearly quantatatively. Precipitates are often hydroxides or hydrated oxides but can also be oxalates, carbonates, and so forth.

The preparation of ferrites by coprecipitation of hydroxides or hydrated oxides is usually quite straightforward. However, the anion of the original soluble salt can apparently affect the sinterability of the final product. Nickel ferrite has been prepared by precipitating hydroxides from sulfate, chloride, and nitrate solutions.[52] The powder precipitated from the sulfate solutions was superior for the purposes of high density and fine microstructure. The specific hydroxide used for the precipitation also affected the sintering properties. Using potassium hydroxide as the precipitating agent gave powders that showed inhibited grain growth during sintering; those precipitated using lithium hydroxide showed discontinuous grain growth.

Other examples of mixed hydroxide or hydrated oxide precipitates include the preparation of magnesium aluminate.[53] A spinel phase was formed at 400°C (compared with 1200 to 1600°C required using conventionally mixed powders). A mixed-oxide precipitate as a precursor to $LaMnO_3$ gives a powder that forms the compound at 400°C below the synthesis temperature needed for mixed oxides.[54]

In cases where a single precipitating agent will not precipitate all desired cation species, a second precipitating agent can be used. Since lithium hydroxide is soluble, a stearate–hydroxide combination will precipitate a precursor for lithium ferrite.[55] In cases where a single solution is not readily available for all cations, separate solutions can be added at the proper rate to a single precipitating agent. For the preparation of lead titanate, titanium is most easily used as the chloride, but lead chloride is insoluble. Thus separate solutions can be added to an oxalate to obtain a mixed precipitate.[45]

Special Precipitation Techniques

The hydrolysis of organometallics is a versatile technique for the preparation of highly reactive powders. In this process the organometallic, often an alkoxide, is prepared and mixed with highly purified water for hydrolysis. Examples of oxides made from alkoxides are yttria-zirconia,[56] barium titanate,[57] and alumina.[58]

Hydrothermal conditions at higher temperatures yield precipitates with relatively large and well-formed crystals. This has been demonstrated for various ferrite compositions[59,60] and for ZrO_2.[61] In some cases the hydrothermal treatment[60] gives an oxide product rather than a hydroxide or a hydrated oxide.

For some hydroxide precipitations, prolonged digestion at high pH and at temperatures near boiling will form an oxide phase directly that can be filtered and used without needing a separate decomposition step. This technique is particularly useful for ferrites,[62,63] but other materials, such as $BaTiO_3$[64] and even complex chalcogenides,[65] have been prepared by similar techniques.

While most precipitations are affected by the addition of a precipitating ion, the addition of an aqueous solution to a large excess of a second solvent in which the original solute is insoluble will also cause precipitation to occur. This principle forms the basis for a powder preparation technique termed "liquid drying."[66] For the preparation of some spinels the appropriate sulfates were dissolved in water and this solution was atomized and sprayed into a large excess of acetone. The small droplet size caused rapid precipitation and minimized the amount of segregation that took place. Other investigators have also reported the preparation of ferrite[1] and transition metal sulfides[67] using similar methods.

In cases where a single salt is being precipitated, a well-crystallized precipitate can be made by homogeneous precipitation.[68] This technique slowly raises the concentration of the precipitating ion, such as hydroxide or oxalate, which is formed by the slow hydrolysis of urea or ethyl oxalate. This provides a precipitation scheme where there is minimal supersaturation and relatively few nuclei form.

ADVANTAGES

The greater degree of homogeniety possible with solution-prepared powders compared with conventional ground, mixed, and calcined powders provides a higher reactivity. The diffusion paths necessary to achieve the desired solid phases are short. This allows processing in a single step or at lower temperatures, which also gives greater control of powder particle size. Consequently, improvements in sinterability are the result. The solution preparation also offers: (1) the advantages of precise control of stoichiometry through gravimetric or volumetric combination of analyzed single cation solutions; (2) an ability to disperse trace additives uniformly; and (3) freedom from the contamination inherent in most powder grinding and mixing operations.

REFERENCES

1. Y. D. Tretyakov, I. Y. Kosinskaya, N. N. Oleinikov, and Y. G. Sakonov, "Synthesis of Ferrites from Solid Solutions of Salts," *Izv. Akad. Nauk. SSSR Neorg. Mater.*, **5** (7), 1255–1258 (1969); translation in *Inorg. Mater.* **5** (7), 1067–1070 (1969).
2. A. G. Elliot and R. A. Huggins, "Phase Relations in the System $NaAlO_2$–Al_2O_3," *J. Amer. Ceram. Soc.*, **58** (11–12) 497–500 (1975).
3. J. G. M. DeLau, "Preparation of Ceramic Powders from Sulfate Solutions by Spray Drying and Roasting," *Amer. Ceram. Soc. Bull.* **49** (6), 572–574 (1970).
4. A. Lagrange, J. Nicolas, and M. Hildebrandt, "Preparation and Properties of Hot-Pressed Ni–Zn Ferrites for Magnetic Head Application," *IEEE Trans. Mag.*, **8** (3), 494–497 (1972).
5. T. Akashi, T. Tsuji, and Y. Onoda, *Sintering and Related Phenomena,* G. C. Kuczynski, N. A. Hooton, and C. F. Gibbon, eds., 747–756, Gordon and Breach, New York, 1967.
6. M. J. Ruthner, H. G. Richter and I. L. Steiner, *Proc. Int. Conf. Ferrites Jap.*, 1970, Y. Hoshino, S. Iida, and M. Sugimoto, eds. University Park Press, Baltimore, 1971, pp. 75–78.
7. A. A. Jonke, E. J. Petkus, J. W. Loeding, and S. Lawroski, "The Use of Fluidized Beds for the Continuous Drying and Calcination of Dissolved Nitrate Salts," *Nucl. Sci. Eng.*, **2,** 303–319 (1957).
8. O. M. Todes, V. A. Seballo, Y. Y. Kaganovich, A. P. Goltsiken, S. P. Nalimov, and O. M. Rozanov, "Design Calculation of Processes for the Dehydration of Solutions in a Fluidized Bed with Selective Discharge," *Khim. Prom.* **46** (8), 612–615 (1970). *Chem. Abstr.* **74,** 5030x (1971).
9. R. Roy, "New Ceramic Materials Produced by Novel Processing Techniques," *International Journal of Powder Metallurgy,* **6** (1), 25–28 (1974).
10. A. L. Stuijts, *Proc. Int. Conf. Ferrites, Japan,* **1970.** Y. Hoshino, S. Iida, and M. Sugamoto, eds., University Park Press, Baltimore, 1971, pp. 108–113.
11. D. W. Ferguson, O. C. Dean, and D. A. Douglas, "Sol–Gel Process for the Remote Preparation and Fabrication of Recycle Fuels," *Proc. interm. Conf. Peaceful Uses At. Energy, 3rd, Geneva, 1964,* **10,** 307–315 (1965).
12. "Sol–Gel Processes for Ceramic Nuclear Fuels," Proceedings of a Panel Held in Vienna, May 1968, International Atomic Energy Agency, Austria, 1968.

13. J. M. Fletcher and C. J. Hardy, "Applications of Sol–Gel Processes to Industrial Oxides," *Chem. Ind.* (*London*) **1968** (2), 48–51.

14. D. M. Roy and R. Roy, "Experimental Study of Formation and Properties of Synthetic Serpentines and Related Layer Silicate Minerals," *Amer. Miner.*, **39** (11,12), 957–975 (1954).

15. P. W. D. Mitchell, "Chemical Method For Preparing $MgAl_2O_4$ Spinel," *J. Am. Ceram. Soc.*, **55** (9), 484 (1972).

16. D. R. Messier and G. E. Gazza, "Synthesis of $MgAl_2O_4$ and $Y_3Al_5O_{12}$ by Thermal Decomposition of Hydrated Nitrate Mixtures," *Amer. Ceram. Soc. Bull.*, **51** (9), 692–697f (1972).

17. C. Marcilly, P. Courty, and B. Delmon, "Preparation of Highly Dispersed Mixed Oxides and Oxide Solid Solutions by Pyrolysis of Amorphous Organic Precursors," *J. Amer. Ceram. Soc.*, **53** (1), 56–57 (1970).

18. L. J. Koppens, "Improved Ferrite Memory Cores Obtained by a New Preparation Technique," *IEEE Trans. Mag.*, **8** (9), 303–305 (1972).

19. Th. J. A. Popma and A. M. Van Diepen, "Magnetization and Mössbauer Spectra of Non-Crystalline $Y_3Fe_5O_{12}$," *Mater. Res. Bull.*, **9** (9), 1119–1128 (1974).

20. F. J. Schnettler, F. R. Monforte, and W. W. Rhodes, *Science of Ceramics*, Vol. 4, G. H. Stewart, ed., The British Ceramic Soc., 1968, pp. 79–90, Stoke-on-Trent.

21. V. V. Merkovich and T. A. Wheat, "Use of Liquid Nitrogen in Spray Freezing," *Amer. Ceram. Soc. Bull*, **49** (8), 724–725 (1970).

22. H. A. Sauer and J. A. Lewis, "Freezing Droplets of Aqueous Solutions for the Cryochemical Process," *Amer. Inst. Chem. Eng. J.*, **18** (2), 435–437 (1972).

23. D. W. Johnson and F. J. Schnettler, "Characterization of Freeze-Dried Al_2O_3 and Fe_2O_3," *J. Amer. Ceram. Soc.*, **53** (8), 440–444 (1970).

24. Y. S. Kim and F. R. Monforte, "Theoretically Dense (99.9%) Polycrystalline Alumina Prepared from Cryochemically Processed Powders," *Amer. Ceram. Soc. Bull.*, **50** (6), 532–535 (1971).

25. F. J. Schnettler and D. W. Johnson, *Proc. Int. Conf. Ferrites: Jap.* **1970** Y. Hoshino, S. Iida, and M. Sugamoto, eds., University Park Press, Baltimore, 1971, pp. 121–124.

26. D. W. Johnson, Jr., P. K. Gallagher, D. J. Nitti, and F. Schrey, "Effect of Preparation Technique and Calcination Temperature on the Densification of Lithium Ferrites," *Amer. Ceram. Soc. Bull.*, **53** (2), 163–167 (1974).

27. A. C. C. Tseung and H. L. Bevan, "Preparation and Characterization of High Surface Area Semiconducting Oxides," *J. Mater. Sci.*, **5**, 604–610 (1970).

28. R. E. Jaeger, T. J. Miller and J. C. Williams, "Effects of Ammonium Hydroxide on Phase Separation in the Cryochemical Processing of Salt Solutions," *Amer. Ceram. Soc. Bull.*, **53** (12), 850–852 (1974).

29. M. C. Tinkle, "Cryochemical Method for Forming Spherical Metal Oxide Particles from Metal Salt Solutions," U.S. Patent 3,776,988 (Dec. 4, 1973).

30. J. T. Wenckus and W. Z. Leavitt, "Preparation of Ferrites by the Atomizing Burner Technique, *Conf. Mag. Mag. Mater., Boston,* **1956,** T-91, American Institute of Electrical Engineers, New York, 1957.

31. W. W. Malinofsky and R. W. Babbit, "Fine-Grained Ferrites I. Nickel Ferrite," *J. Appl. Phys.* **32** (3) (supplement), 237s–238s (1961).

32. S. DiVita and R. J. Fischer, "Barium Titanates," U.S. Patent 2,985,506 (May 23, 1961).

33. M. L. Nielson, P. M. Hamilton, and R. J. Walsh, Ultrafine Particles, W. E. Kuhn, ed., Wiley, New York, 1963, pp. 181–195.

34. K. S. Mazdiyasni, C. T. Lynch, and J. S. Smith, "Preparation of Ultra-High-Purity Submicron Refractory Oxides," *J. Amer. Ceram. Soc.*, **48** (7), 372–375 (1965).

35. G. J. Duffy, "Vapor-Phase Production of Colloidal Silica," *Ind. Eng. Chem.*, **51** (3), 232–238 (1959).

36. K. A. Loftman, Ultrafine Particles, W. E. Kuhn, ed., Wiley, New York, 1963, pp. 196–205.

37. M. Formenti, F. Juillet, P. Meriaudeau, S. J. Teichner and P. Vergnon, "Preparation in a Hydrogen Oxygen Flame of Ultrafine Metal Oxide Particles," *J. Colliod Interface Sci.*, **39** (1), 79–89 (1972).

38. F. I. Kukoz and E. M. Feigina, "Effect of Ultrasound and other Precipitation Conditions on the Particle Size Composition of Nickeleous Hydroxide Precipitates," *Zh. Prikl. Khim.* (*Leningad*), **42** (9), 1978–1983 (1969); *Chem. Abstr.* **72,** 4736d (1970).

39. A. W. Adamson, *Physical Chemistry of Surfaces*, Wiley, New York, 1967.

40. W. S. Clabaugh, E. M. Swiggard, and R. Gilchrist, "Preparation of Barium Titanyl Oxalate Tetrahydrate for Conversion to Barium Titanate of High Purity," *J. Res. Natl. Bur. Stand.*, **56,** 289–291 (1956).

41. P. K. Gallagher and F. Schrey, *Proc. ICTA, 3rd,* **2,** H. G. Wiedemann, ed., 623–634, Burkhauser Verlag, Basel, 1972.

42. P. K. Gallagher, "A Simple Technique for the Preparation of R. E. FeO_3 and R. E. CoO_3," *Mater. Res. Bull.*, **3,** 225–232 (1968).

43. G. Wagner and H. Binder, "The Binary Systems BaO–SnO_2 and BaO–PbO_2. I. Phase Analysis," *Z. Anorg. U Allgen. Chem.*, **297,** 328–346 (1958).

44. P. K. Gallagher and D. W. Johnson, Jr., "Kinetics of the Formation of $BaSnO_3$ from Barium Carbonate and Tin (IV) Oxide or Oxalate Precursors," *Thermochim. Acta,* **4,** 283–289 (1972).

45. P. K. Gallagher, F. Schrey, and F. V. DiMarcello, "Preparation of Semiconducting Titanates by Chemical Methods," *J. Amer. Ceram. Soc.*, **46** (8), 359–365 (1963).

46. D. W. Johnson, Jr. and P. K. Gallagher, "Studies of Some Perovskite Oxidation Catalysts Using DTA Techniques," *Thermochim. Acta,* **7,** 303–309 (1973).

47. P. K. Gallagher, D. W. Johnson, Jr., and F. Schrey, "Studies of Some Supported Perovskite Oxidation Catalysts," *Mater. Res. Bull.*, **9** 1345–1352 (1974).

48. F. Schrey, "Effect of pH on the Chemical Preparation of Barium-Strontium Titanate," *J. Amer. Cer. Soc.*, **48** (8), 401–405 (1965).

49. P. K. Gallagher and F. Schrey, "Preparation and Thermal Analysis of Mixed Magnesium–Manganese–Iron Oxalates and Hydrated Oxides," *J. Amer. Ceram. Soc.*, **47** (9), 434–437 (1964).

50. W. J. Schuele, "Preparation of Fine Particles from Bimetal Oxalates," *J. Phys. Chem.*, **63** (1), 83–86 (1959).

51. P. K. Gallagher, H. M. O'Bryan, Jr., F. Schrey, and F. R. Monforte, "Preparation of Nickel Ferrite from Coprecipitated $Ni_2Fe_8C_2O_4 \cdot 2H_2O$," *Amer. Ceram. Soc. Bull.*, **48** (11), 1053–1059 (1969).

52. H. M. O'Bryan, Jr., P. K. Gallagher, F. R. Monforte, and F. Schrey, "Microstructure Control in Nickel Ferrous Ferrite," *Amer. Ceram. Soc. Bull.*, **48** (2), 203–208 (1969).

53. R. J. Bratton, "Coprecipitates Yielding $MgAl_2O_4$ Spinel Powders," *Amer. Ceram. Soc. Bull.*, **48** (8), 759–762 (1969).

54. S. A. Prokudina, Y. S. Rubinchik, and M. M. Pavlyuchenko, "Kinetics of Low Temperature Synthesis of $LaMnO_3$," *Isv. Akad. Nauk SSSR, Neorg. Mater.* **10** (3), 488–492 (1974); translation in *Inorg. Mat.*, **10** (3), 416–419 (1974).

55. A. L. Micheli, "Preparation of Lithium Ferrites by Co-precipitation," *IEEE Trans. Magnet.*, **6** (3), 606–608 (1970).

56. K. S. Mazdiyasni, C. T. Lynch, and J. S. Smith, II, "Cubic Phase Stabilization of Translucent Yttria-Zirconia at Very Low Temperatures," *J. Amer. Ceram. Soc.*, **50** (10), 532–537 (1967).

57. K. S. Mazdiyasni, R. T. Dolloff, and J. S. Smith, II, "Preparation of High-Purity Submicron Barium Titanate Powders," *J. Amer. Ceram. Soc.*, **52** (10), 523–526 (1969).

58. D. F. Saunders and A. Packter, "Effects of Rate and Temperature of Mixing on the Surface Area of Precipitated Alumina Hydrates," *Chem. Ind.* (*London*) **18,** 594 (1970).

59. A. A. Van Der Giessen, "Hydrothermal Preparation of Ceramic Powders," *Klei Keram.*, **20** (2), 30–38 (1970); *Chem. Abstr.* **72** 114604s (1970).

60. F. Schrey, "Hydrothermal Preparation of a Manganese–Zinc Ferrite," paper 17-EI-67F, 1967 Annual Meeting of the American Ceramic Soc.; *Bull. Amer. Ceram. Soc.* **46** (4), 788 (1967).

61. N. T. Okopnaya, V. I. Zelentsov, V. M. Chertov, and B. N. Lyashkevich, "Regulation of the Particle Size of Zirconium Dioxide by a Hydrothermal Method," *Adsorbtsiya Adsorbentry*, **2,** 108–109 (1974); *Chem. Abstr.*, 111783z (1974).

62. P. E. D. Morgan, "Direct Aqueous Precipitation of Lithium Ferrite and Titanate," *J. Amer. Ceram. Soc.*, **57** (11), 499–500 (1974).

63. M. Kiyama, "Conditions for the Formation of Fe_3O_4 by the Air Oxidation of $Fe(OH)_2$ Suspensions," *Bull. Chem. Soc. Jap.*, **47** (7), 1646–1650 (1974).

64. S. S. Flaschen, "Aqueous Synthesis of Barium Titanate," *J. Amer. Chem. Soc.*, **77** (12), 6194 (1955).

65. P. W. D. Mitchell and P. E. D. Morgan, "Direct Precipitation Methods for Complex Crystalline Chalcogenides," *J. Amer. Ceram. Soc.*, **57** (6), 278 (1974).

66. R. E. Jaeger and T. J. Miller, "Preparation of Ceramic Oxide Powders by Liquid Drying," *Amer. Ceram. Soc. Bull.*, **53** (12), 855–859 (1974).

67. R. J. Bouchard, "The Preparation of Pyrite Solid Solutions of the Type $Fe_xCo_{1-x}S_2$, $Co_xNi_{1-x}S_2$, and $Cu_xNi_{1-x}S_2$." *Mater. Res. Bull.*, **3,** 563–570 (1968).

68. L. Gordon, M. L. Salutsky, and H. H. Willard, *Precipitation from Homogeneous Solution,* Wiley, New York, 1959.

13

Lattice Strain in Alumina Powders

J. P. Page

E. A. Metzbower

D. J. Shanefield

D. P. H. Hasselman

The fundamental factors underlying the densification kinetics of powder materials during sintering appear to be well understood. The effect of surface energy, particle size, and temperature, as well as the various mechanisms of material transport, such as viscous and plastic flow, evaporation and condensation, volume, and surface diffusion, all have been the object of a number of experimental and theoretical studies.[1-4] The shrinkage behavior of a powder compact, as theoretically derived, can be expressed as:

$$\frac{\Delta L}{L_0} = \mathrm{N}\left[\frac{\gamma D a^3_0}{R^p K T}\right]^n t^m \qquad (1)$$

where L is the linear shrinkage of the powder compact after time t, L_0^3 is the original length of the compact, N is a constant, γ is the surface energy, D is the diffusion coefficient, a_0 is the vacancy volume, R is the particle radius, K is the Boltzman constant, T is the temperature, and p, n, m are exponents whose values depend on the particular mechanism for material transport. Equation 1 is a simplified form of an equation given by Kingery.[5]

In practice we have observed that different batches of alumina made by the same chemical process and similar in surface area, particle size distribution, morphology, and purity can exhibit marked differences in their sinterability. The differences in sinterability cannot be explained in terms of surface energy, particle size, and purity.

One possible explanation for this behavior, suggested in Chapter 9, is the presence of transition aluminas (nonalpha phase) in the powder. Another possibility, presented in this chapter, is that the kinetics of material transport during sintering are different in each powder because of different levels of defects in the powders.

Defects can change the rate-controlling mechanism of material transport (e.g., substitution of surface diffusion for bulk diffusion). Also, defects can suppress or enhance the rate of material transport for a given sintering mechanism.

Enhancement of the rate of material transport can be due to increased concentrations of lattice defects, such as vacancies.[6] In ionic compounds excess vacancy concentrations can be created by the addition of impurities.[7] To conserve charge neutrality, impurities of a higher positive valence than the host material create cation vacancies. For example, TiO_2 impurities have been observed to increase densification rates of MgO.[8] Similar effects have been observed for TiO_2 impurities in ZnO.[9] However, additions of Ga^{3+} and Al^{3+} decreased sintering rates for ZnO as the result of decreases in cation-vacancy concentration.[9]

High concentrations of dislocations or subgrain boundaries can also significantly affect sintering rates, particularly in those materials for which sintering is controlled by bulk diffusion. This latter hypothesis is substantiated by the observations of Pines and Sirenko,[10] who noted that the sinterability of cold-worked copper powders showed a positive correlation with the dislocation density. Also, Fedorchenko[11] ascribed the superior sintering of carbonyl nickel compared to electrolytic nickel to its having higher densities of dislocations and higher internal lattice strain. Additionally, it is well recognized that mechanically worked powders, such as ball-milled powders, generally show a greater sintering activity than annealed powders with the same particle size.[12,13]

MEASURING LATTICE STRAIN

In the search for a suitable technique to establish the validity of this hypothesis, it should be noted that lattice strain can cause a distinct broadening of the lines obtained in X-ray analysis. As a result, if excess lattice defects and internal strain have an effect on the sintering behavior of alumina powders, a correlation between sinterability and X-ray line broadening should be found.

The magnitude of average internal strain and coherently diffracting domain size can be measured by X-ray analysis using the Warren-Averbach method.[15] Using nickel-filtered copper radiation, the (012) and the (113) peaks and their second-order reflections were step-scanned at intervals of $2\theta = 0.05°$ using a time period of 20 sec. Each peak required a total of 81 measurements. The as-recorded X-ray data were corrected for polarization and geometric factors, atomic scattering factors, and the K_α doublet with the aid of a previously developed computer program.[16] The corrected profiles were then further corrected for instrument broadening by the Stokes method,[17] followed by the separation of the effects on internal strain and domain size by the Warren-Averbach method. The average ($\bar{\epsilon}$) of the mean strain perpendicular to the (012) and (113) planes at a depth of 50 Å was calculated.

Lattice-strain measurements were carried out on five alumina powders. These powders, prepared by the Bayer process and dry ball milled for a period of approximately 24 hr, were obtained from commercial sources and were designated A, B, C, D, and E. These powders were selected on the basis of previously established differences in sintering behavior. The powder characteristics and sintering behavior (described later) were measured in the as-received condition.

Powder purity was measured by optical emission and spark ion-mass spectroscopy and the results are recorded in Table 13.1. Powder surface area and particle size were determined by the B.E.T. and sedimentation methods, respectively. Particle shape and the presence of dislocations and internal strain were determined qualitatively by scanning electron microscopy and transmission electron microscopy.

SINTERABILITY OF POWDERS

The five powders were cold-pressed into compacts approximately ⅞ in. in diameter by 0.1 in. thick in a circular steel die at 3000 psi and were fired at 1475°C for 7 hr. Prior to firing, green density was determined from the specimens's weight and dimension. After sintering, density was measured by ASTM C20-70 using toluene as the suspending fluid. The sinterability of the powders was expressed by the Hirschhorn densification parameter (D.P.)[18] defined by:

$$\text{D.P.} = \frac{\text{final density} - \text{green density}}{\text{theoretical density} - \text{green density}}$$

Table 13.1. Impurities in alumina samples (ppm)

Impurity	Sample A	Sample B	Sample C	Sample D	Sample E
Iron	90	70	61	82	144
Magnesium	525	350	365	450	<250
Silicon	399	516	500	407	402
Titanium	<50	<50	<50	<50	<50
Calcium	250	<100	<100	250	360
Sodium	520	400	450	550	530
Gallium	25	23	18	20	72
Potassium	<50	<50	<50	<50	<50
Lithium	68	40	38	45	39
Barium	70	<50	<50	90	<50
Boron	7	12	NA	12	4
Fluorine	67	102	NA	69	109
Sulfur	93	106	NA	64	41

The parameter D.P. can vary from 0 to 1 and defines the ratio of the actual densification achieved under the sintering conditions selected to the total densification that could have been achieved theoretically.

To establish the time dependence of the densification, additional specimens measuring approximately ½ in. diameter by 0.500 ± 0.002 in. long were isostatically pressed at 7500 psi, followed by sintering at 1475°C in a dilatometer.

Table 13.2 gives the experimental data for the median particle size, particle-size range,* surface area, strain domain size, and green density together with the data for fired density, densification parameter, and sintering constant.

MULTIPLE REGRESSION ANALYSIS

A multiregression analysis showed that the fired density can be expressed by:

$$\rho_f = 1.692 + 0.949\,\rho_g + 8.0 \times 10^{-2}\,\bar{\epsilon} - 9.65 \times 10^{-3}\,A_s \quad (2)$$

where ρ_f = fired density in g/cc
ρ_g = green density in g/cc
$\bar{\epsilon}$ = mean strain
A_s = surface area in m^2/g.

* Particular-size range between particle sizes corresponding to the 10 and 90% deciles, as determined by sedimentation.

Table 13.2. Alumina powder characteristics, X-Ray data and sintering behavior

Sample	Median Particle size	Band width (μ)	Surface Area (m^2/g)	Mean strain $\times 10^3$	Domain size (Å)	Green Density (g/cc)	Fired Density (g/cc)	Densification Parameter	Sintering Constant (*m* in Eq. 1)
A	0.45	1.08	14.20	3.143	393	2.265	3.931	0.96	—
B	0.39	0.92	10.24	2.369	376	2.126	3.793	0.89	—
C	0.54	1.15	10.75	3.175	428	2.127	3.843	0.92	—
D	0.38	1.06	11.28	1.917	393	2.090	3.691	0.84	0.34
E	0.61	1.35	9.59	2.630	352	2.148	3.851	0.90	0.43

The analysis showed that the variation in fired density could be accounted for by 80.1, 13.2, and 6.7% of the variations in green density, strain, and surface area, respectively, with a multiple correlation coefficient of .994. The fired density was not affected by differences in median particle size, particle size range, or domain size.

Similarly, the densification parameter (D.P.) could be expressed by:

$$\text{D.P.} = -3.962 \times 10^{-2} + 0.397\,\rho_g + 4.656\,\bar{\epsilon} - 3.33 \times 10^{-3} A_s \qquad (3)$$

The variations in densification parameter could be accounted for by 81.7, 11.8, and 2.6% of the variations in green density, strain, and surface area, respectively, with a multiple correlation coefficient of .993. As for the fired density, the present results for the densification parameter were not affected by the variation in median particle size, particle-size range, and domain size.

As expected, of all the characteristics determined in the present study, green density has the major effect on final density values to be obtained. However, in agreement with the original hypothesis and as indicated by the percentage contributions, the internal strain also plays a vital role in the densification process. The positive coefficients of the internal strain terms in equations 2 and 3 indicate that high internal strain is desirable in achieving high fired densities.

The negative coefficients in Equations 2 and 3 suggest that the high final density requires low surface area, which in turn implies the requirement of large particle size. This appears contradictory, since high densification rates require small particle size, as indicated by Equation 1. In addition, the statistical analysis of the present data showed that particle size had no effect. A regression analysis of green density in terms of surface area, particle size, and particle-size range showed that green density can be expressed by:

$$\rho_g = 1.650 + 0.0387A_s + 0.816a - 0.287\,\Delta a$$

in which the relative contribution by A_s, a, and Δa to ρ_g are 63, 22, and 4%, respectively. These results suggest that, in spite of the negative coefficients for the effect of surface area on fired density, high surface area is desirable in view of its greater effect on fired density by way of the green density compared to its direct effect on fired density.

LATTICE STRAIN AND SINTERABILITY

Since the firing conditions for all samples were identical, the significant effect of internal strain on final density must have been to accelerate the rate of densification. In the absence of externally applied loads, any permanent internal strains in the individual particles can only result from the presence of lattice imperfections. In monocrystalline powder particles, lattice imperfections can exist in the form of dislocations, stacking faults, subgrain boundaries, and perhaps nonuniform vacancy concentrations. These lattice imperfections can affect material transport during sintering in two ways. First, the strain field associated with these imperfections can provide an additional driving force for mass transport as they are being annealed out at higher temperatures. This mechanism is expected to be effective primarily only during the initial stage of the densification process and is not thought to be sufficient to explain the observed results. More likely, as suggested earlier, lattice imperfections in the form of immobile dislocations serve as diffusion pipes, thereby accelerating material transport by lattice diffusion. This mechanism is expected to be effective over the total period of sintering and, in the opinion of the authors, presents the most likely explanation for the observed results. Further confirmation of this conclusion can be found by comparing the sintering curves shown in Figure 13.1, in which the slope of the relative density versus time curve corresponds to the value m in Equation 1. The numerical values for m are

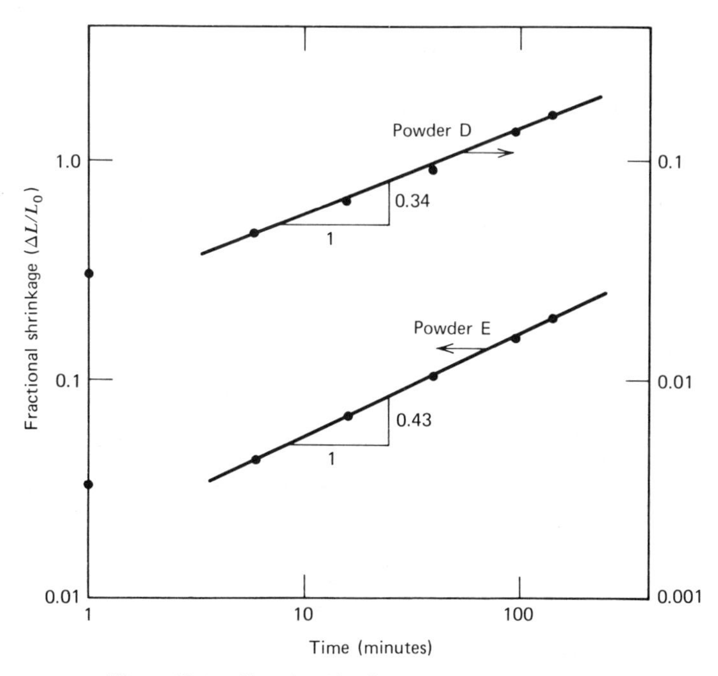

Figure 13.1. Sintering kinetics of two alumina powders.

given in Table 13.2. Sample D with the lower value of strain energy has a value for the sintering constant near ⅓, which suggests that for this powder, grain-boundary diffusion was the primary mechanism for material transport. For powder E, however, the value of m approaches 0.5, which is representative of material transport by volume diffusion.[19]

The detection of the presence of dislocation within the particles by transmission electron microscopy presented difficulties. The majority of particles were too large for transmission of the electron beam so that only less representative platelike particles could be examined. Although the presence of dislocations could not be determined unambiguously, numerous particles showed evidence of the presence of internal strains in the form of extinction patterns, as shown in Figure 13.2 for powder sample E.

The dislocation densities within the particles can be estimated from the X-ray data. From the known value of domain size (D), the dislocation density (ρ_D) can be calculated from[20]:

$$\rho_D = \frac{3n}{D^2} \tag{4}$$

where n is the number of dislocations per domain face. The dislocation density (ρ_S) can be determined from the strain ($\bar{\epsilon}$) by[20]:

$$\rho_S = \frac{K(\epsilon)^2}{Fb^2} \tag{5}$$

where F = dislocation interaction factor
b = Burger's vector
K = 20.6, calculated by the procedure of Williamson and Smallman.[20]

Values for ρ_D and ρ_S calculated by means of Equation 4 and 5 are given in Table 13.3. The values listed may be in error by as much as a factor of 5 in

Table 13.3. Dislocation densities in alumina powders

Powder Sample	Dislocation Calculated from Strain	Densities Calculated from Domain Size
A	1.36×10^{11}	1.94×10^{11}
B	0.77×10^{11}	2.21×10^{11}
C	1.39×10^{11}	1.64×10^{11}
D	0.51×10^{11}	1.94×10^{11}
E	0.95×10^{11}	2.42×10^{11}

Figure 13.2. Extinction pattern showing microstrain and possible dislocations in powder sample E.

view of the simplifying assumptions that underlie Equations 4 and 5. Nevertheless, it is of interest to note that the values for the dislocation densities approach those found for cold-worked metals.[20]

It is most probable that these large dislocation densities were introduced during the ball-milling stage of the powder production process. Under these conditions, it is quite likely that during particle impact the stress exceeds the critical stress required for dislocation generation or multiplication. Indeed, the abrasion of aluminum oxide[21] is known to introduce high densities of dislocations in the surface.

Of interest to the present study is the difference in dislocation densities for the various powders. It appears unlikely that dislocation densities on the order of those given in Table 13.3 could have been introduced during the chemical stage of the powder production process. An alternative explanation is based on the general observation that in the growth of single crystals, high rates of growth generally result in lattice imperfections. Undoubtedly, such lattice imperfections do exist in the present powders, even in relatively low concentrations. For the various powder samples differences in these concentrations of lattice, imperfections may be the result of even slight variations in temperature during the growth process. It is well known that lattice imperfections in the form of properly oriented and pinned dislocations can act as Frank-Read sources for the multiplication of additional dislocations during mechanical impact. During the ball-milling process, the multiplication of dislocations could result in the dislocation densities given in Table 13.3, with any differences between the powder samples related to initial differences in lattice imperfections prior to the ball-milling stage. It is suggested here that relating the particular growth conditions of the crystallites to the final densities of internal stress generated during ball milling may constitute a fruitful area for future research.

As is pointed out above, the accelerated sintering could also be due to excess vacancy concentrations as the result of certain types of impurities. In this present study, however, no correlation among impurity content, final density, and densification parameter could be found. As a result, it is felt that the present observation cannot be attributed to excess vacancy concentrations.

Regardless of the validity of the above explanations, the present results show that a high degree of internal strain is beneficial in achieving high final densities in a sintering operation. The results also suggest that a general internal strain be considered an important parameter for the characterization of powder samples.

REFERENCES

1. J. Frenkel, *J. Phys.* (*USSR*), **9**, 385 (1945).
2. G. C. Kucyzinski, *J. Appl. Physics*, **21**, 632 (1945).
3. D. L. Johnson, *Phys. Sintering*, **1** (Y3), (July 1969) B1.
4. B. Ya. Pines, *Uspekhi Fiz. Nauk*, **52**, 501 (1954); translation AES-TR 5563, 2/15, 1963.
5. W. D. Kingery, *Introduction to Ceramics*, Wiley, New York, 1963, p. 376.
6. P. Reyner, *Reactivity of Solids*, Wiley-Interscience, New York, 1969.
7. F. A. Kroger, *The Chemistry of Imperfect Crystals*, Wiley, New York, 1964.
8. D. N. Polyboyarinov, *Refractories* (*Moscow*), **27**, 137 (1962).
9. W. A. Weyl, *Ceram. Age*, **60**, 28 (1952).
10. B. Ya. Pines and A. F. Sirenko, *Fiz Tverd. Tela*, **7**, 687 (1965).
11. I. M. Fedorchenko and V. V. Skorokhod, *Sov. Powder Metall.*, **10** (58) 805 (1967).
12. A. L. McLaren and P. W. M. Atkinson, *J. Nucl. Mater.* **17**, 332 (1965).
13. V. A. Bron and N. V. Semkina, *Sov. Powder Metall.*, **5**, 332 (1962).
14. W. H. Gitzen, "Alumina as a Ceramic Material," *Amer. Ceram. Soc.*, Columbus, Ohio, 1970, p. 7.
15. B. E. Warren and B. L. Averbach, *J. Appl. Phys.*, **21**, 595 (1950).
16. E. A. Metzbower, NRL Report 7253, 1971.
17. A. R. Stokes, *Proc. Phys. Soc., London*, **B61**, 382, (1948).

18. J. L. Hirschorn, *Introduction to Powder Metallurgy,* Colonial Press, New York, 1969, p. 206.
19. F. Thummler and W. Thomma, *Metal Rev.,* **12,** 69 (1967).
20. G. K. Williamson and R. E. Smallman, *Philos. Mag.,* **1,** 34 (1956).
21. B. J. Hockey, "*Science of Ceramic Machining,*" S. J. Schneider and R. W. Rice, eds., National Bureau of Standards, Washington, 1972, p. 333.

14

Agglomeration Effects on the Sintering of Alumina Powders Prepared by Autoclaving Aluminum Metal

R. T. Tremper
R. S. Gordon

Aluminum oxide can be sintered to translucency with a density near theoretical by the addition of a small amount of magnesia and by firing in an appropriate atmosphere. This technique, first described by Coble,[1] has been studied intensively and the properties of high-density, sintered alumina allow it to be used as an arc chamber in high-pressure, sodium vapor lamps. For this application the sintered body must be as dense as possible. To achieve a total light transmission comparable to that which is possible with a single crystal, the porosity level in sintered Al_2O_3 must lie below 10^{-4} to 10^{-5} volume fraction for pore sizes in the range of 0.5 to 5μm. It is well established[2] that voids and second-phase particles are primarily responsible for the diffuse light transmission characteristics of polycrystalline Al_2O_3 and not grain boundaries. The optical anisotropy in Al_2O_3 is too small for grain boundaries to influence, to any significant degree, the total transmission through a dense polycrystalline body.

Most of the previous studies on the sintering kinetics of polycrystalline alumina focused on alum or Bayer-process powders that are frequently used in the production of commercial products. These powders are normally obtained by means of the high-temperature (1100 to 1200°C) thermal decomposition of an aluminum hydroxide (e.g., gibbsite and boehmite) or an alum salt. A new low-temperature process has recently been developed in this laboratory for the production of high-purity aluminum oxide. It consists of converting in an autoclave high-purity (99.99%) aluminum metal directly to the oxide at low temperatures (300 to 500°C) and moderate steam pressures (500 to 5000 psi).

By autoclaving, alumina powders can be prepared with a wide range of characteristics (e.g., surface areas, agglomerate sizes and distributions, and phase compositions). The purpose of the present study was to determine quantitatively the effects of these characteristics on "powder sinterability," with the ultimate goal of specifying the properties of an alumina powder that are necessary and sufficient for it to be sintered to densities up to and higher than 99.99% of theoretical, a level that is attractive for application in a sodium vapor lamp.

EXPERIMENTAL PROCEDURE

Synthesis of Autoclave Alumina

Using the previous results of Krischner and Torkar[3] as a starting point, an extensive study was undertaken of the conversion of aluminum metal to an aluminum oxide powder by reaction with high-pressure steam. Since this work will be reported extensively elsewhere,[4] only a brief description of the process is given here. In essence, by placing aluminum metal in any form (splatter, ingot, chopped rods) in a high-pressure steam chamber (autoclave) and subjecting it to the proper combination of temperature (350 to 450°C), steam pressure (1000 to 4500 psi), and time (up to 100 hours), alumina powders can be synthesized that contain up to 100% of the alpha phase and possess surface areas between 1 and 20 m^2/g.

Two basic types of autoclave processes were used in synthesizing the alumina powders for this study. The first technique used was direct steam transfer. In this procedure the aluminum metal, which had been preheated in the reaction autoclave, was subjected directly to the final operating pressure by the transfer of steam from the reservoir autoclave. Pressures employed in this type of experiment were usually between 1500 and 4500 psi. By use of a double-chamber autoclave in this manner, the temperature in the reaction autoclave could be maintained at a point equal to or above the saturation temperature. The specific experiment that produced 100% alpha alumina involved the direct transfer of steam at the saturation temperature and pressure. This variation is referred to as the "transfer-saturation" method.

The second general type of process used in the production of alumina powders was the two-stage process. In this method the aluminum is subjected initially to steam at a relatively low pressure (1000 to 1500 psi) and temperatures above saturation (approximately 400°C); this is followed by a second steam transfer to pressures between 2200 and 3200 psi and temperatures of no more than 30°C over saturation. During the first stage, the reaction product is essentially all gamma alumina, which is converted to 100% alpha alumina during the second steam transfer. This technique allows for the production of an alumina powder with an agglomerate-size distribution different from that encountered in the single-stage, steam-transfer procedure.

Both the steam-transfer and two-stage processes are indicated schematically in Figure 14.1. Another possible reaction path is to heat the metal and steam together along the saturation curve and then into the

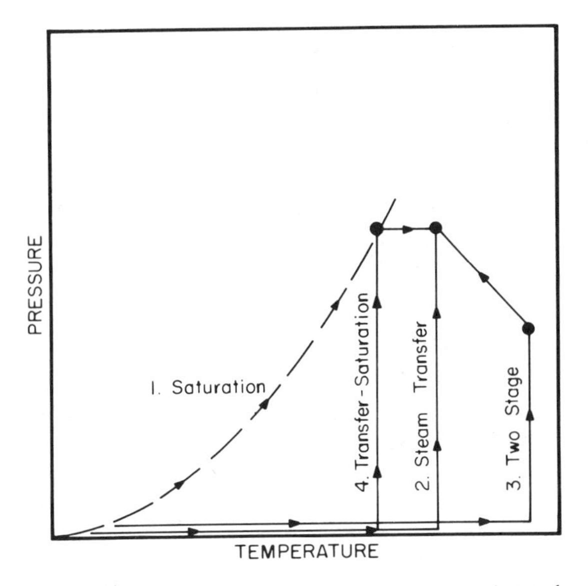

Figure 14.1. Schematic diagram of autoclave reaction path.

superheated region. This method is undesirable because of the formation of boehmite[4] as a precursor.

Powder Characterization

The reader is referred elsewhere[4] for a complete description of the technique used to characterize the autoclave alumina powders, namely, ignition loss, thermal gravimetric analysis, surface area measurement by BET, absolute density determination, X-ray diffraction, and transmission electron microscopy. In addition to these methods, the powders were further characterized by scanning electron microscopy* and for agglomerate-size distribution was determined by use of a Coulter Counter.

In addition to routine agglomerate-size studies at 1000 to 4000+, high-resolution micrographs of the agglomerate structure at 60,000+ were obtained.

Median agglomerate sizes and distributions were determined by Coulter Counter particle-size analysis.† Prior to analysis, each sample was dispersed in an electrolyte of 4% $Na_4P_2O_7$ by subjecting the suspension to a 5 minute ultrasonic treatment. In all measurements a 50 μm aperature with a 0.8 μm minimum detectable point was used. Agglomerates below 1 μm of course, were undetected in this analysis.

Sample Protection

Upon removal from the autoclave, the powders were screened through a 90 mesh nylon screen, and the fines of this process were called the as-prepared powder. Portions of the powders were doped with the appropriate amount of $Mg(NO_3)_2 \cdot 6H_2O$ to give 0.1 wt % MgO in the sintered ceramic. An aqueous slurry technique was used, followed by drying at 600°C and screening through a 100 mesh nylon screen. The sample to be milled (10 g lots) were placed in 500 ml polypropylene centrifuge bottles along with 125 g of alumina grinding media and a few drops of isopropanol as a milling aid. These were placed on an adapted laboratory vibratory mill and milled for 2 hours. The powders were then rescreened through 100 mesh. One gram portions of the processed powders were pressed into pellets (0.667 in diameter) in a tungsten carbide die in a hydraulic laboratory press with no binders or lubricants. A pressure of 15,000 psi was used yielding pellets of 40 to 55% theoretical density.

* JEOL 50 A Electron Probe Microanalyzer.
† Coulter Counter, Model B and Model T.

Sintering

The furnace used for this study was a tungsten mesh element vacuum furnace.* In all runs the temperature was programmed† to conform to the following schedule: (1) heat to 1200°C at 500°C/hour, (2) heat to the soak temperature at 125 to 150°C/hour, (3) soak for 6.7 hours, and (4) cool to room temperature at 350°C/hour. The vacuum was maintained at 1 to 5 $\times$ 10^{-6} torr throughout each sintering cycle. Typically, three to four pellets were sintered simultaneously, each separated by squares of tungsten foil.

Density and Ceramographic Preparation

The bulk density of the sintered pellets was determined by a normal weight and measurement technique. Samples whose density was > 90% had virtually all closed porosity, and their densities were redetermined by the fluid immersion technique in a temperature-controlled bath of monobromobenzene. The density of a single crystal of aluminum oxide was determined with each set of samples, and the relative densities (sample to single crystal value of that set) are reported as percent of theoretical. Cross sections of many of the sintered samples were prepared by normal ceramographic techniques, followed by thermal etching for 1 hour at 1600°C in hydrogen.

RESULTS

Effect of Autoclaving Conditions on Powder Sinterability

The principal objective of the sintering studies in the early stages of the program was to discover the powder properties that have the biggest influence on powder sinterability. In the synthesis of autoclave alumina, several powder characteristics were determined as a function of time in the autoclave at several steam pressures, using steam-transfer conditions at temperatures above saturation. Two of these were the relative amounts of α-Al_2O_3, KI–Al_2O_3, and γ-Al_2O_3 in the powder and the powder surface area (BET). The KI phase is unique to autoclaving conditions. Several autoclaving conditions were chosen for study. These are summarized in Table 14.1

* Astro Industries Model 1100 V.
† Data-Trak 5300 Programmer.

along with the equilibrium compositions (long autoclaving times) and the range of surface areas. For short autoclaving times the surface areas of the powders were high (> 25 m^2g) and γ-Al_2O_3 was the predominant phase. The approach to an equilibrium phase composition was rapid at the highest pressure (~ 5 hours) and fairly slow at the lower pressures (> 50 hours). From these four conditions, powders were available for sintering with a range of α-Al_2O_3 contents and surface areas. In general, as the α-Al_2O_3 content increased with autoclaving time, the surface area of the powder decreased, presumably because of the conversion of the fine γ-Al_2O_3.

From these four conditions several powders were selected at different autoclaving times for each pressure. These powders (as prepared) were pressed into pellets and sintered at 1625°C, a temperature suitably high for testing the powder's sintering characteristics and yet not too high to encounter problems with excessive and discontinuous grain growth. The results of these sintering studies are presented in Figure 14.2, in which bulk densities are plotted versus autoclaving time at each of the four pressures. The variations in α-Al_2O_3 content and surface area of these powders with autoclaving time are also shown in Figure 14.2. (Of course, complete conversion to α-Al_2O_3 occurs during sintering.)

This series of experiments indicates conclusively that the sinterability of autoclave alumina powders increases with increasing α-Al_2O_3 content. In all cases in which the powder contained significant amounts of γ-Al_2O_3, poor sintering characteristics were observed. The surface area becomes an important variable after conversion to α-Al_2O_3 (or an α-KI mixture) is complete. In some cases at 3100 and 4500 psi, converted powders with low surface areas (~ 2 m^2/g at long autoclaving times) either sintered to low densities or did not sinter at all. The presence of the KI phase (~ 40% in the 2400 and 3100 psi powders) appeared not to be deleterious to sintering in these initial experiments. Subsequent experiments, however, with powders containing essentially 100% KI–Al_2O_3 revealed rather poor sintering characteristics.

Table 14.1. Typical autoclaving conditions

Pressure (psi)	Temperature (°C)	Equilibrium Phase Composition	Range of Surface Area (m^2/g)
1500	400	70% α, 30% γ	27–14
2400	420	60% α, 40% KI	26–2
3100	440	60% α, 40% KI	31–2
4500	420	90% α, 10% KI	25–2

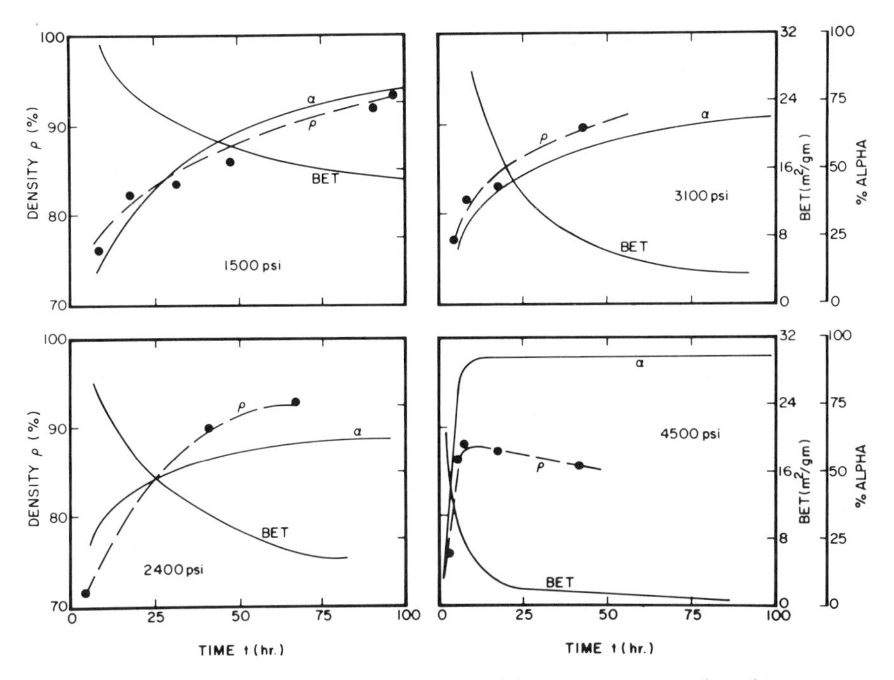

Figure 14.2. Powder properties and sintered densities versus autoclave time.

Final-Stage Sintering of Autoclave Powders

Based on the initial screening experiments (Figure 14.2), six autoclave powders were chosen for a systematic sintering study. Each powder, because of its phase composition (50 to 80% α-Al_2O_3) and reasonably high surface area (> 6 m^2/g) was believed to have a reasonable chance to be sintered in four conditions; as-prepared, milled only, doped (MgO) only, and milled and doped with MgO. Milling was introduced because of the highly agglomerated nature of the powders in the as-prepared (or autoclaved) condition. Pressed pellets of these powders in the four conditions were sintered in the temperature range between 1700 and 1900°C. For comparison, two commercial, high-purity (99.98 to 99.992%) powders were prepared in the doped condition.* One of these powders (CA) was highly agglomerated, while the other (CAF) was reported to be in an agglomerate-free state. After autoclaving, powder F was screened through a 400 mesh nylon screen with the fines retained for sintering in the four conditions. A summary of autoclaving conditions and powder characteristics for powders in this series is given in Table 14.2.

* Adolph Meller Co., Providence, Rhode Island.

Table 14.2. Autoclave conditions and properties of final stage powders

Powder	Source of Al Raw Material	Pressure (psi)	Temperature (°C)	Time (hours)	Surface Area (m^2/g)	Approximate Phase Composition
A	Ingot	1500	400	94	35	50% α, 50% γ
B	Chopped rod	1500	400	113	14	60% α, 40% γ
C	Chopped rod	3100	445	32	11	60% α, 40% KI
D	Ingot	3100	440	69	11	60% α, 40% KI
E	Chopped rod	4500	420	5.5	6	80% α, 20% KI
F[a]	Ingot	4500	425	19.5	8	70% α, 30% KI

[a] Screened through 400 mesh.

In Figure 14.3*A* the sintered densities (percent of theoretical) at different sintering temperatures are given for the powders in Table 14.2 in the as-prepared condition. All powders exhibited rather poor sintering characteristics, as might be expected for undoped and agglomerated powders. Powder F, which had been screened to break up large agglomerates, sintered to the highest density (~ 97.5%). The relative ranking of the powders in terms of their densities was essentially independent of temperature. A typical microstructure of a specimen sintered in the "as-prepared" condition is shown in Figure 14.4*A*. In all these specimens the grains were quite small (10 to 24 μm) and large pores were present, both on the grain boundaries and within the grains.

Sintering data for the milled-only series of powders are presented in Figure 14.3*B*. Two important features are to be noted. First, the relative differences in sinterability of the powders was significantly reduced by the milling procedure. However, even these small differences in reactivity appear to be significant, since the "ranking" of the powders remains roughly the same across the entire temperature range. The second important feature is that there is no increase in density with temperature, indicating that these specimens had reached their ultimate density at temperatures below 1700°C, and that sintering at higher temperatures resulted in essentially no further pore annihilation or removal. In support of this conclusion, a typical microstructure of a specimen sintered in the milled-only condition is shown in Figure 14.4*B*. In the entire series of samples, extremely large grains (100 to 300 μm), each entrapping a large number of small pores, were encountered. Milling leads to the breakdown of large agglomerates and hence to the removal of large pores from the green powder compact. Small, well-dispersed pores are less effective in inhibiting discontinuous grain growth than the larger pores present in the agglomerated structures. Furthermore, milling, by broadening the distribution of particle sizes, can lead to a larger driving force for discontinuous grain growth.[5]

The sintering data for the doped-only powders and the doped commercial powders (CA and CAF) used as a reference are shown in Figure 14.3*C*. Again, as in the as-prepared series, the trend of a gradual increase in sintered density with temperature was observed. Two autoclave powders (D and F) could be sintered to densities over 99% at 1900°C, while the remaining powders exhibited rather poor sintering characteristics. The commercial powders, especially the agglomerate-free powder, exhibited excellent sintering properties with the production of translucent pieces above 1800°C. It is perhaps significant that milling (Figure 14.3*B*) produced a greater improvement in sintered densities than did doping (Figure 14.3*C*), although doping alone did permit powders D and F to be sintered to a high density. These particular powders possessed the smallest agglomerates. These results indicate the effect of agglomerates and associated large pores in inhibiting the densification process.

A typical microstructure of a specimen sintered in the doped-only condition is shown in Figure 14.4*C*. As expected from the data in Figure 14.3*C*, a large number of pores remain both on the grain boundaries and within the grains. These large pores probably originate from large voids between the agglomerates in the powder compacts. Again the grains were fairly small, ranging from 15 to 35 μm depending on the temperature. The inability of most of the doped-only powders to sinter to high densities is related to the agglomeration characteristics of the powders and possibly to a nonuniform dispersal of the dopant.

The sintering data from the doped and milled autoclave powders are presented in Figure 14.3*D* along with the results for the commercial powders (note that the density scale is greatly expanded). By doping and milling, virtually all the powders sintered to $>99\%$ of theoretical density. In fact, at 1740 and 1800°C, the sinterability of powder B approached that of the commercial agglomerate-free powder, while powder D did the same at 1900°C. Virtually all the autoclave powders sintered to higher densities than the agglomerated commercial powder. Trends among the powders, although not immediately obvious from the graph, can be detected with careful scrutiny and are discussed later. The best of this series was very good indeed, with translucency comparable to that of the doped agglomerate-free commercial powder.

A typical microstructure of a 99.8% dense specimen from this series is given in Figure 14.4*D*. It is a classic equiaxed, nearly pore-free microstructure, with the few remaining pores trapped within grains (sometimes in

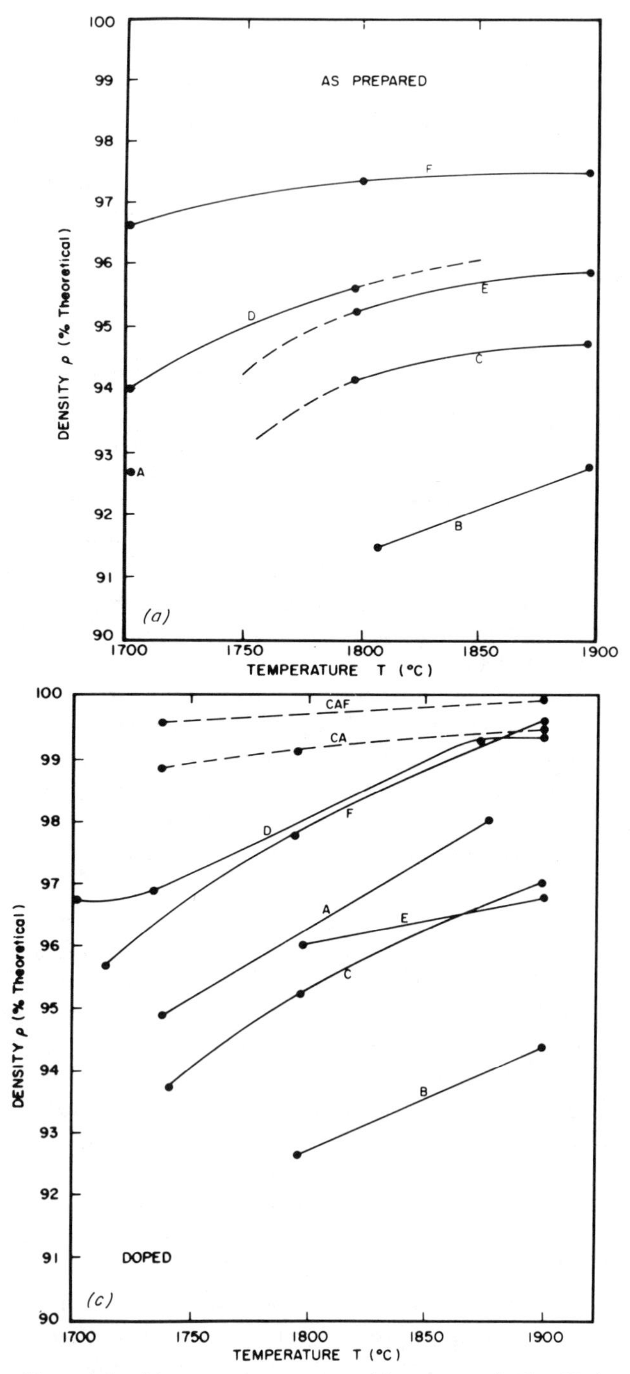

Figure 14.3. Sintered densities versus temperature: (*a*) as prepared, (*b*) milled only, (*c*) doped only, (*d*) doped and milled.

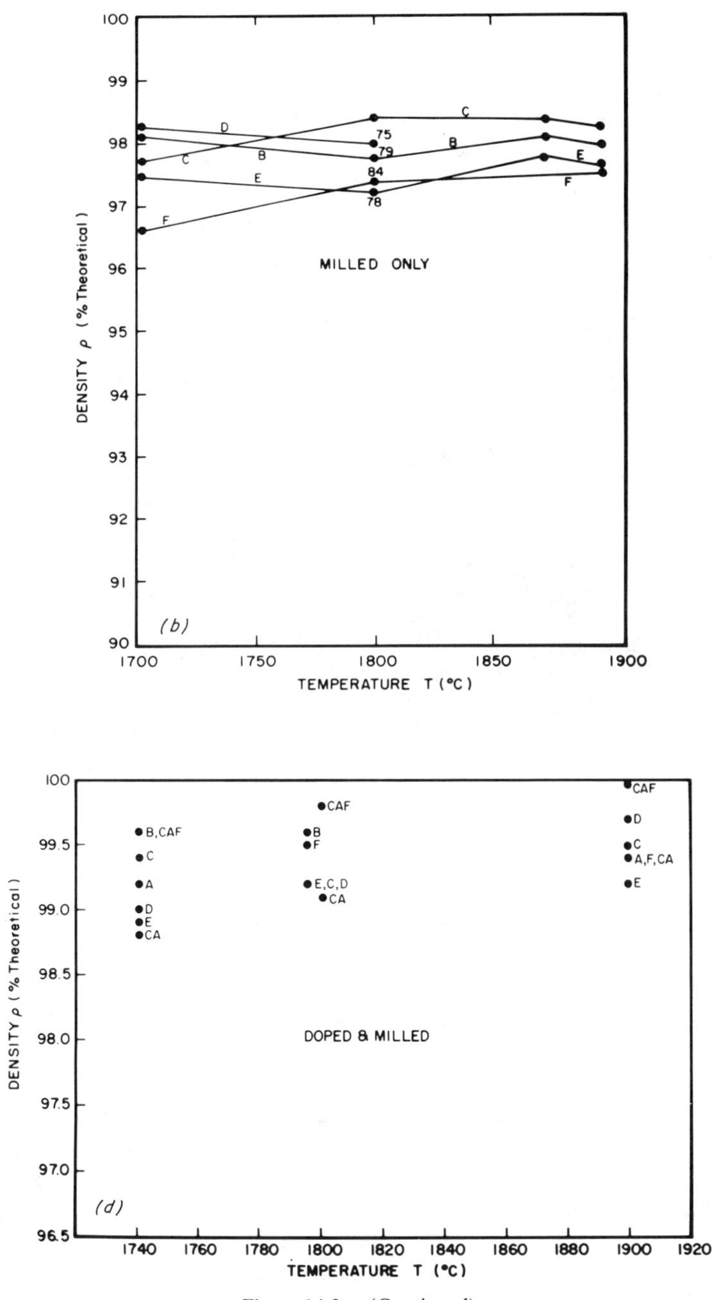

Figure 14.3. (Continued)

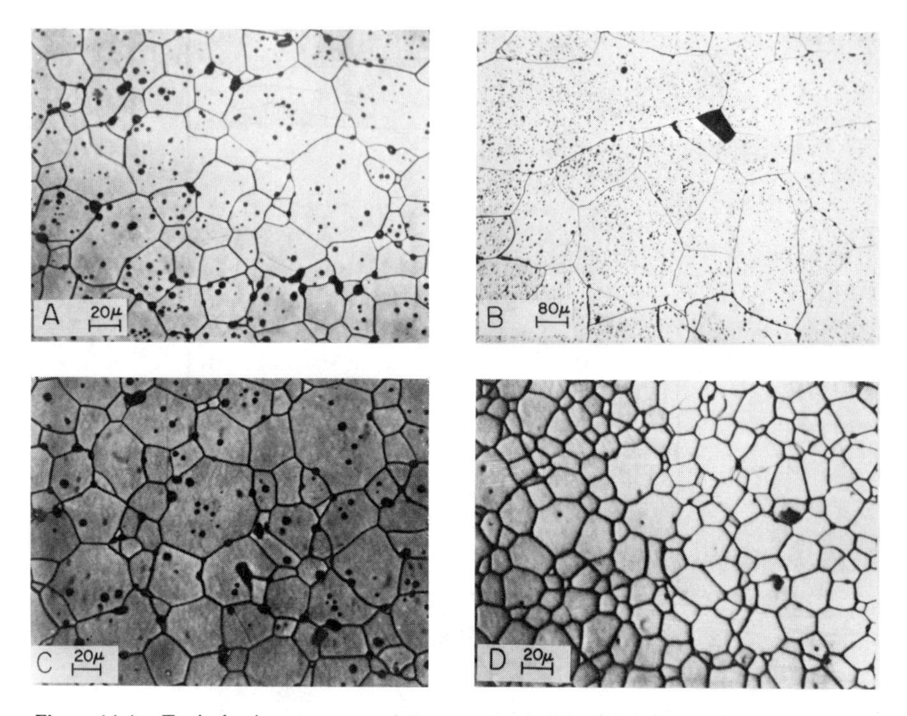

Figure 14.4. Typical microstructures; (*A*) as prepared, (*B*) milled only, (*C*) doped only, (*D*) doped and milled.

clusters). The milling step has probably done two things to promote final density: breakdown of agglomerates and dispersion of the dopant. The grain sizes of the samples in this series were between 20 and 40 μm.

Agglomeration Characteristics of Autoclave Powders

Early in the sintering program it became apparent that alumina powders prepared by autoclaving Al metal could only be sintered to theoretical density providing the powders were milled after synthesis. Preliminary examination of the autoclave powders by low-magnification scanning electron microscopy revealed that the powders were highly agglomerated, with individual agglomerates up to 20 μm in size.

Consequently, it was decided to make a thorough investigation into the nature of agglomerate formation by taking into account the variables of different Al metal reactants, autoclave temperature, steam pressure, and reaction conditions in the autoclave. In addition, a preliminary study was conducted to determine what happens to the agglomerates during vibratory milling.

In Figure 14.5 Coulter Counter cumulative plots of agglomerate sizes are given for powders synthesized from an ingot reactant at four different pressures (1500, 2400, 3100, and 4500 psi) in steam-transfer experiments. Distribution curves for the other reactant metals were similar in shape to those at the same autoclave pressures and temperatures. At 1500 psi powders were prepared at 400 ± 10°C for reaction times exceeding 90 hours to insure 70 to 80% conversion to α-Al_2O_3. For the 2400 and 3100 psi conditions the reaction times at 425 to 440°C were over 30 hours to insure 60 to 70% conversion to α-Al_2O_3, with 30 to 40% KI–Al_2O_3 making up the balance. At 4500 psi the reactions times at 425°C were 7.5 hours to achieve at least a 90% conversion to α-Al_2O_3. Included also in Figure 14.5 for comparison is the distribution curve for the deagglomerated commercial powder (CAF). The powder synthesized at 1500 psi was the "best" as-prepared autoclave powder in that its cumulative Coulter plot was closest to that of the commercial deagglomerated powder and nearly log-normal.

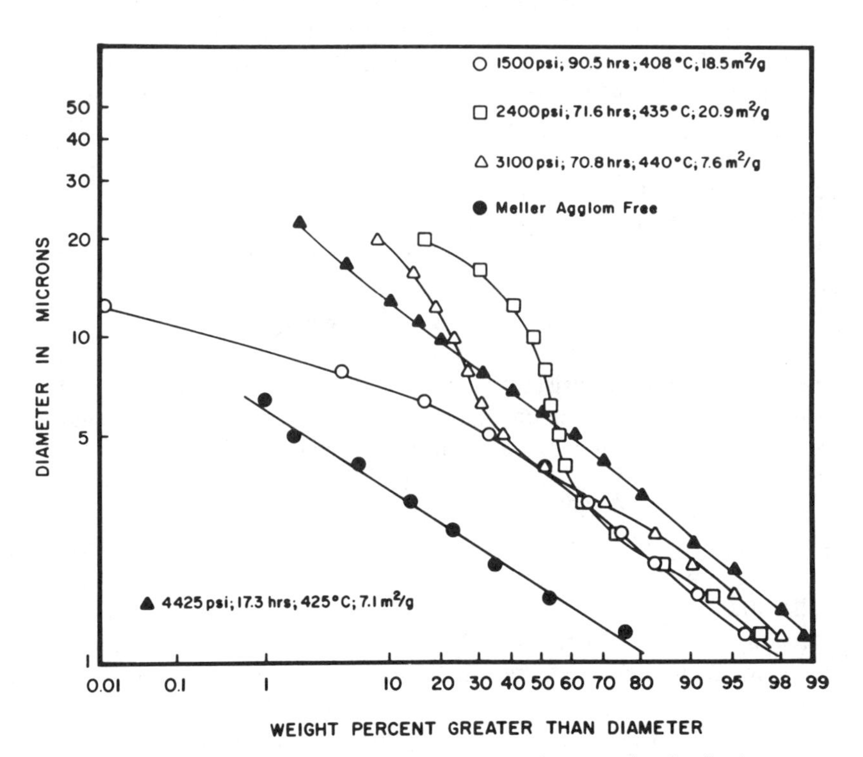

Figure 14.5. Effect of autoclave pressure on agglomerate-size distribution.

All the plots in Figure 14.5 are similar at small agglomerate sizes (< 5 μm). However, powders synthesized in the intermediate pressure regime (2400 to 3100 psi) under supersaturated conditions where KI–Al_2O_3 is a significant secondary phase (30 to 40%) to α-Al_2O_3 possess sigmoidal distribution curves, and significant fractions (20 to 60%) of the agglomerates exceed 10 μm in size. This tendency for coarse-agglomerate formation is most pronounced in the powders synthesized at 2400 psi. Powders prepared at these intermediate pressures under steam-transfer conditions at saturation temperatures show only a slight tendency towards a duplex or sigmoidal agglomerate distribution. These powders are essentially pure α-Al_2O_3, while the powders prepared under supersaturated conditions contain appreciable amounts of the KI phase. Powders prepared at 4500 psi possess a nearly log-normal agglomerate distribution curve (similar to powders prepared at 1500 psi), except that the agglomerates are much larger than those formed at the lower pressure.

To investigate the effect of Al raw material, three powders, each prepared from a different reactant (splatter, chopped rod, and ingot), were investigated. These powders were prepared by autoclaving at 1500 psi and 400 $\pm$ 10°C for reaction times in excess of 90 hours to insure 70 to 80% conversion to α-Al_2O_3. Examination of these powders by low-magnification (4000$\times$) scanning electron micrographs revealed the presence of agglomerates of various sizes up to 10 to 20 μm. Ultimate particle sizes, which can be estimated from the surface areas of the powders (12 to 35 m^2/g), are on the order of 0.05 to 0.1 μm. Thus the agglomerates are considerably larger than the individual crystallite sizes. The largest agglomerates are present in the powders prepared from the splatter and chopped-rod reactants, while the smallest agglomerates form from the bulk ingot reactants.

In Figure 14.6 cumulative agglomerate-size plots from the Coulter measurements are given for these three powders. All the curves are of the same general shape. Consistent with the scanning microscope observations, powders prepared from the chopped rod possess agglomerates considerably larger (about two times as large) than those prepared from the ingot.

In Figure 14.7 high-resolution photomicrographs (40,000 to 60,000$\times$) are presented to illustrate the fine structure of the agglomerates. Two distinct types of crystallites are present within the agglomerates: (1) relatively large (0.5 to 1.5 μm) rhombohedral (hexagonal) crystals (Figure 14.7*d*) and (2) smaller (0.02 to 0.1 μm), equiaxed crystals, where hexagonal facets or forms are readily apparent (Figure 14.7*c*). It appears that the micron-size crystals form from agglomerates of the smaller crystallites at an intermediate stage of the process.

Investigation of these powders by transmission electron microscopy and diffraction also revealed two forms of α-Al_2O_3: (1) large rhombohedral

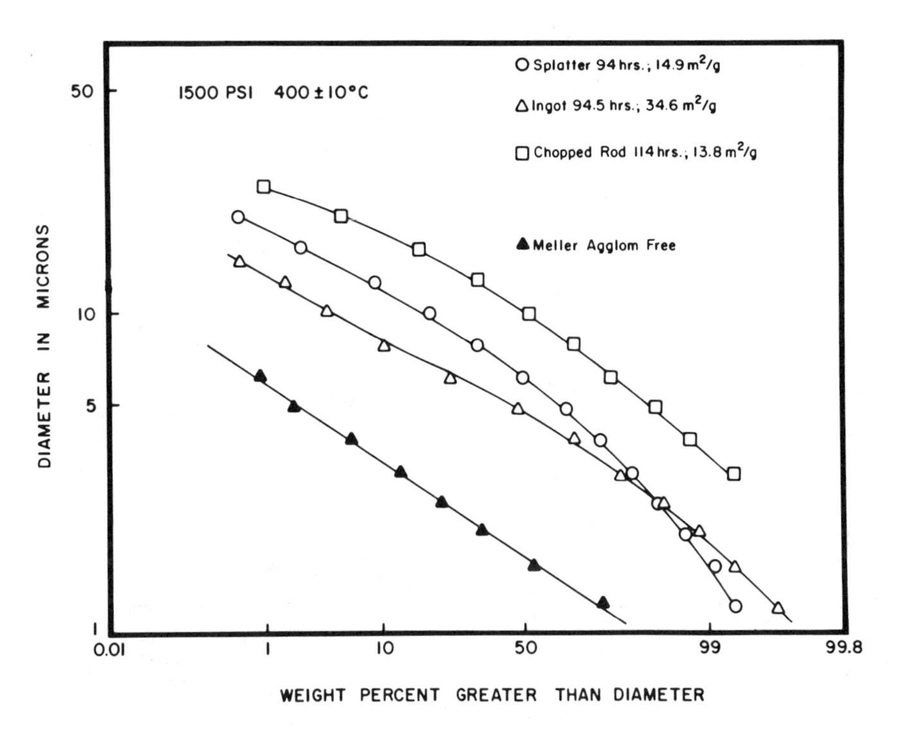

Figure 14.6. Effect of aluminum raw material on agglomerate-size distribution.

crystals ~1 μm and (2) smaller crystallites ~0.1 μm. These observations are consistent with those obtained by high-resolution, scanning electron microscopy. Evidently the small crystallites (~0.1 μm) are the "active" form of α-Al_2O_3 reported by Krischner and Torkar.[3] This phase is predominant at 1500 psi. For long reaction times agglomerates (~1 μm) of these small crystals transform into larger single crystals of well-defined crystal habit.

For powders synthesized at higher pressures (≥ 2400 psi), very little effect was obtained by changing the initial reactant, and the data in Figure 14.5 are representative for all reactant types for these steam-transfer conditions.

The best agglomeration characteristics (i.e., log-normality and small median agglomerate sizes) were discovered in powders that had been synthesized in the two-stage, steam-transfer process. This process involved the prior formation of γ-Al_2O_3 at either 2400 or 3100 psi. The Coulter curves for these powders were nearly always log-normal and agglomerates were usually smaller than those shown for the 1500 psi powders in Figure 14.5.

As mentioned earlier, milling was found to be a necessary operation

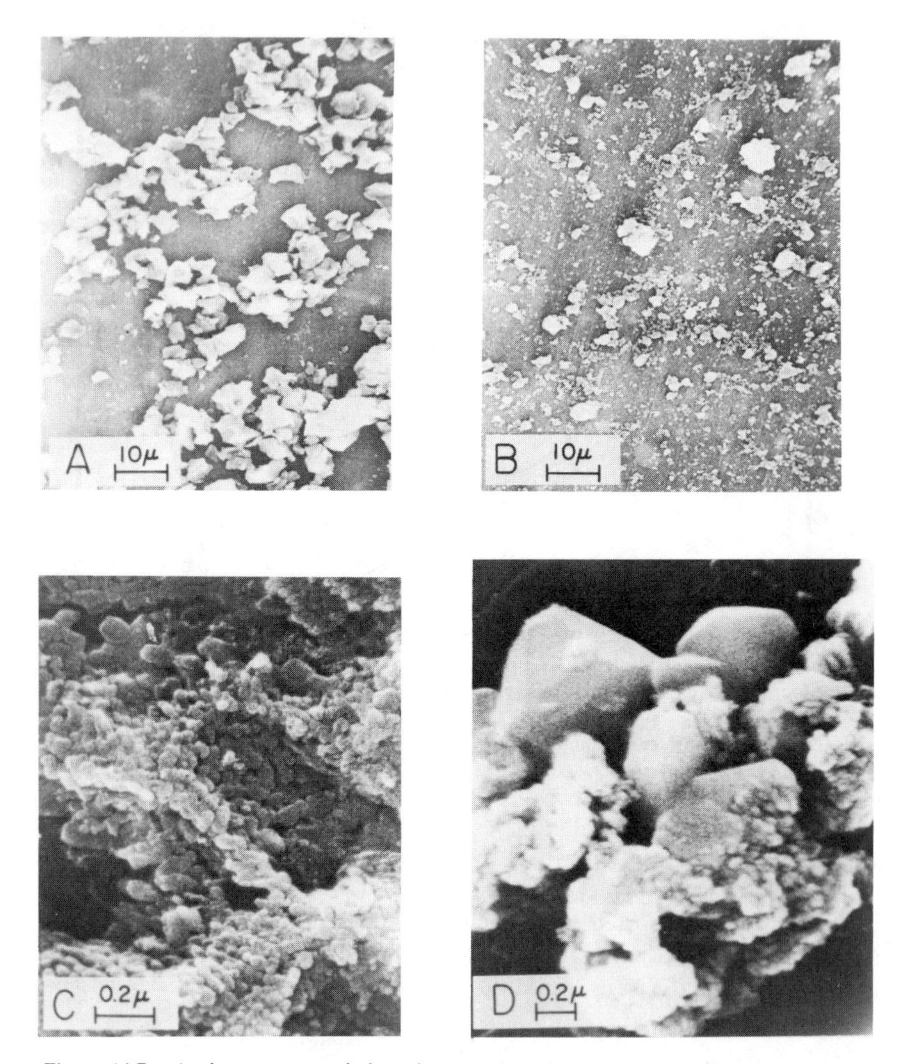

Figure 14.7. Agglomerate morphology by scanning electron microscopy: (*A*) as-prepared powder, (*B*) milled powder, (*C*) "active" alpha alumina, (*D*) transformation from "active" to "nonactive" alpha alumina.

before the autoclave powders could be sintered to theoretical density. As a result an exploratory study on the effects of milling on agglomerate size and distribution was initiated. In Figures 14.7*a* and 14.7*b*, scanning electron photomicrographs (1000×) show the effect of vibratory milling on a typical autoclave powder. While the effect of milling is clearly apparent, some large agglomerates (5 to 15 μm) still persist in the powders even after comminution. The results of the milling study are summarized in Figure 14.8, in which the range of agglomerate sizes for the milled autoclave powders is indicated schematically. Powders on the low end of the indicated range exhibited the best sintering characteristics. The distribution curve for the commercial deagglomerated powder is included in Figure 14.8 for comparison.

Several facts become clear from the figure. (1) Vibratory milling reduces the median agglomerate size by about two to four times, depending on the extent of initial agglomeration. (2) All the milled powders have similar agglomerate sizes and cumulative distribution curves. The agglomerate distributions are slightly sigmoidal and indicate a pronounced coarse fraction over 5 μm. (3) The lowest median agglomerate sizes (~2.3 μm) obtained by the laboratory milling procedure are still almost a factor of 2 higher than those in milled or deagglomerated commercial powder.

Correlation of Sintering Data with Powder Agglomeration Characteristics

The singlemost decisive change in sinterability of the autoclave powders was noticed at the time that milling was introduced as a process variable. Consequently, a method was sought to relate quantitatively the stage of agglomeration in an autoclave powder to its sintering characteristics. To this end, the agglomeration characteristics of the autoclave powders have been described by the use of an "agglomeration factor" described by

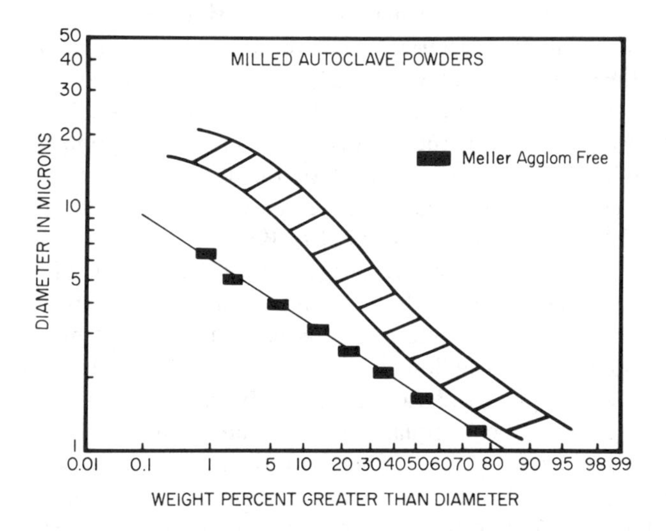

Figure 14.8. Range of agglomerate-size distributions for milled powders.

Johnson *et al*[6] and originally proposed by Balek.[7] It is defined as the ratio of the median agglomerate diameter, as measured by the Coulter technique, to the equivalent spherical diameter of the individual crystallites, as measured by gas adsorption (BET). This agglomeration factor, AF(50), is defined as

$$AF(50) = \frac{\text{median agglomerate diameter (Coulter Counter)}}{\text{equivalent spherical diameter of crystallites (BET)}} \quad (1)$$

An agglomeration factor on the order of 1 corresponds to complete deagglomeration, while higher values indicate increased agglomeration. Since a number of the powders used in this study possessed a bimodal distribution of agglomerate sizes (particularly those synthesized at 2400 and 3100 psi) and since the authors agree with Cutler (Chapter 3) that it is the coarse fraction of the powder that limits its ability to be sintered to theoretical density, another agglomeration factor [AF (10)] has been defined as the ratio of the agglomerate diameter that 10% of the powder is larger than, to the equivalent spherical diameter of the individual crystallites, that is:

$$AF(10) = \frac{\text{diameter that 10\% of agglomerates are larger than (Coulter Counter)}}{\text{equivalent spherical diameter of crystallites (BET)}} \quad (2)$$

In Table 14.3 the two agglomeration factors [AF (50) and AF (10)] are given for the six powders studied in the final-stage sintering studies. The powders are ranked according to their overall sinterability in both the doped and undoped conditions for the unmilled and milled categories. The rankings were made subjectively using sintered densities and visually observed translucency as selection criteria. (Relative rankings for doped and undoped samples of the same powders were nearly always the same.)

An examination of the data in Table 14.3 reveals that, in general, the correlation between increased sinterability and decreased agglomeration factor (smaller agglomerates) is very good. In the unmilled category, the best powder was easily F, which had been screened through a 400 mesh screen; this sample had the second lowest AF(10). Powder D, which sintered not as well as powder F, had a lower AF(50) (25 to 29) but higher AF(10) (132 to 64), indicating that perhaps the assumption that the coarse fraction of agglomerates controls the sinterability is correct. The rest of the powders sintered in order of their agglomeration factors except powder B, which was made from the chopped rod raw material. It was our experience that all measurable quantities being equal, powders made from the chopped rod did not sinter as well as those synthesized from the more massive raw materials (ingot). It was noted also that powders synthesized from chopped rods of Al possess higher agglumeration factors than powders prepared from Al

Table 14.3. Correlation of sinterability with agglomeration factor

Rank	Sample No.	Agglomeration Factor (50)	Agglomeration Factor (10)	% Alpha	Surface Area (m^2/g)
Unmilled					
1	F	29	64	60	8
2	D	25	132	60	11
3	E	22	62	80	6
4	C	46	124	60	11
5	A	120	202	50	35
6	B	91	164	60	14
Milled					
1	D	17	56	60	11
2	C	16	55	60	11
3	B	23	64	60	14
4	A	62	200	50	35
Commercial					
CAF (agglomerate free) ~12			~25	100	12–30
CA (agglomerated) ~50			~110	100	~25

ingots. The top three powders in the unmilled category were ingot-derived powders.

In the milled category, the same general trends can be seen, with the most sinterable powders having the lowest agglomeration factors.

In Table 14.3 the agglomeration factors are given for the two commerical high-purity aluminas used in this study as standards. All through this study the best autoclave powders sintered better than the commercial agglomerate-free powder. Examination of the relative agglomeration factors shows this result to be reasonable.

Since a method of synthesis was developed in which both the surface area and the agglomerate-size distribution of an alumina powder could be obtained by specifying a particular autoclaving condition, a series of four powders, each with different agglomeration factors, AF(10), were synthesized and prepared for sintering studies. Pressures of 2400 and 3100 psi were chosen. At each pressure powders were synthesized by two techniques: (1) steam transfer at saturation and (2) two-stage steam transfer. The autoclaving times were chosen so that the surface areas of the powders would be similar. Under these autoclaving conditions the synthesized powders were pure α-Al_2O_3. The powders were doped and milled for sinter-

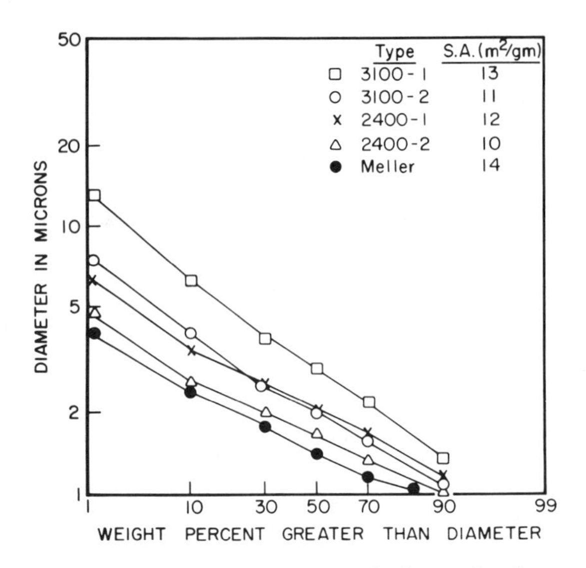

Figure 14.9. Agglomerate-size distributions for final-stage sintering experiments.

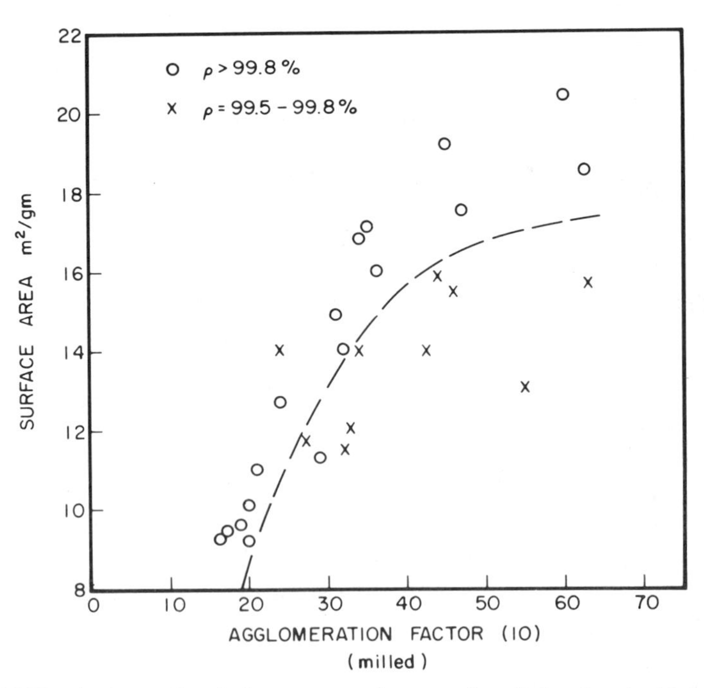

Figure 14.10. Agglomeration factor versus surface area for all doped and milled sintered specimens.

Table 14.4. Effect of agglomeration factor on final density

Sample	Surface Area (m^2/g)	AF(10)	Percent Theoretical Density
3100 (one-stage)	13	53	99.60
3100 (two-stage)	11	30	99.85
2400 (one-stage)	12	27	99.90
2400 (two-stage)	10	17	99.95
CAF	14	22	99.98

ing studies in the normal manner. The agglomerate-size distributions for the powders after milling are shown in Figure 14.9. Pellets of each powder were sintered at 1900°C in vacuum and the results are summarized in Table 14.4.

The results of this set of experiments clearly indicate the importance of the degree of agglomeration, as quantified by the agglomeration factor, in determining the final density of a sintered alumina ceramic. The autoclave powder with the lowest (AF(10)) agglomeration factor (2400 two-stage) sintered to the highest density (99.95%). The agglomerate characteristics of this powder were nearly indentical to those of the commercial agglomerate-free powder, which also sintered to a very high density (99.98%).

Finally, for use in determining what agglomeration factors are necessary to sinter an autoclave powder of a given surface of high density, the data from a large number of sintering experiments are summarized in Figure 14.10. The density of 99.98% of theoretical was chosen arbitrarily to separate the powders into two groups. It is clear from these data that the agglomeration factor [AF(10)] alone cannot predict whether or not a given powder can be sintered to a high density. Powders with higher surface areas can evidently tolerate larger agglomeration factors than powders with smaller surface areas and still be sintered to high densities (>99.8%).

Optical Properties of Sintered Autoclave Alumina Bodies

As stated in the introduction, the objective of this study was to determine the powder characteristics that are necessary and sufficient for the achievement of high sintered densities in polycrystalline Al_2O_3. From the experiments to date, the following necessary powder properties have been identified:

1. Nearly complete conversion to α-Al_2O_3.
2. Doping with a requisite amount of MgO.

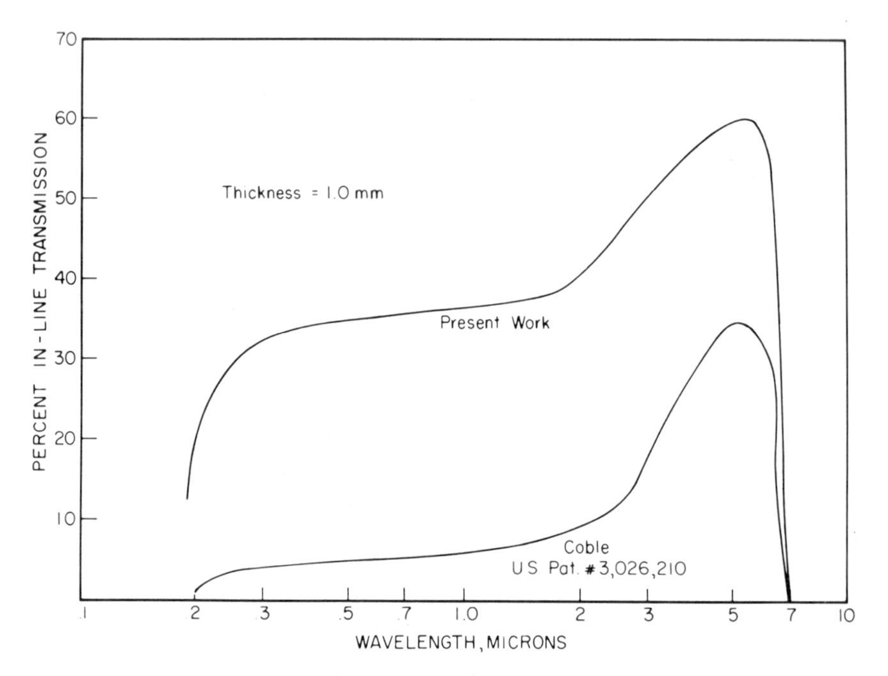

Figure 14.11. Comparison of optical transmission properties.

3. Milling to reduce the median agglomerate size below 1 to 2 μm and the AF(10) agglomeration factor below 10 to 20.
4. A powder surface area that is reasonably high (> 8 m^2/g).

To achieve all these conditions, autoclave powders were prepared by the two-stage process at 2400 and 3100 psi (refer to Table 14.4). Typical in-line, optical transmission data for translucent polycrystalline Al_2O_3 (99.85 to 99.95% theoretical density) prepared from these two-stage powders are given in Figure 14.11. These data, particularly in the visible spectrum for the highest density (99.95%) specimen, are a significant improvement (two to five times) over the optical transmission data of conventionally prepared translucent alumina,[5] shown also in Figure 14.11 for comparison.

To achieve further improvements, porosity levels will have to be reduced by approximately one order of magnitude (i.e., sintered densities over 99.99%). This result will probably require more efficient powder preparation procedures so that median agglomerate sizes are under 1 μm and no agglomerates over 2 to 3 μm exist in the powder.

CONCLUSION

The use of alumina powders synthesized hydrothermally in an autoclave has proven extremely valuable in determining which powder characteristics are important in sintering aluminum oxide to theoretical density. By choosing the appropriate autoclaving conditions, powders with varying α-Al_2O_3 contents, surface areas, and agglomerate-size distributions were formed. A study of the sintering properties of these alumina powders has led to the conclusion that 100% α-Al_2O_3 in the powder is necessary for doping with magnesium oxide and milling it to break up agglomerates has been reaffirmed. The sinterability of an alumina powder was also related to the relative number of ultimate particles in the agglomerates by an agglomeration factor. It was shown that the agglomeration factor required for sintering a powder to translucency varied with the surface area of the powder. The highest sintered densities (99.95%) and best optical-transmission characteristics were obtained by sintering autoclave powders synthesized by the two-stage process. These powders were completely converted to α-Al_2O_3 and were reasonably fine (10 to 15 m^2/g). They possessed median agglomerate sizes after milling of under 2 μm and essentially a log-normal distribution of agglomerate sizes.

ACKNOWLEDGMENTS

The authors express their gratitude to I. Cisan, T. King, and W. L. Taylor of this laboratory for valuable help in sample preparation. E. S. Palik of the Quartz and Chemical Products Department and W. H. Thresh of this laboratory performed the many BET and Coulter analyses required for this work. E. Lifshin of Corporate Research and Development is responsible for the excellent SEM micrographs. The help of J. Cooper, D. McClellan, and D. C. Fries in performing the quantitative X-ray analyses is acknowledged.

REFERENCES

1. R. L. Coble, "Sintering Crystalline Solids. I. Intermediate and Final State Diffusion Models," *J. Appl. Phys.*, **32,** (5), 787–792 (1961); R. L. Coble, "II. Experimental Test of Diffusion Models in Powder Compacts," *J. Appl. Phys.*, **32,** (5), 793–799.
2. J. G. J. Peelen and R. Metselaar, "Light Scattering by Pores in Polycrystalline Materials: Transmission Properties of Alumina," *J. Appl. Phys.*, **45,** (1), 216–220 (1974).

3. (a) K. Torkar and H. Krischner, "Untersuchungen uber Aluminium-hydroxyde und-oxyde," *Monatsh. Chem.*, **91**, 658–668. (1960); (b) H. Krischner and K. Torkar, "Microcrystalline Active Corondum," Vol. 1, G. H. Stewart, ed., *Science of Ceramics*, 1962, pp. 63–76, Academic Press, London.
4. R. S. Gordon and R. T. Tremper, unpublished work
5. G. R. Chol, "Influence of Milled Powder Particle Size Distribution on the Microstructure and Electical Properties of Sintered Mn-Zn Ferrites," *J. Amer. Ceram. Soc.*, **54,** 34–39 (1971).
6. (a) L. Berrin, D. W. Johnson, and D. J. Nitti, "High Purity Reactive Alumina Powders: I, Chemical and Powder Density," *Bull. Amer. Ceram. Soc.*, **51,** (11), 840 (1972); (b) L. Berrin, D. W. Johnson, and D. J. Nitti, "II. Particle Size and Agglomeration Study," *Bull. Amer. Ceram. Soc.*, **51,** (12), 896 (1972).
7. V. Balek, "Temperature Dependence of Characteristic Properties of α-Fe_2O_3 Powders," *J. Mater. Sci.*, **5,** (8), 714 (1970).
8. R. L. Coble, U. S. Patent 3,026,210. (1962).

15
Layer-Silicate Minerals

H. Heystek

Through the centuries the potter has always known that different clays and shales have different plastic, dry, and fired properties, but he did not know why. The dramatic advances in the development of research tools and techniques, especially in the last 25 years, allowed the ceramist to develop a basic understanding of the structure and composition of ceramic raw materials. Rapid and important changes have occurred in the manufacturing processes as a result of increased automation, production,and firing rates, and demands for improved products and quality control. This in turn requires the more accurate control and, in some cases, specific tailoring of raw material that can only be done when these materials are well characterized.

Clay mineralogists have evolved the modern concept of layer silicates (phyllosilicates) as small crystalline hydrous aluminum silicate particles, with magnesium and iron substituting wholly or in part for aluminum in some minerals, and with lithium, calcium, sodium or potassium in others. Ample evidence exists that the unique characteristics that predetermine the ceramic use and application of the layer silicates are as follows:

1. Composition and structure of the layer silicates.
2. Particle shape and size distribution.
3. Nature and amount of nonlayer-silicate constituents.

Characterization of a raw materials system requires a qualitative and quantitative mineralogical analysis and chemical analysis of all constituents, including gangue minerals, organics, and soluble salts. Therefore, knowing the mineralogy and classification scheme of layer silicates in general is particularly valuable. The behavior and properties of the raw material should also be characterized. This includes particle-size distribution, particle shape, exchangeable ions, rheological properties, workability, drying behavior, shrinkage, and strength.

In this chapter the classification scheme for layer silicates and methods for characterizing these materials are described. In 1969 the Clay Minerals Society Nomenclature Committee submitted to the Association International Pour L'Etude Des Argilles (AIPEA) the latest version of a classification scheme[1] for the layer silicates. This classification (Table 15.1) represents a general agreement among clay scientists of 32 countries and is the result of many discussions at International and national clay mineral meetings and with the Commission on New Minerals and Mineral Names of the IMA (International Mineralogical Association).

LAYER-SILICATE CLASSIFICATION

The basic structural and compositional units in layer silicates are Si—O tetrahedral sheets and Mg—OH or Al—OH octahedral sheets combined to form the so-called 1:1, 2:1, and 2:1:1 layer types.

The 1:1 Types

As shown in Table 15.1, the combination of one silica tetrahedral and one octahedral sheet results in the kaolinite–serpentine group. The difference between kaolinite and serpentine is that in the former only two-thirds of the octahedral cation positions are filled by aluminum ions (dioctahedral), whereas in the latter, all the positions are filled by magnesium ions (trioctahedral).

The kaolinite group, with the general composition $Al_2Si_2O_5(OH)_4$, includes the minerals kaolinite, dickite, nacrite, and metahalloysite. This classification proposes that the term "halloysite" be used for the fully hydrated ($4H_2O$) variety and that is a layer-stacking notation be used for the structural varieties kaolinite, dickite, and nacrite analogous to the mica polymorphs. As yet no decision has been made on the appropriate symmetry symbols to be used.[2]

In the dioctahedral 1:1 layer silicates, there is no charge per formula unit, and very little if any substitution occurs within the structural unit.

Table 15.1. Classification scheme for layer silicates

Type	Group (x = charge per formula unit)	Subgroup	Species
1:1	Kaolinite-serpentine $x \sim 0$	Kaolinites	Kaolinite, halloysite
		Serpentines	Chrysotile, lizardite, antigorite
	Pyrophyllite talc $x \sim 0$	Pyrophyllites	Pyrophyllite
		Talcs	Talc
	Smectite $x \sim 0.25$–0.6	Dioctahedral smectites	Montmorillonite, beidellite, nontronite
		Trioctahedral smectites	Saponite, hectorite, sauconite
2:1	Vermiculite $x \sim 0.6$–0.9	Dioctahedral vermiculites	Dioctahedral vermiculite
		Trioctahedral vermiculites	Trioctahedral vermiculite
	Mica $x \sim 1$	Dioctahedral micas	Muscovite, paragonite
		Trioctahedral micas	Biotite, phlogopite
	Brittle mica $x \sim 2$	Dioctahedral brittle micas	Margarite
		Trioctahedral brittle micas	Clintonite
2:1:1	Chlorite x variable	Dioctahedral chlorites	Donbassite
		Di,trioctahedral chlorites	Cookeite, sudoite
		Trioctahedral chlorites	Pennine, clinochlore, prochlorite

However, some reports [3,4] have suggested that iron and titanium can occur within the kaolinite structure. This is not the case with the trioctahedral 1:1 serpentine minerals listed below, where nickel, ferrous and ferric iron, and aluminum are substituted for magnesium in the octahedral positions, and aluminum is substituted for silicon in the tetrahedral sheet.

Chrysotile	$Mg_6Si_4O_{10}(OH)_8$—tubular, monoclinic
Lizardite	$Mg_6Si_4O_{10}(OH)_8$—platy, orthohexagonal
Antigorite	$Mg_6Si_4O_{10}(OH)_8$—platy, monoclinic

Amesite	$(Al_2Mg_4)(Si_2Al_2)O_{10}(OH)_8$
Berthierine	$(Al, Fe^{3+}, Fe^{2+}, Mg)_6(Si_3Al)O_{10}(OH)_8$
Garnierite	$(Ni, Mg)_6Si_4O_{10}(OH)_8$

The 2:1 Types

The 2:1 type of layer silicate refers to the fact that there is one octahedral sheet sandwiched between two tetrahedral sheets in the unit structure and includes the pyrophyllite talc, smectite, and mica groups. The different groups are a result of varying relative surface-charge densities ranging from an essentially zero electron change per formula unit for pyrophyllite and talc to a charge of about 2 for the brittle micas.

Pyrophyllite is a dioctahedral hydrous aluminum silicate, and talc is a trioctahedral hydrous magnesium silicate with limited substitution by aluminum possible.[5] Commercial "talcs" may or may not be predominantly talc, as associated minerals such as calcite, dolomite, quartz, magnesite, and chlorite are common. However, it is the commercial label of "talc" for material containing appreciable quantities of the chain silicates tremolite and anthophyllite that concerns the Health Division of the Mining Enforcement and Safety Administration (MESA) in the establishment of safety standards.[6] Ceramists are aware of the presence of talc or tremolite $Ca_2Mg_5(Si_4O_{11})_2(OH)_2$ because of the different body-pressing characteristics and fired mineralogy[7] imparted by the particle shape or chemistry of these minerals.

For the 2:1 minerals with a layer charge of 0.25 to 0.6, there is still general national and international disagreement between smectite and montmorillonite-saponite as a group name. The present approach is to leave the question open and await the evolvement of the most acceptable term through usage.

In the so-called expanded 2:1 minerals, substitution for silicon may occur in the tetrahedral sheet, while the ions commonly substituting for one another in the octahedral sheet are aluminum, magnesium, lithium, and ferrous and ferric iron. Frequently these substitutions are by ions of similar ionic radius rather than by those of the same charge. The subsequent deficit or excess charge on the layer structure is compensated by loosely held interlayer potassium, sodium, calcium, and magnesium ions and organic molecules.[8] This explains the relatively high cation-exchange capacity of the smectites and their expansion in the *c*-direction when water enters between the unit layers. The interlamellar distances are variable, depending on water content and the exchangeable ion present. The nature of the initially absorbed water layers is considered to have a partially ordered structure[9] that is affected specifically by the surface forces and structure of the smectites and the different exchangeable ions that may be present.

The dioctahedral smectites are montmorillonite, $Al_{2-x}Mg_xSi_4O_{10}(OH)_2$, with all the substitution in the octahedral sheet; beidellite, $Al_2Si_{4-y}Al_yO_{10}(OH)_2$, with all the substitution in the tetrahedral sheet; and nontronite, $(F3, Al)_2\ Si_{4-y}Al_yO_{10}(OH)_2$, with substitution dominant in the tetrahedral sheet. The trioctahedral smectites are represented by the clay minerals hectorite, $Mg_{3-x}Li_xSi_4O_{10}(OH)_2$; saponite, $Mg_{3-x}Al_xSi_{4-y}Al_yO_{10}(OH)_2$, and sauconite, $Zn_{3-x}Al_xSi_{4-y}Al_yO_{10}(OH)_2$.

The vermiculites have a higher charge per formula unit (0.6 to 0.9) than the smectites, and this charge deficiency is balanced by interlayer exchangeable magnesium ions. The magnesium ions are hydrated, resulting in an interlayer double sheet of water molecules and a *c*-axis dimension of 14 Å that collapses to 10 Å on heating. Most vermiculites are trioctahedral, $Mg_{3-x}Fe_XSi_{4-y}Al_yO_{10}(OH)_2$, although the dioctahedral analogue has been reported to be common in soils.[10] The industrially known macrocrystalline vermiculite, which has exfoliation properties when heated, is actually a hydrobiotite, which is an interstratified biotite and vermiculite.

The mica minerals have a 2:1 layer structure similar to the smectites except that the charge deficiency (the substitutions) is typically concentrated in the tetrahedral sheet. As counterions, potassium is tenaciously held in the interlayer space, resulting in a characteristic 10 Å basal spacing. Much isomorphous substitution occurs in addition to many possible variations in the manner of stacking of the structural units above each other. Some typical dioctahedral micas are muscovite, $KAl_2Si_3A10_{10}(OH)_2$, and paragonite, $NaAl_2Si_3A10_{10}(OH)_2$. Trioctahedral micas are phlogopite, $K(Mg, Fe)_{2+x}(Fe, Al)_{2-x}Si_{2+x}(Al,Fe)_{2-x}O_{10}(OH)_2$.

The micaceous minerals common in soils, clays, shales, and slates have been called illite as originally defined by Grim *et al.*[11] It has become clear, however, that the terms illite or hydromica have been used to cover a fairly wide range of minerals. These terms are considered useful field terms and presently are not included in the classification table.

The brittle micas represented by the dioctahedral variety margarite, $CaAl_2Si_2Al_2O_{10}(OH)_2$, and the trioctahedral mineral clintonite, $CaMg_3Al_2Si_2O_{10}(OH)_2$, have not been reported as occurring in clays. Although not of interest to ceramists, they are layer silicates and are included in the classification scheme.

The 2:1:1 Types

In chlorites the fundamental unit layer is a 2:1 structure plus an interlayer hydroxide sheet resulting in a 2:1:1 layer. The common chlorite minerals occurring in clays are trioctahedral, such as pennine, clinochlore, and donbassite, has been described by Lazarenko[12] (dioctahedral in both the 2:1 layer and the interlayer hydroxide sheet). Cookeite and sudoite[13] are examples of di,trioctahedral chlorites that are dioctahedral in the 2:1 layer but trioctahedral in the interlayer hydroxide sheet. It is known that the various chlorites $(Mg, Fe, Al)_6(Si, Al)_4O_{10}(OH)_8$ differ from each other in the kind or amount of substitution within the hydroxide sheet or the tetrahedral or octahedral sheets in the 2:1 layer. In addition, variations not only occur in the orientation of the sheets within a 2:1 layer, but also in the stacking sequence of chlorite units.

Interstratified-Layer Silicates

Considering the similarities in crystalline structure and chemical composition of the various layer silicates, it is not surprising that random and regular intimately interstratified units of two distinct species are quite common. In addition, the layer silicates usually originate as the weathered products of primary minerals, and alternation sequences such as biotite, vermiculite, chlorite, kaolinite; and mica, illite, vermiculite, montmorillonite in soils have been reported frequently[14] and are supported by synthetization in laboratory experiments.[15] It is to be expected, therefore, that interlaying of these minerals in various combinations would occur as a result of preferential weathering of some layers with respect to others.

No general agreement has yet been reached on the preferred terminology for interstratified minerals, but it has been suggested that with the random or irregularly interstratified-layer silicates the material should be described in terms of the component layers with the dominant component listed first, for example, "irregular chlorite–mica interstratification." Furthermore, if the minerals are regularly interstratified, distinctive names, such as corrensite (1 chlorite/1 vermiculite), should not be used but rather the prefix "regular", for example, "regular chlorite–vermiculite interstratification." Reported in the literature are occurrences of interstratifications such as illite–montmorillonite, vermiculite–biotite, vermiculite–chlorite, mica–chlorite, montmorillonite–chlorite, saponite–chlorite, berthierite–chlorite, montmorillonite–kaolinite, and muscovite–kaolin.

LAYER-SILICATE CHARACTERIZATION

When considering the use of layer silicates in ceramic processing, we find that particle shape, fine particle size (frequently 2 μm or less), large surface area, and ion-exchange capacity of the materials are usually the principal factors that control viscosity, plasticity or workability, pressing behavior, green and dry strength, and drying shrinkage. Characterizing layer silicates as to their structural and chemical composition, particle morphology and size distribution, and cation-exchange capacity is therefore absolutely necessary to the ceramist in designing bodies and controlling processing (Table 15.2).

The most valuable instrumental characterization techiques are X-ray diffraction, thermal methods, electron optical methods, and infrared spectroscopy. No one technique, however, is adequate on its own but should be used in combination with another and with total chemical analysis, cation-exchange capacity determinations, and surface-area measurements.

X-Ray Diffraction

Since the proposed layer-silicate classification is based on the structure and composition of these minerals, it is not surprising that X-ray diffraction has become indispensable in the identification of clay minerals. Much has been written[16-18] on sample preparation and treatment techniques (orientation, glycolation, heating), so only brief mention is made here of the basic diagnostic parameters used to identify the different layer silicates.

The interplanar spacings normal to the (001) cleavage as shown in Table 15.3 are the most significant criteria used in X-ray differentiation between the layer silicates. The 7 Å minerals, kaolinites and serpentines, are further distinguished on the basis of their 060 reflections, which are 1.48 to 1.49 and 1.52 to 1.59 Å, respectively. This also applies to the 9.3 Å minerals, pyrophyllite talc. X-Ray diffraction patterns of the smectites (expanded 2:1 minerals) yield typical 12 to 15 Å basal spacings that change to 17 Å spacings after treatment with ethylene glycol. To separate the 14 Å chlorites and vermiculites, heat treatment followed by X-ray diffraction is used. The 14 Å spacing of chlorites persists up to 550°C, whereas at 350°C, dioctahedral and trioctahedral vermiculites show a shift of the spacing to 12 Å and 9 Å, respectively.

Frequently, samples for X-ray investigation are given extensive preparation to aid in the identification of natural mixtures of layer silicates. The material is ground, fractionated to obtain the minus 2 μm fraction, and sometimes chemically treated to remove iron or organic phases. This should be kept in mind when the bulk characterization of the material is considered, because the amount and type of the nonlayer silicate admixtures, such as free quartz, feldspar, carbonates, sulfates, oxides and hyroxides of iron, anatase, and amorphour material, could play a dominant role in its ceramic properties.

The problems of analyzing clay mineral mixtures quantitatively by X-ray diffraction are principally due to preferential orientation effects, variations in crystallinity,[19] chemical composition, and particle size as well as to intimate layer interstratification. Most methods attempt to relate the diffrac-

Table 15.2. Simplified characterization of layer silicates

Mineral	Composition (%)			Cation exchange capacity	Particle		Ceramic Properties	Industrial Uses
	SiO_2	Al_2O_3	Other	(meq/100 g)	Size (μm)	Shape		
Kaolinites	45	39		3–15	1–10	Hexagonal plates	White firing Refractory, c/32 +	Refractories; Whitewares; Structural clay products
Halloysite				10–40	0.2 × 1	Tubes	Low to medium: Plasticity Drying shrinkage Green strength	Refractories; Whitewares; Structural clay products
Pyrophyllite	67	28		0		Plates		Refractories; Whitewares; Structural clay products
Talc	63		32 MgO	0		Plates	White firing	Whitewares; Structural clay products
Smectites	55	20	Fe_2O_3 MgO Li_2O	75–150	1	Plates	Usually red burning	Structural clay products
Vermiculite	35	15	25 MgO	100–150	2	Plates	c/04 to c/7	Structural clay products
Mica	55	25	5 K_2O	10–40	1	Plates	High: Plasticity	Structural clay products
Chlorites	29	20	30 MgO	10–40	2	Plates	Drying shrinkage Green strength	Structural clay products
Interstratified layer silicates	combinations of above			25–77	2	Plates		Structural clay products

Table 15.3. Interplanar spacings of layer silicates

Type	Group	Basal Spacing (Å)
1:1	Kaolinites–serpentine	7
2:1	Pyrophyllitetalc	9.3
	Smectite	12–15
	Vermiculite	14
	Mica	10–11
2:1:1	Chlorite	14

tion intensity from a given compound to its percentage in artificial mixtures[20] or use internal standards such as $Zn(OH)_2$,[21] pyrophyllite,[22] and aluminum powder.

If other techniques, such as selective dissolution analyses,[23] thermal gravimetric analyses, and cation-exchange capacities, are used in conjunction with X-ray diffraction, reasonably good quantitative mineralogical estimates can be made. Hussey[24] recently optimized chemical allocation to minerals within prescribed limits using a simultaneous linear equations program.

Thermal Methods

In conjunction with X-ray diffraction, thermal techniques such as differential thermal analysis,[25] thermogravimetry, and differential thermogravimetry are extremely useful in identifying such things as amorphous or poorly crystalline phases, organic material, pyrite, carbonates, and gibbsite. Improvements in commercially available equipment, such as increased sensitivity, atmosphere control, and registration of evolved gases, have made thermal techniques more effective in quantitative determinations. A committee on standardization established by the First International Conference on Thermal Analysis in 1965 recommended[26] a uniform reporting system for authors so that thermal data could be critically assessed.

Absorption Spectroscopy

Infrared absorption spectroscopy is now widely used to distinguish structural hydroxyl groups from structural and sorbed water[27] and is effective in investigating poorly crystallized and amorphous materials. It is a technique complementary to X-ray diffraction and thermal investigations in obtaining structural data.

Mossbauer spectroscopy[28] has become increasingly useful in the study of the coordination (octahedral or tetrahedral) or valence (Fe^{3+} or Fe^{2+}) of iron in layer silicates. In addition, siderite and goethite, common gangue minerals in clays and shales, can also be detected.

Ultraviolet visible spectroscopy[29] also shows promise as a quick method of determining the presence, location, and valance state of iron in a layer-silicate structure.

Electron-Optical Methods

Electron microscopy is of value in supplying morphological data and in the case of kaolinite (plates) and halloystie (tubes) results in conclusive identification. The use of high-resolution electron microscopy on samples that were ultramicrotomed[30] made possible direct photographs of the (001) atomic planes and representation of 7, 10, and 14 Å structures. This technique allows for the identification of interstratified clay minerals[31] and the textural or fabric study of undisturbed samples of clays, shales, and slates. The scanning electron microscope with its great depth of focus is also uniquely suited to textural and fabric investigations of the fine-grained layer silicates.

A combination of selective-area electron diffraction with electron microscopy allows no one to obtain diffraction diagrams of single crystals with diameters of about 1000 A. In addition, electron-microscope microprobe analysis[4] can be carried out to determine the elemental composition of individual clay-mineral particles. Finally, direct electron-microscope measurements of dispersed clay particles provide a valuable check on results of standard[32] particle-size-distribution methods.

Surface Area

In a clay–water system the surface area of the clay particles and the ion population in the aqueous media are the major factors affecting the properties of a slip; the extrusion or pressing of a body; and the strength, porosity, and drying characteristics of the processed ware. The most widely used method to determine surface area of fine dry particles involves the adsorption of gases as described by Brunauer *et al.*[33] (see Chapter 6). However, this technique does not record the so-called internal surfaces of the expandable 2:1 layer silicates. Simple and rapid procedures have been devised to overcome this problem, that is, measurement of the quantity of glycerol, dodecylamine hydrochloride,[34] or methylene blue[35] that is adsorbed by clays. In the last case, the cation-exchange capacity of the clay is also determined.

Cation-Exchange Capacity

Many procedures for determining the cation-exchange capacity (C.E.C.) of clays have been described in the literature, including the conventional Kjeldahl method and, most recently, the use of an ammonia electrode.[36] The C.E.C. of kaolinite, which is given in Table 15.2, is low[37] compared with that of other clay minerals but varies among different kaolinites. A detailed study of some Georgia kaolinites[38] has shown that their exchange capacity, surface area, and iron and titanium content increased as their crystalline stacking order decreased.

In addition to total C.E.C., it is important to determine the amount and type of naturally occurring exchangeable ions. These ions play an important role in the rheological properties and drying characteristics[39] of a raw material, especially in structural clay products, in which very little or no beneficiation is utilized.

NEED FOR COMPLETE CHARACTERIZATION

In all layer-silicate deposits, horizontal and vertical variability occurs to a greater or lesser degree, creating a problem for the ceramic manufacturer intent on producing products of uniform quality. In the past the practice was to use a large amount of different raw materials in a body in an attempt to offset changes in the properties of any one component. Today the trend is to use materials guaranteed by suppliers to conform to specified compositions and particular ranges of physical properties.

Much has been said and written about the characterization of "pure" layer silicates. But pure systems do not exist even with beneficiated materials such as kaolins, ball clays, and talcs used in the whiteware industry. Therefore, a suppliers data sheet should, in addition to such information as chemical analysis, percent of plus 325 mesh material, viscosity curve, and percent drying shrinkage, include a total quantitative mineralogical analysis, screen analysis, particle-size distribution to less than 1 μm, particle-shape and surface-area data, C.E.C., and exchangeable ions and soluble salts present. As an example, it is necessary to know (1) if a "talc" contains fibrous tremolite in addition to the platy talc; (2) the quantity of free quartz, calcite, or dolomite that may be present, or (3) in the case of a ball clay, how much quartz, smectite, mica, or organic material occurs with the predominant mineral kaolinite.

To the brick, clay pipe, or lightweight aggregate manufacturer, a careful mineralogical evaluation of his particular clay or shale deposit is imperative. If, for example, kaolinite, illite, chlorite, quartz, calcium carbonate, iron hydroxide, and organic material are present, it is important to determine their quantitative distribution pattern. With additional information, such as soluble salts present, particle morphology, chemical analysis, and total size distribution available as presented in Table 15.4, a planned mining, stockpiling, blending, and quality-control program becomes feasible. In years past it was stated that this approach was too expensive for the industrial clay products industry—today it is too expensive not to do it. Once a deposit is completely characterized, it is possible to determine what compositional variables are the most critical and to concentrate on them for control purposes.

A complete detailed qualitative and quantitative characterization is also necessary of material used in rheological and viscosity studies. This must be the initial step toward developing precise methods for evaluating workability or plasticity and obtaining data that can be used in predicting or controlling the process behavior of ceramic bodies.

An aid toward a better understanding and correlation of characterization techniques and data from different laboratories should result from the Source Clay Minerals Project recently activated by the Clay Minerals Society.[40] A clay-minerals repository has been established at the Geology Department of the University of Missouri. Unbeneficiated, well-blended samples of natural clays are available for study. It is hoped that all data obtained by investigators will be accumulated in the repository for future evaluation.

Finally, it is necessary to stress again that to obtain quantitative characterization data, more than one investigative technique must be used. Also, it is imperative that the sample used for characterization be representative of the shipment, stockpile, or deposit considered for use by a consumer.

REFERENCES

1. S. W. Bailey, "Summary of National and International Recommendations on Clay Mineral Nomenclature," 1969–1970 CMS Nomenclature Committee, *Clays and Clay Minerals*, **19**, 129–132 (1971).
2. S. W. Bailey, "Polymorphism of the Kaolin Minerals," *Amer. Miner.*, **48**, 1196–1209 (1963).
3. D. L. Dolcater, J. K. Syers, and M. L. Jackson, "Titanium as Free Oxide and Substituted Forms in Kaolinites and Other Soil Minerals," *Clay Clay Miner.*, **18**, 71–79 (1970).
4. W. B. Jepson and J. B. Rowse, "The Composition of Kaolinite—An Electron Microscope Microprobe Study," presented at 23rd Annual Clay Minerals Conference, Cleveland, Ohio, 1974.

5. I. S. Stemple and G. W. Brindley, "A Structural Study of Talc and Talc–Tremolite Relations," *J. Amer. Ceram. Soc.*, **43,** 34–42 (1960).
6. A. Goodwin, ed., *Proc. Symp. Talc,* **1974,** Bureau of Mines IC 8639.
7. H. Heystek and E. Planz, "Mineralogy and Ceramic Properties of Some California Talcs," *Bull. Amer. Ceram. Soc.*, **43,** (8), 555–561 (1964).
8. J. Moum, C. N. Rao, and T. S. R. Ayyar, "A Natural 17 A Montmorillonite–Organic Complex from Alleppey, Kerala State—India," *Clays Clay Miner.* **21,** 89–95 (1973).
9. P. F. Low, "Physical Chemistry of Clay—Water Interaction," Adv. Agronomy, **13,** 269–327 (1961).
10. G. Brown, "The Dioctahedral Analogue of Vermiculite," *Clay Miner. Bull.*, **2,** 64–70 (1953).
11. R. E. Grim, R. H. Bray, and W. F. Bradley, "The Mica in Argillaceous Sediments," *Amer. Miner.*, **22,** 813–829 (1937).
12. E. K. Lazarenko, "Donbassites, A New Group of Minerals from the Donetz Basin," *C. R. Acad. Sci., USSR,* **28,** 519–521 (1940).
13. G. Muller, "The Identification of Dioctahedral Four Layer Phyllosilicates (Sudoit, in the series Sudoit-Chlorite)," *Int. Clay Conf., 1st, London,* **1963**, Pergamon, pp. 121–130.
14. M. J. Wilson, "The Weathering of Biotite in Some Aberdeenshire Soils," *Mineral. Mag.*, **35,** 1080–1093 (1966).
15. R. Roy and L. B. Sand, "A Note on Some Properties of Synthetic Montmorillonites," *Amer. Miner.*, **41,** 505–509 (1956).
16. G. Brown, ed., *The X-Ray Identification and Crystal Structure of Clay Minerals,* Mineralogical Society, London, 1961.
17. D. Carroll, *Clay Minerals: A Guide to Their X-ray Identification,* The Geological Society of America, Special Paper 126, 1970.
18. K. G. Sansom and D. White, "Aggregation and Dispersion in Clays with Particular Reference to Montmorillonites, " *Trans. Br. Ceram. Soc.*, **70,** 163–165 (1971).
19. D. N. Hinckley, "Variability in Crystallinity Values Among the Kaolin Deposits of the Coastal Plain of Georgia and South Carolina," *Clays Clay Miner., 11th, Proc. Natl. Conf.*, **1963** pp. 229–235.
20. A. D. Buck, "Quantitative Mineralogical Analysis by X-ray Diffraction," U.S. Army Engineer Waterways Experiment Station Miscellaneous Paper C-72-2, February 1972.
21. M. H. Mossman, D. H. Freas, and S. W. Bailey, "Orienting Internal Standard Method for Clay Mineral X-Ray Analysis," *Proc. Natl. Conf., Clays Clay Miner., 15th,* **1967** pp. 441–453.
22. J. M. Roberts, "X-Ray Diffraction and Chemical Techniques for Quantitative Soil Clay Mineral Analaysis," Ph.D. Thesis, Dept. Civil Eng., Pennsylvania State University, June 1974.
23. C. A. Alexiades and M. L. Jackson, "Quantitative Clay Mineralogical Analysis of Soils and Sediments," *Proc. Natl. Conf., Clays Clay Miner., 14th,* **1966**, pp. 35–52.
24. G. A. Hussey, "Use of a Simultaneous Linear Equations Program for Quantitative Clay Analysis and the Study of Mineral Alteration During Weathering," Ph.D. Thesis, Department of Agronomy, Pennsylvania State University, August 1972.
25. R. C. Mackenzie, ed., *The Differential Thermal Investigation of Clays,* MIneralogical Society, London, 1957.

26. H. G. McAdie, "Recommendations for Reporting Thermal Analysis Data," *Anal. Chem.*, **39,** 4 (1967).
27. V. C. Farmer, ed., *The Infrared Spectra of Minerals,* The Mineralogical Society, London, 1974.
28. C. E. Weaver, J. M. Wampler, and T. E. Pecuil, "Mossbauer Analysis of Iron in Clay Minerals," *Science,* **156,** 504–508 (1967).
29. S. W. Karickhoff and G. W. Bailey, "Optical Absorption Spectra of Clay Minerals," *Clay Clay Miner.* **21,** 59–70 (1973).
30. J. L. Brown and M. L. Jackson, "Chlorite Examination by Ultramicrotomy and High Resolution Electron Microscopy," *Clays Clay Miner.* **21,** 1–7 (1973).
31. T. Yoshida, "Elementary Layers in the Interstratified Clay Minerals as Revealed by Electron Microscopy," *Clays Clay Miner.,* **21,** 413–420 (1973).
32. O. Lauer, Grain Size Measurements on Commercial Powders," issued by Alpine Ag, Augsburg, 1966.
33. S. Brunauer, P. H. Emmett, and E. Teller, "Adsorption of Gases in Multimolecular Layers," *Amer. Chem. Soc.,* **60,** 309–319 (1938).
34. G. W. Kalb and R. B. Curry, "Determination of Surface Area by Surfactant Adsorption in Aqueous Suspension, I. Dodecylamine Hydrochloride," *Clays Clay MIner.,* **17,** 47–57 (1969).
35. P. T. Hang and G. W. Brindley, "Methylene Blue Absorption by Clay Minerals, Determination of Surface Areas and Cation Exchange Capacities (Clay-Organic Studies XVIII)," *Clays Clay Miner.* **18,** 203–212 (1970).
36. E. Busenberg and C. V. Clemency, "Determination of the Cation Exchange Capacity of Clays and Soils Using an Ammonia Electrode," *Clays Clay Miner.* **21,** 213–217 (1973).
37. R. E. Grim, *Clay Mineralogy,* McGraw-Hill Book Co., Inc., New York, 1953.
38. H. H. Murray and S. C. Lyons, "Further Correlations of Kaolinite Crystallinity with Chemical and Physical Properties," *Proc. Natl. Conf., Clays Clay Miner., 8th,* **1960** pp. 11–16.
39. R. T. Bailey, "Measurement of Strength of Unfired Clays and Ceramic Bodies," *Trans. Br. Ceram. Soc.,* **71,** 272–277 (1972).
40. W. F. Moll, "Source Clay Minerals Project," *Clays Clay Miner.,* **21,** 71–73 (1973).

PART THREE

PARTICULATE–WATER SYSTEMS

Most ceramic processes at some stage involve a particulate–liquid state. Water is the most commonly used liquid for such purposes. The interactions of the aqueous phase with ceramic particles, the properties of slips, the rheology of solutions, and the phenomena associated with mixing and drying are important areas covered in the chapters that follow.

16

The Structure of Water and its Role in Clay–Water Systems

W. G. Lawrence

Many papers have been published on the effects of adding different ions on the properties of clay–water systems. However, little has been written from the viewpoint of the structure of water and how it relates to the effects in clay–water systems. In this chapter a review is presented of water structure and how ions affect this structure. This information is used to discuss possible structural effects in the interaction of aqueous solutions with clay surfaces.

WATER STRUCTURE

The water molecule has the configuration shown in Figure 16.1.[1] The H—O—H bond angle is 104° 40′, and therefore the molecule is a dipole with a dipole moment of 1.71 to 1.97×10^{-18} esu. Because of its dipole nature, a water molecule attracts ions and other molecules, forming combinations that are not related to electron transfer or sharing even though

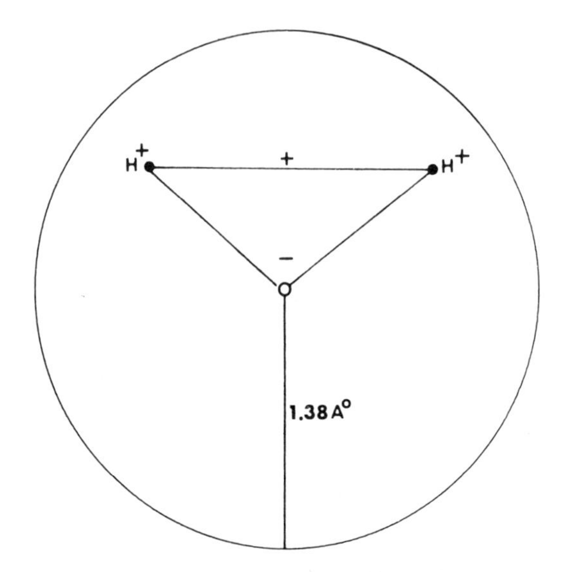

Figure 16.1. Structure of the water molecule.

these combinations may be stoichiometric. Among the types of interactions possible,[2] two that are important for our consideration are: (1) dipole–dipole attractions between water molecules, resulting in some type of structure found in associated liquids and (2) ion–dipole attractions, which are responsible for the fact that ions in water solutions are not separate entities but are associated with and travel with a certain number of water molecules.

With water and other liquids having angular intensity distributions of the molecular field, the arrangement of molecules deviates from a close-packed system. The molecules arrange in a structure that best satisfies the bond angles determined by the angular charge distribution.[3]

The details of the structure of water have been extensively investigated and many theories have evolved. However, this subject is still controversial and no single theory is widely accepted. Two general approaches have been to view water as association complexes and as defective crystalline systems. Association complexes are disconnected clusters of water molecules. By means of X-ray diffraction studies, Morgan and Warren[4] showed that the average number of nearest neighbors around a water molecule is 4.6. This is close to the tetrahedral coordination of ice. However, they cautioned that their results cannot be interpreted in terms of certain number of neighbors at certain distances. At any given time, any given molecule may have more or less than four neighbors, and these neighbors may be at a continuous variety of distances. The authors therefore suggest the possibility of having disconnected association complexes.

Lennard-Jones and Pople[5] maintain that the presence of broken intermolecular bonds necessary for "disconnected association complexes" must be rejected because of thermodynamic considerations. They maintain that the weak, diffuse second maxima in the radial distribution curve is due to distortion rather than breaking of intermolecular bonds.

Eyring[6] and Frenkel[7] have introduced the concept of lattice defects to the water structure. Erying considers the water molecules to be associated with regular lattice positions, but some of those lattice sites may be left vacant ("hole" theory of liquids). These holes or vacancies can then diffuse or move by the process of a molecule jumping into the vacant position and leaving behind another hole.

Figures 16.2 represents the idealized water lattice. Each water molecule has tetrahedral coordination, having four nearest neighbors at a distance of 2.85 Å at 1.5°C and 3 Å at 83°C.[4] Megaw[8] indicates the hexagonal axial ratio $c/a = 1.6289$ at 0°C. This is somewhat lower than the ideal ratio for the tetrahedral structure, which is $c/a = \sqrt{8}/3 = 1.6330$. The unit cell of ice at 0°C is $a_0 = 4.5135$ Å and $c_0 = 7.3521$ Å, with each unit cell containing four molecules.

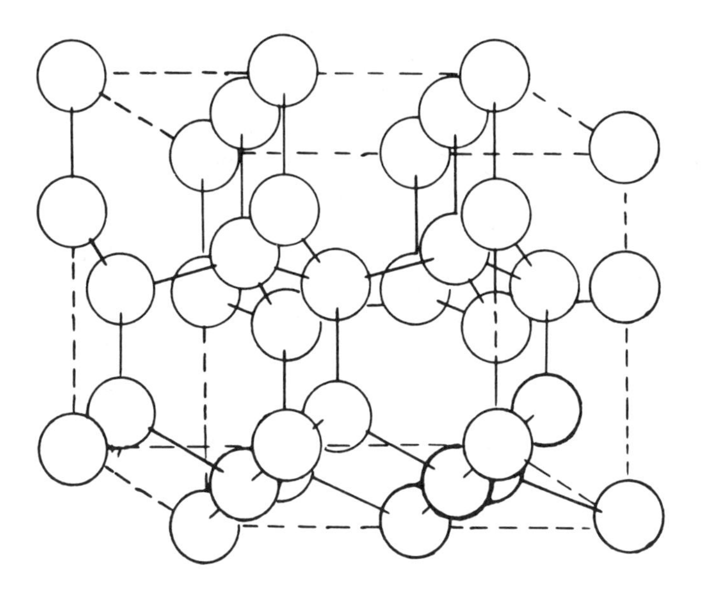

Figure 16.2. The ideal water lattice.

ROLE OF THE HYDROGEN BOND

Forslind[3] has discussed in detail the process of molecular association and the hydrogen bond. For a hydrogen bond to form between two electronegative atoms, it is first necessary that there be a close approach of these atoms. The dipole–dipole interaction between water molecules themselves is apparently a slow, sluggish type based on the difficulty of nucleation of the condensed vapor phase. It appears that the initial approach between water molecules is due to the van der Waals and molecule dipole attraction until interpenetration of electron clouds is achieved. At this point the attraction between the hydrogen nuclei and the two unpaired electrons of the oxygen provides a significant orienting influence.

It is known that the hydrogen atom may be simultaneously attracted to two or more electronegative atoms, such as fluorine, oxygen, chlorine, and nitrogen. The hydrogen atom may thus act as a bond between oxygens. When a hydrogen is bonded to one electronegative element, such as oxygen in the case of H_2O, the strong pull exerted by the atom on the bonding electrons leaves an effective positive charge on the hydrogen sufficient to cause attraction for a second electronegative atom.

Frank and Wen[9] present a resonance scheme for the hydrogen bond in water. This considers resonance between three types of structures. It recognizes that hydrogen bond formation is an acid–base interaction. When a bond is formed, one molecule becomes more acidic and the adjacent molecule more basic than the unbonded water molecule. Thus a cooperative-type bonding is pictured such that when one bond is formed the tendency is for several to form and when one bond is broken a cluster breaks down. This gives a picture of flickering clusters of various sizes and shapes first bonded or associated and then breaking down. The half-life of such clusters is 10^{-10} to 10^{-11} seconds, which corresponds to the dielectric and bulk relaxation time of water but is long enough (10^2 to 10^3 times a molecular vibration period) to be meaningful.

Tanford[10] presents facts of interest in consideration of the water structure. The heat of sublimation of ice is 12,200 cal/mole at 0°C. Of this, 1400 cal/mole represents the heat required to give the water molecules the random motion they possess in the gaseous state (translational enthalpy). The remainder, 10,800 cal/mole, must be the energy required to break the hydrogen bonds holding the structure together. The heat of fusion of ice is 1400 cal/mole, only about 15% of the energy required to break the hydrogen bonds in the crystal. It is therefore obvious that liquid water must retain a considerable amount of structure, which is maintained by the hydrogen bonding mechanism.

These observations and the previously cited data that show the O—O distance to be approximately 2.85 Å and the O—O bond angle to be tetrahedral indicate that the water molecules are hydrogen bonded, with the hydrogen located 1.00 Å from one oxygen and 1.76 Å from the other. It can jump from one equilibrium position to the other, which can temporarily create H_3O^+ and OH^- ions. A small number of these ions are always present in ice crystals.

EFFECTS OF NONPOLAR MOLECULES

Frank and Evans[11] present data on the temperature coefficient of solubility of simple nonpolar gas molecules in water. From such data the entropy loss in the solution process is calculated. For nonpolar gases in nonpolar solvents the entropy loss is in the 10 to 15 cal/degree mole range. In water, however, it is in the 25 to 40 cal/degree mole range. Since the entropy of a system is an indication of the degree of disorder present, the extra entropy loss in aqueous solutions of nonpolar gases means the water structure has become more ordered under the influence of the dissolved nonpolar molecules. As Frank and Evans describe it "the water builds a microscopic iceberg round the nonpolar molecule." This writer prefers to say that the water structure is immobilized adjacent to such molecules.

If one considers the flickering-cluster type of water structure in which the half-life of the cluster (10^{-10} to 10^{-11} seconds) depends on the resonance scheme of the hydrogen bonding, it seems reasonable that such a cluster having one side or a portion of its surface adjacent to a nonpolar molecule with which hydrogen bonding is not possible would have a longer half-life. It is being protected. This concept is borne out by the observation that the entropy loss is greatest for the larger solute molecules, such as radon and chloroform.

EFFECT OF IONS ON WATER STRUCTURE

If nonpolar molecules show such effects as previously described, one can imagine the greatly exaggerated effects that might be produced by ionic solutions. In addition to the possible effect of surface alone, there is the intense electric field due to the ionic charge. When small distances are involved, the field intensity is around 1 million V/cm, assuming the bulk dielectric constant of water (80) gives a field of 0.5×10^6 V/cm at a distance of 6 A from the center of a monovalent ion. This value is certainly minimal, since the dielectric constant of water has been shown to be approximately 2 to 4 around a point charge such as an ion.

Frank and Wen[9] present the simple model shown in Figure 16.3 for the structure modifications in water produced by a small ion. From previous discussions regarding the dipole nature of the water molecule, dipole–dipole interactions, ion–dipole interactions, and the cooperative type of bonding in water resulting in association of the liquid in clusters having a certain half-life, a picture of the effect of such ions emerges.

The high field strength of the ion will lead to immobilization of the nearest-neighbor water molecules as a result of the ion–dipole attraction (*A*). This type of attraction is shown in Figure 16.4. The second region (*B*) is one in which the arrangement is more random than normal, and the outer region (*C*) contains normal water having the normal water structure as a result of dipole–dipole attraction and hydrogen bonding. The inner region (*A*) will contain water molecules highly oriented with the negative side of the water molecule towards the positive side pointing away from the positive ion. The outer region (*C*) will maintain the tetrahedral arrangement as previously described. The intermediate region between *A* and *C* may be of finite width and will contain no more orientation disorder than either *A* or *C*.

Frank and Evans[11] believe that region *A* is composed of nearest-neighbor water molecules, with the extent of region *A* being increased by a small or multiply charged ion. Such ions induce additional structure (entropy loss)

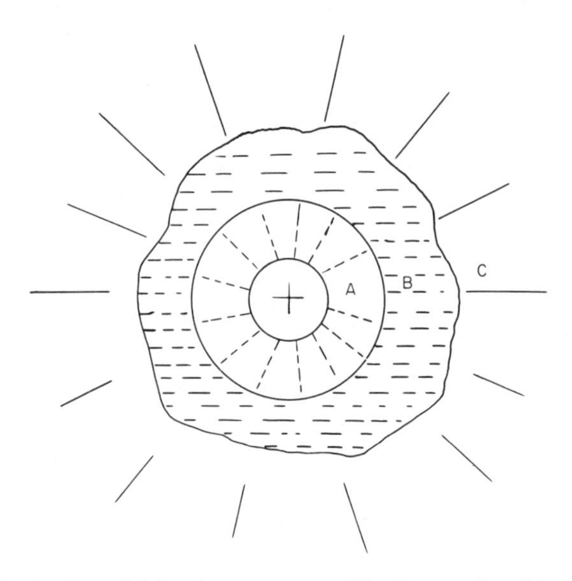

Figure 16.3. A simple model for the structure modifications produced by a small ion: (*A*) region of immobilization of water molecules due to ion–dipole attraction, (*B*) region of structure breaking, (*C*) structurally "normal" water.

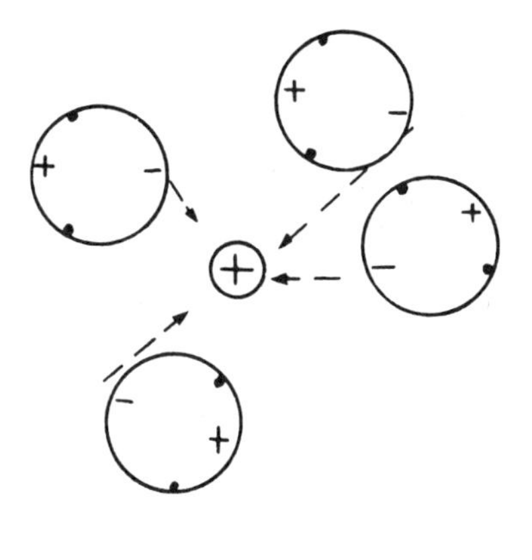

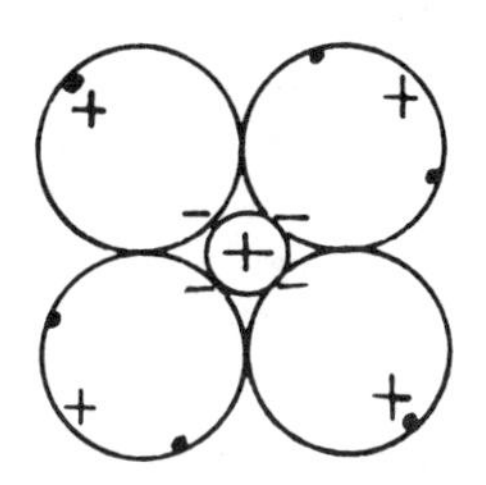

Figure 16.4. Ion–dipole attraction.

beyond the first water layer. Thus region *A* may be enlarged at the expense of region *B*. The large single charged ions, such as Cs^+ and 1^-, seem to have enough structure-breaking effects to enlarge region *B* at the expense of *A* and may in some cases eliminate it entirely. This effect would involve entropy gain.

Thus it is possible to predict the structure-altering properties of a considerable number of ions from entropy data. Cations smaller or more highly charged than K^+ are structure formers, becoming more so the smaller and more highly charged they are. K^+ itself is slightly structure-breaking, with this tendency increasing through Rb^+ and Cs^+. F^- is a structure former, but the other halide anions, which are larger, are structure breakers. OH^- is a structure former, and NO_3^-, $C10_4^-$, and SO_4^{-2} are structure breakers.

In any concept pertaining to the effect of ions on water structure, the structure formed about any ion is not a rigid, firmly held layer resembling ice; instead it is a water structure, which on the average has a higher degree of order than does normal water. Although hydration numbers have been assigned to various ions to indicate how many water molecules are associated with an ion, the values are quite variable and meaningless. Thus one may find hydration numbers for the Na^+ ion varying from 71 to 1.

Samoilov[12] has provided an excellent dynamic picture of the hydration process. His concept is based on the logical approach that the water molecules are in constant motion; thus continual exchange may take place between the water molecules closest to the ion and those farther away. The magnitudes defining the frequency of exchange of water molecules near the ions are the quantitative characteristics of ion hydration in solutions.

If a water molecule is in the immediate vicinity of other water molecules, then it spends an average time γ in the immediate vicinity of a certain selected water molecule. Time γ is the average time the two molecules remain as neighbors. The activation energy of exchange or the energy required to separate them as neighbors is E. The time a water molecule is in the immediate vicinity of an ion will not be γ, because an ion is not energetically equivalent to a water molecule. The time a molecule is associated with the ion will be γ_i, where $\gamma_i \neq \gamma$ mainly because of the difference in the activation energy of exchange. This energy is no longer E but $E + \Delta E$.

γ_i/γ and ΔE are quantitative characteristics of ion hydration in solutions. If an ion firmly holds the water molecules it means the time a molecule spends in the vicinity of the ion is much greater than the time it spends associated with another water molecule and γ_i/γ is large. For permanent bonding to an ion $\gamma_i/\gamma = \infty$. Any decrease in γ_i/γ indicates weakening of the ion bond with the water molecules of the solution. γ_i/γ is related to ΔE by the relation $\gamma_i/\gamma = \exp(\Delta E/RT)$; thus finding ΔE for the ion results in a quantitative description of its hydration.

Samoilov[12] worked out the method for calculating ΔE for individual ions from experimental data on self-diffusion in water and the temperature coefficients of ion mobility in solutions and arrived at the following relationship

$$\frac{1}{U_i}\frac{dU_i}{dT} + \frac{1}{T} - \frac{1}{D}\frac{dD}{dT} = \frac{\Delta E}{RT^2}\,\frac{1}{1 + \alpha \exp(\Delta E/RT)}$$

where U_i = ion mobility
D = self-diffusion coefficient of water
α = numerical coefficient with a value of 0.0655

The values of ΔE and γ_i/γ calculated from the above equation are given in Table 16.1.

Table 16.1. ΔE and γ_i/γ for some monatomic ions (21.5°C)

Ion	Li^+	Na^+	K^+	Cs^+	Cl^-	Br^-	I^-	Mg^{2+}	Ca^{2+}
ΔE(kcal/mole)	0.73	0.25	−0.25	−0.33	−0.27	−0.29	−0.32	2.61	0.45
γ_i/γ	3.48	1.46	0.65	0.57	0.63	0.61	0.58	86.3	2.16

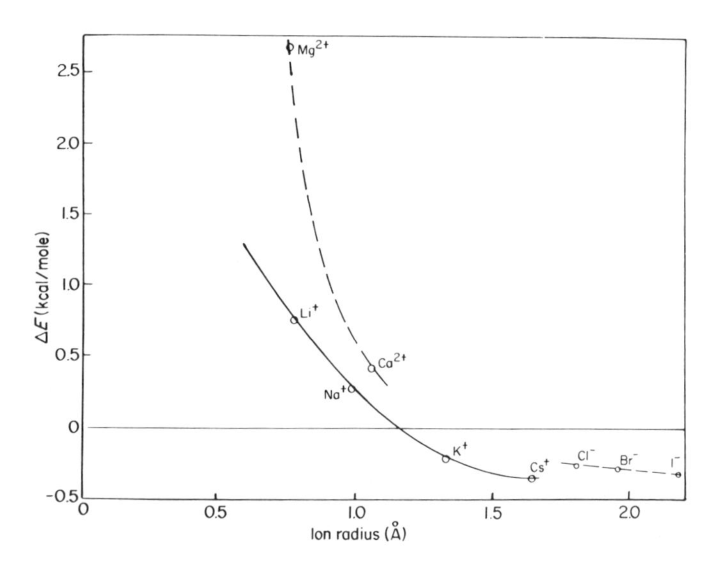

Figure 16.5. ΔE versus ion radius for some cations and anions.

The values for ΔE given in Table 16.1 are plotted against ionic radius in Figure 16.5. It is seen that ΔE decreases with increasing radius and that at approximately 1.20 Å, radius changes from positive to negative. In this case $\gamma_i/\gamma \sim 1$ and a water molecule in the vicinity of an ion is more mobile than in pure water. This provides a concept termed "negative hydrogen." Wang,[13] in studying self-diffusion of water in solutions of KCl and KI, showed them to be higher than in pure water, which is supporting evidence of the negative hydration concept.

RELATIONS BETWEEN ION CHARACTERISTICS AND IMMOBILIZATION OF WATER MOLECULES

An increase in the ordered structure of water involves entropy loss, while a decrease in structure involves entropy gain. It is thus possible to use the entropy of solution ΔS which is the entropy change (calories per degree per mole) in passing the hypothetical mole fraction of unity in solution, as a basis for correlating structure effects of ions. It is thus possible to calculate ΔS^{st}, which is the contribution to the entropy change ΔS due to the effect of ions on the structure of water. Robinson and Stokes[14] present data for a number of ions (Table 16.2).

For all alkali and halide ions except the smallest (Li^+ and F^-) ΔS^{st} indi-

Table 16.2. Entropy of solution of monatomic ions in water[14]

Ion	ΔS	ΔS^{st}
F^-	−40.9	−3.5
Cl^-	−26.6	+10.2
Br^-	−22.7	+13.9
I^-	−18.5	+17.9
H^+	−38.6	
Li^+	−39.6	−1.1
Na^+	−33.9	+4.0
K^+	−25.3	+12.0
Rb^+	−23.1	+14.1
Cs^+	−21.3	+15.7
Mg^{2+}	−84.2	−27.6
Ca^{2+}	−65.5	−10.7
Al^{3+}	−133.	−42.0

cates an increase of disorder in the structure of water, being the greatest for the largest ions (Cs^+ and I^-). The negative ΔS^{st} for Mg^{2+}, Ca^{2+}, and Al^{3+}, on the other hand, indicates these ions would promote a long-range order in the water structure.

There is obviously a size effect involved as well as a charge effect. A plot of the ionic field strength z/a^2 for the ions versus ΔS^{st} from Tables 16.2 and

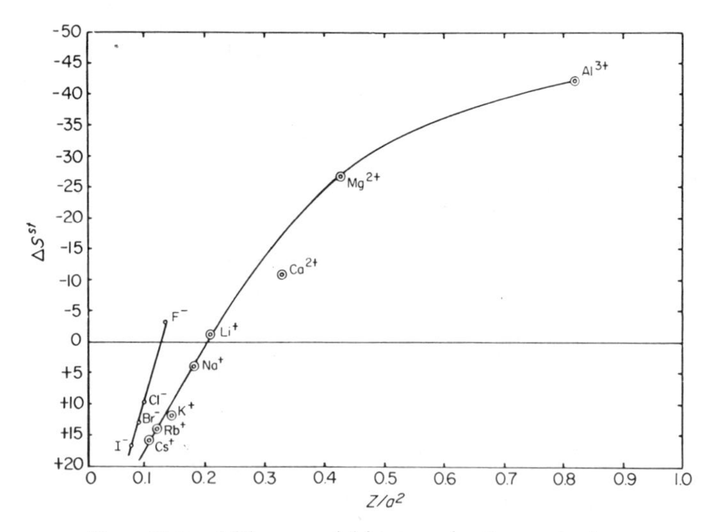

Figure 16.6. ΔS^{st} versus z/a^2 for several cations and anions.

16.3 is shown in Figure 16.6. This shows good correlation for cations and anions if they are considered separately. One would expect some difference between the two because of the difference in bonding mechanism between a cation-H_2O and an anion-H_2O molecule. In the case of the anion-H_2O molecule interaction, the role of the hydrogen bonding must be considered. The hydrogen bonding mechanism is most effective between anions having high electronegativity. Thus F^-, being the most electronegative of the anions in this series, has an abnormally high effect as far as increasing the ordered structure of water. This effect is reduced as the electronegativity of the anion is in the series of F^-, Cl^-, Br^-, and I^- decreased.

Bingham[15] presents data (Table 15.3) on the ionic elevations of fluidity for several cations. The plot of this data versus z/a^2 (Figure 16.7) for the ions results in a curve very similar to that of Figure 16.6. The correlation of this data is not as good as shown by the plot of the entropy ΔS^{st} data.

Table 16.3. Ionic elevations, Δ, and z/a^2 for selected cations[15]

Cation	R_c	R_{H_2O}	$R_c + R_{H_2O}$	a^2	z/a^2	Δ
Cs^+	1.69	1.38	3.07	9.42	0.106	2.59
Pb^{2+}	1.21		2.59	6.72	0.299	2.6
Rb^+	1.48		2.86	8.20	0.122	1.86
NH_4^+	1.43		2.81	7.9	0.126	0.44
K^+	1.33		2.71	7.35	0.136	0.28
H^+	1.38		2.76			−6.41
Ag^+	1.26		2.64	6.96	0.144	−8.91
Na^+	0.95		2.33	5.44	0.184	−9.60
Hg^{2+}	1.10		2.48	6.25	0.32	−12.2
Li^+	0.60		1.98	3.92	0.255	−14.0
Ba^{2+}	1.35		2.73	7.19	0.279	−25.3
Sr^{2+}	1.13		2.51	6.30	0.318	−28.4
Ca^{2+}	0.99		2.37	5.61	0.356	−31.3
Co^{2+}	0.72		2.10	4.42	0.450	−34.4
Cu^{2+}						−34.7
Mn^{2+}	0.80		2.18	4.75	0.421	−34.8
Cr^{3+}	0.65		2.03	4.12	0.730	−35.4
Zn^{2+}	0.74		2.12	4.50	0.444	−35.6
Mg^{2+}	0.65		2.03	4.12	0.484	−36.5
Cd^{2+}	0.97		2.35	5.52	0.361	−37.2
Ni^{2+}	0.69		2.07	4.30	0.465	−39.1
Be^{2+}	0.31		1.69	2.86	0.70	−45.0
Cr^{3+}	0.65		2.03	4.12	0.730	−49.6
Fe^{3+}	0.60		1.98	3.92	0.77	−52.2
Al^{3+}	0.50		1.88	3.53	0.855	−70.5

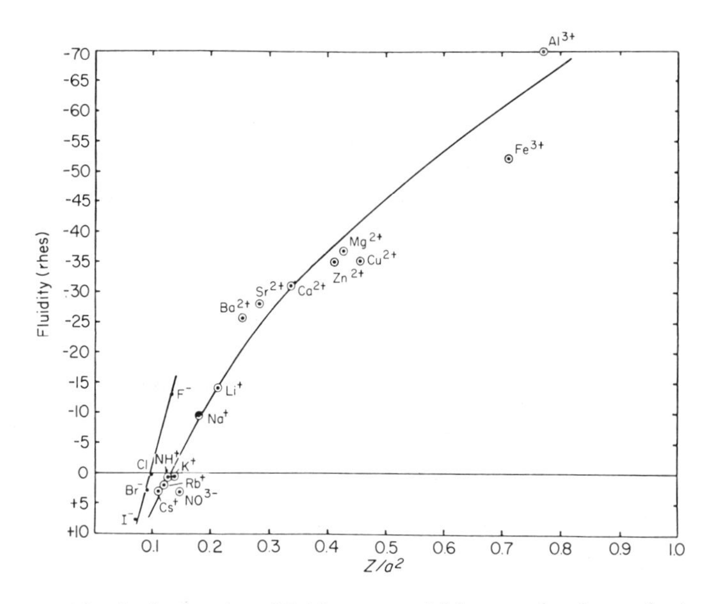

Figure 16.7. Ionic elevation of fluidity versus z/a^2 for several cations and anions.

Bingham's data indicate the Na^+ ion to be a structure promoter, while the entropy data show it to be a structure modifier. The structure-promoting tendency of ions is related to the field strength of the ion; the higher the field strength the greater the surface promotion due to the ion. This involves the size and charge of the ion, as the smaller the size and larger the charge, the greater the field strength.

THE "WATER HULL"

The concept of immobilization of water induced by ions of high charge and small size provides a basis for the mechanism of formation of a "water hull" or "solvated layer" surrounding a clay particle. Since the surface of a clay crystal is composed of positive and negative sites resulting from the exposure of such ions as Al^{3+}, Si^{4+}, O^{2-}, and OH^-, one would expect such active sites to exert an influence on the surrounding water. This influence might be even more pronounced, since these ions at the surface are part of a rigid crystal lattice. They have the capability of attracting the dipole water molecule or other positive or negative ions to satisfy their charge deficiency. The extent or thickness of the water layer built up will depend on the types of ions associated with the liquid. Those ions that are structure promoters, such as Ca^{2+}, Mg^{2+}, and Al^{3+} as shown in Figure 16.8, will result in a larger, thicker solvated layer being built up around the particle. This would be comparable to enlargement of area *A* in Figure 16.3. This concept has been verified in part by East[16] who has calculated the water-film thickness from shrinkage measurements and volume water loss. For a kaolinite in the size range 0.4 to 0.8 μ, the water-film thickness was 80 Å for H-kaolin, 86 Å for Na-kaolin, and 106 A for Ca-kaolin. The experimental techniques involved in the precise measurement of these thicknesses are extremely difficult and these values are only qualitative indications of size that serve to confirm the theory.

The influence of the size of ions on their ability to enhance the development of the water structure has been pointed out by many investigators.

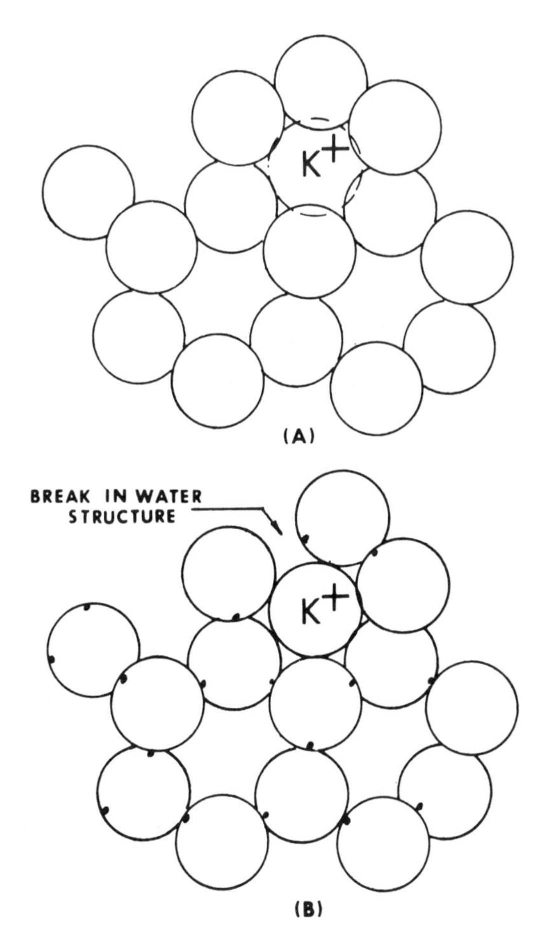

Figure 16.8. Water structure showing size of K^+ ion with respect to hole in hexagonal ring.

Forslind[3] suggests that small ions that fit into the water structure without disrupting it would enhance its development. Larger ions would be expected to retard its development. From the data presented, the critical size radius for a monovalent cation is approximately 1.36 Å. This is in close agreement with the size of the "hole" in the hexagonal water structure. Ions such as K^+ or Cs^+, being larger than this critical size, would disrupt this structure (Figure 16.8) unlike those ions that are smaller or more highly charged such as Ca^{2+}, Mg^{2+}, and Al^{3+} (Figure 16.9), which would retain the structure.

The extent or thickness of the water layer built up around a clay particle depends on the type of ions associated with the water, with such properties

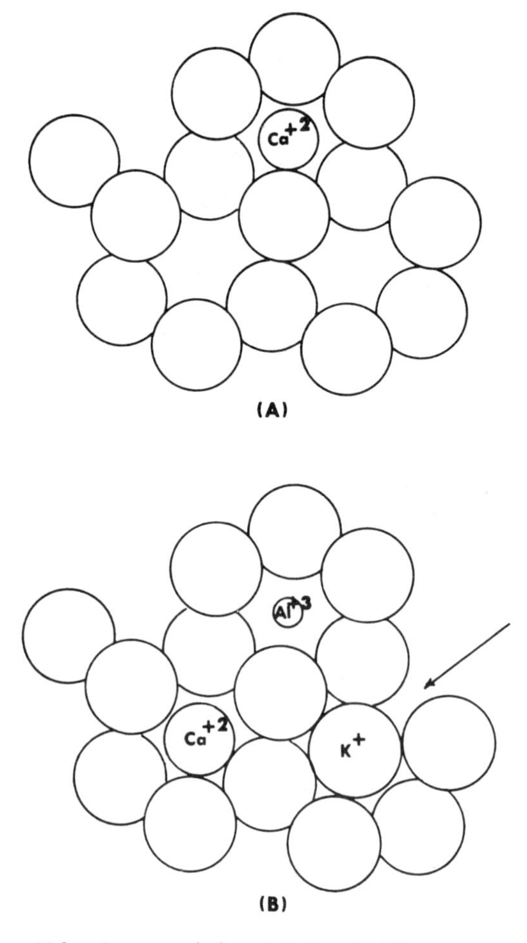

Figure 16.9. Accommodation of Ca^{+2} and Al^{+3} in water structure.

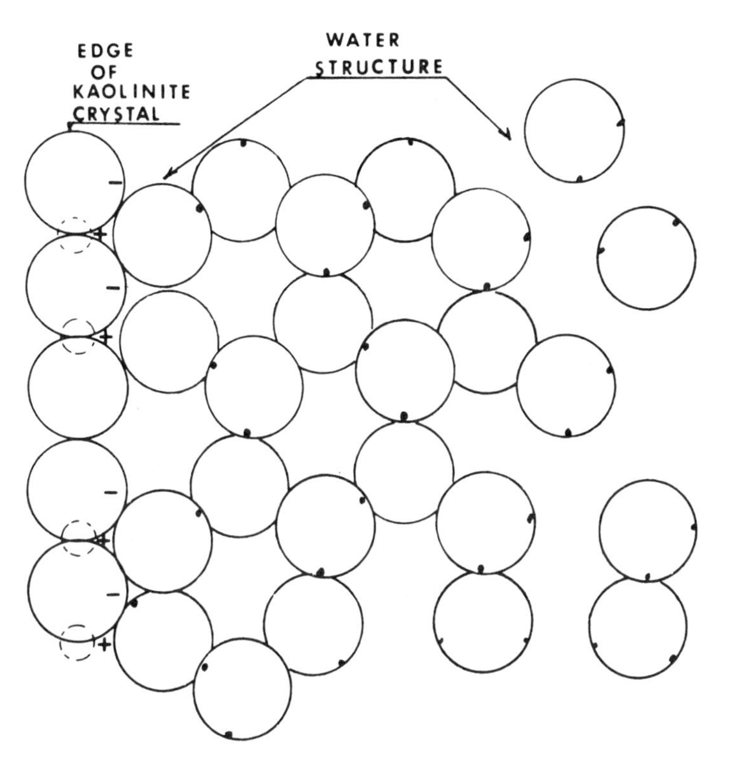

Figure 16.10. Kaolinite surface (left) and buildup of hexagonal water structure on this surface.

as plasticity and viscosity being greatly affected. In systems having small particle size, large surface area, and high solids content, the separation distance between particles may approach the water-film thickness. The properties of this interparticle water determine the behavior of the system.

The excellent geometric fit between the kaolinite crystal and the ice structure has been pointed out by Mason.[17] Because of the small (−1.1%) misfit with ice, kaolinite is a natural effective nuclei for ice formation. Thus one would expect very little disorder at the kaolinite–water interface and good bonding between the two. Figure 16.10 shows the buildup of the water structure on the edge of a kaolinite crystal and the good fit and the hexagonal ring arrangement of the water molecules. Figure 16.11 shows the effect of the large K^+ ion on the structure. It is obvious that the addition of such ions disrupts the structure and prevents its buildup to any great distance, and it can be seen that it would take little force to shear such a system.

Figure 16.12 shows the case for the Ca^{2+} ion as the impurity. This strengthens the structure and enhances its development to larger distances,

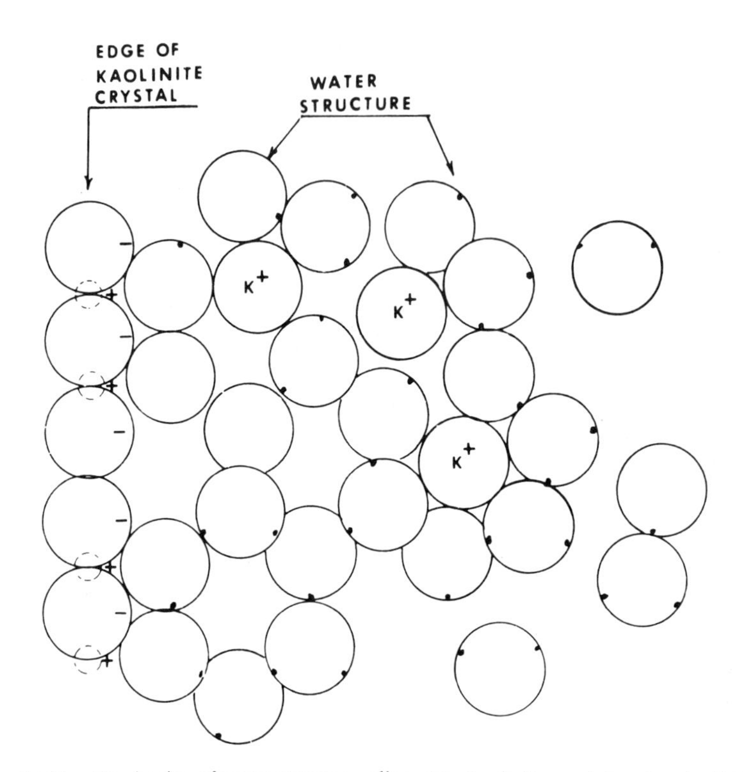

Figure 16.11. Weakening of water structure adjacent to kaolinite crystal surface by K^+ ion.

and if such a structure reaches dimensions approaching the interparticle distance, high forces would be required to shear the system.

CONCLUSIONS

It is obvious from this discussion that there is no one theory of water structure that completely explains its behavior. The theory proposed to explain the effect of ions on the structure as related to the behavior of clay–water systems is certainly not applicable in all instances. If size is a controlling influence, with the suggested 1.36 Å being critical, it does not explain the fact that the viscosity of clay suspensions is reduced equally by Li^+, Na^+, K^+, and Cs^+ as shown by Johnson and Norton.[18]

As a ceramist studies the water structure it becomes evident that there are similarities between the water and glass structures. Both have tetrahedral coordination with oxygens bonded by a cation in a random network having only short-range order. The addition of the large monovalent ions disrupts or weakens the structure in both cases. The addition of the smaller

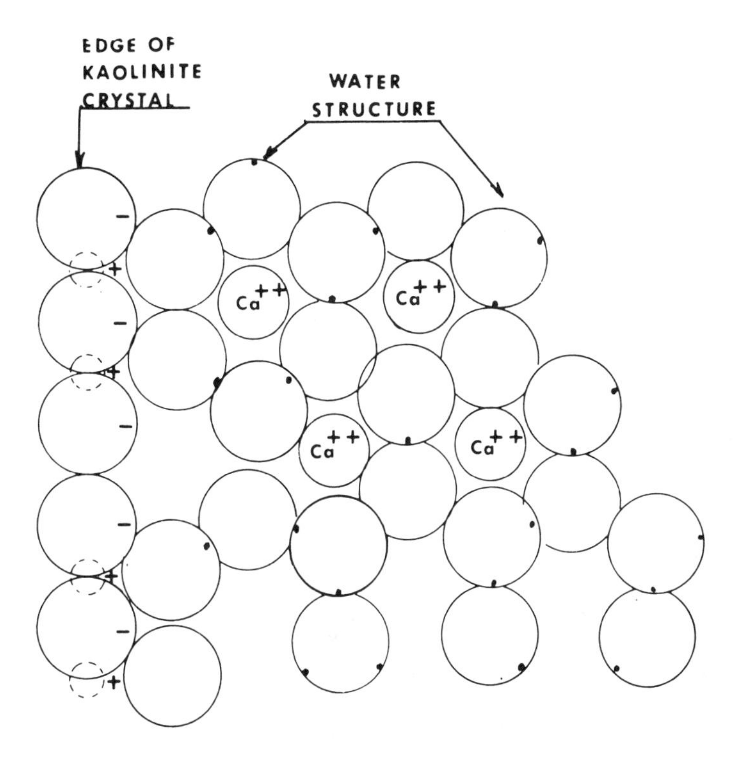

Figure 16.12. Water structure adjacent to kaolinite crystal surface containing Ca^+ ions.

multiply charged ions strengthens the structure of each. Theories have been presented pertaining to the structures of water and glass that have nearly identical concepts, Tilton[19] on glass structure and Pauling[20] on the water structure. Water containing high concentrations of ions, when cooled, becomes increasingly viscous and freezes without crystallization into a glasslike material.

Further clarification must await the development of techniques for more-detailed study of the nature of the water structure in the immediate vicinity of ions and clay-particle surfaces.

REFERENCES

1. J. D. Bernal and R. H. Fowler, *J. Chem. Phys.*, **1**, 515 (1933).
2. T. Moeller, *Inorganic Chemistry*, Wiley, New York, 1952, p. 187.
3. E. Forslind, "A Theory of Water," Swedish Cement and Concrete Research Institute at the Royal Institute of Technology, Stockholm, 1952.
4. J. Morgan and B. E. Warren, *J. Chem. Phys.*, **6**, 666 (1938).

5. J. Lennard-Jones and J. A. Pople, *Proc. Roy. Soc. London, Ser. A.*, **205,** 155 (1951).
6. H. Eyring, *J. Chem. Phys.*, **4,** 283 (1936).
7. J. Frenkel, *Kinetic Theory of Liquids*, Clarendon Press, Oxford, 1946.
8. H. D. Megaw, *Nature*, **134,** 900 (1934).
9. H. S. Frank and Wen-Yong Wen, "Interactions in Ionic Solutions," *Discuss. Faraday Soc.*, **24,** 133–140 (1957).
10. C. Tanford, "The Structure of Water and Aqueous Solutions," *Temp. Measurement Control Sci. Ind.*, **3,** 123–129 (1963).
11. H. S. Frank and M. W. Evans, *J. Chem. Phys.*, **13,** 507 (1945).
12. O. Ya Samoilov, "A New Approach to the Study of Hydration of Ions in Aqueous Solutions," *Discuss. Faraday Soc.*, **24,** 141–146 (1957).
13. J. H. Wang, *J. Phys. Chem.*, **58,** 686 (1954).
14. R. A. Robinson and R. H. Stokes, *Electrolytic Solutions; Measurement and Interpretation of Conductance, Chemical Potential and Diffusion in Solutions of Simple Electrolytes*, Academic, New York, 1955, p. 50
15. E. C. Bingham, *J. Phys. Chem.*, **45,** 885 (1941).
16. W. H. East, "Fundamental Study of Clay: X, Water Films in Monodispersed Kaolinite Fractions," *J. Amer. Ceram. Soc.*, **33, 7,** 211–218 (1950).
17. B. J. Mason, "The Growth of Snow Crystals," *Sci. Amer.*, **1,** 204 (1961).
18. A. L. Johnson and F. H. Norton, "Fundamental Study of Clay: II Mechanism of Deflocculation in the Clay–Water System," *J. Amer. Ceram. Soc.*, **24,** (6) 189–203 (1941).
19. L. W. Tilton, "Noncrystal Ionic Model for Silica Glass," *Jr. Res. Natl. Bur. Stand.*, **59,** (2) 139 (1957).
20. L. Pauling, *The Nature of the Chemical Bond*, Cornell University Press, Ithaca, New York, 1960, p. 464.

17

Particle-Size Distribution and Slip Properties

G. W. Phelps
M. G. McLaren

Casting slips are fluid suspensions of one or more particulate ceramic materials dispersed in a liquid (usually water) at high solids volume by deflocculating agents. A casting operation involves consolidation of suspended solids into a semirigid mass through removal of a portion of the liquid by an absorbent mold. Casting-slip suspension particles are normally predominantly noncolloidal (> 0.5 μm), although some colloidal material must be present to insure slip stability and good rheological qualities.

Deflocculation of clay and clay-based aqueous suspensions by alkali hydroxides and hydrolyzable alkali salts has received detailed attention,[1,2] as has the role of organic[3] and inorganic[4] polyions in promoting deflocculation of clay slips. The potential-determining functions of hydrogen and hydroxyl ions in deflocculation of oxide slips are well established.[5-8] Selective adsorption of organic polyions in deflocculation of metal oxides has also been thoroughly investigated.[9-12]

Although interrelationships between rheological and deflocculation properties and slip casting have been the subject of a great many studies,[13-17] the part played by particle size in general and by particle-size distribution in particular seems to have received little attention in the literature.[18,19] This discussion deals with literature of polydisperse systems and the role of particle-size distribution in governing deflocculation and rheology of ceramic suspensions.

SLIP RHEOLOGY AND PARTICLE-SIZE DISTRIBUTION

Some slips move easily at slower rates of shear but resist stirring at higher speeds, assuming a dry, grainy appearance. When allowed to stand quietly the dry-appearing slip returns to its original shiny, fluid state. This phenomenon is termed dilatancy,[20] although shear hardening has been found to occur in advance of actual volume increase in a number of systems,[21] so some authorities[22] prefer the term "shear thickening." Other kinds of slip become progressively more fluid as shearing rate rises, and upon cessation of stirring these systems thicken more or less rapidly. This phenomenon is usually termed "pseudoplastic," although some observers recommend the more general expression "shear thinning."[22]

Alumina Slips

Calcined alumina powders with different particle-size distributions were compared for their flow behavior as slips. The flow characteristics were measured with a capillary tube viscometer (Cooke-Harrison).[23,24] Size distributions were determined by a sedimentation technique (Numinco Sedigraph 5000, Micromeritics Instrument Co.) after the powder was dispersed with sodium hexametaphosphate in water in a Waring Blendor. For rheology measurements, a powder was deflocculated in water with Darvan 7. A "practical loading limit" was determined by the ability of a 0.1 hp mixer (Lightnin', Mixing Equipment Co., Inc.) to continue stirring the slip. All slips studied were loaded to this limit. Each powder, therefore, was characterized by a different limit.

Three alumina powders were studied individually and in two combinations: a "coarse" alumina, a "fine" alumina, and a commercial casting grade alumina. The combinations were a 4% fine alumina–96% coarse alumina mixture and a calculated body comprised of a mixture of casting-grade alumina, coarse alumina, and fine alumina. The particle-size distribution for the five resulting powders is given in Figure 17.1*a*.

The flow curves for the five slips are shown in Figure 17.1*b*. The coarse alumina was strongly shear thickening and, in fact, would not increase in flow rate at shear stresses above about 400 dynes/cm^2. The casting-grade alumina was slightly shear thickening. The fine alumina, in contrast, was shear thinning.

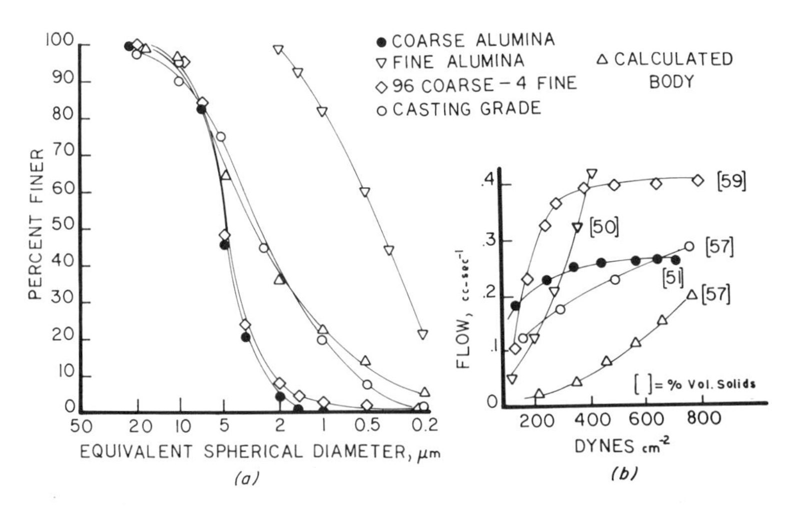

Figure 17.1. Particle-size distribution and rheology of alumina slips: (*a*) particle-size distribution and (*b*) flow behavior for different aluminas deflocculated in water.

Adding 4% fine alumina to the coarse alumina did not change the shear-thickening character possessed by the coarse alumina alone. However, the mixture flowed more rapidly at each shear stress (a decrease in the apparent viscosity). The calculated body was less fluid than the commercial casting alumina slip and showed moderate shear thinning.

The loading limits for the five alumina slips are noted in brackets in Figure 17.1*b*. The fine and coarse powders, having narrow size distributions, had a lower loading limit than the other three powders.

Quartz Slips

Figure 17.2*a* gives distributions for two air-elutriated quartz powders (A and B), each having minimal colloid and narrow size distribution. Slips deflocculated with sodium silicate in water attained volume loadings for quartz A and B of 48 and 50% percent, respectively. The flow rate–stress curves in Figure 17.2*b* show strong shear thickening. An equal parts blend of A and B had an extended distribution and gave a slip with much higher flow rate than did either component powder.

Whiteware Slips

Most whiteware casting bodies are heterogeneous mixtures of clays, ground feldspathic minerals, and powdered silica. Typical of such formulas are the sanitaryware bodies. Figure 17.3 gives particle-size distribution and rheology data for the ball-clay component, the kaolin component, and the

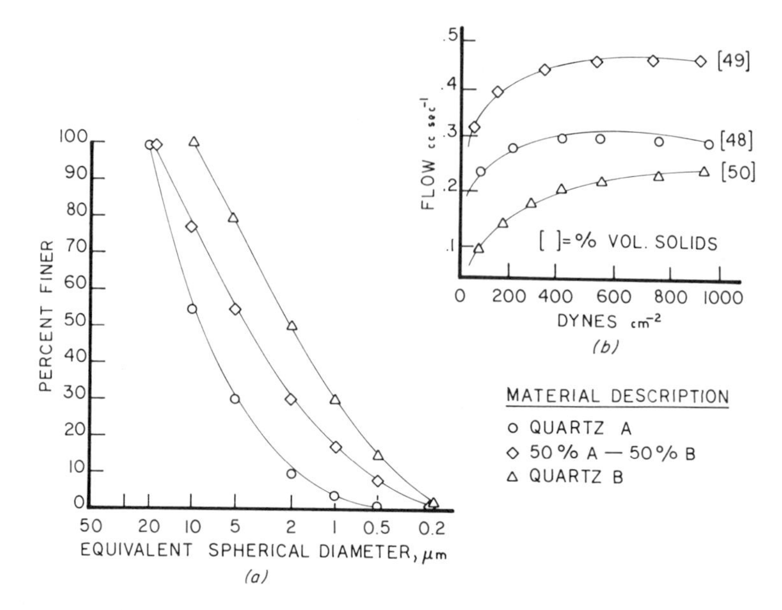

Figure 17.2. The effect on slip fluidity of combining dilatant slips prepared from two different narrow-size-range distributions of quartz (quartz A and quartz B): (*a*) size distribution and (*b*) flow behavior.

nonplastic component of a sanitaryware body. The clay components and nonplastic components each had narrow distributions.

The high-surface ball clay could be loaded only to 43% solids by volume and the slip was typically shear thinning. The low-area kaolin gave shear thickening at 49% solids by volume, while the very-low-area nonclay powder slip was strongly shear thickening at 52% solids.

A body was prepared by mixing the components. This mixing gave the very broad size distributions shown in Figure 17.3. The body exhibited moderate shear-thinning rheology and could be loaded to 54%. This slip could be cast, while none of the individual component slips would qualify as a sanitary-ware slip in terms of rheology and loading.

Occasionally the replacement of a particular low-specific-surface material by another of the same mineral character and surface area but having a different particle distribution results in markedly different deflocculation response and slip rheology.[25] Figure 17.4 describes such an occurrence; in this instance the nonclays component of a sanitary-ware casting body (inverted triangles) was changed from a jointly ground feldspar–quartz (open circles) to a blend of separately ground (diamonds) feldspar and quartz. The distribution of the replacement body (upright triangles) in the intermediate range of sizes showed an apparently modest change. However,

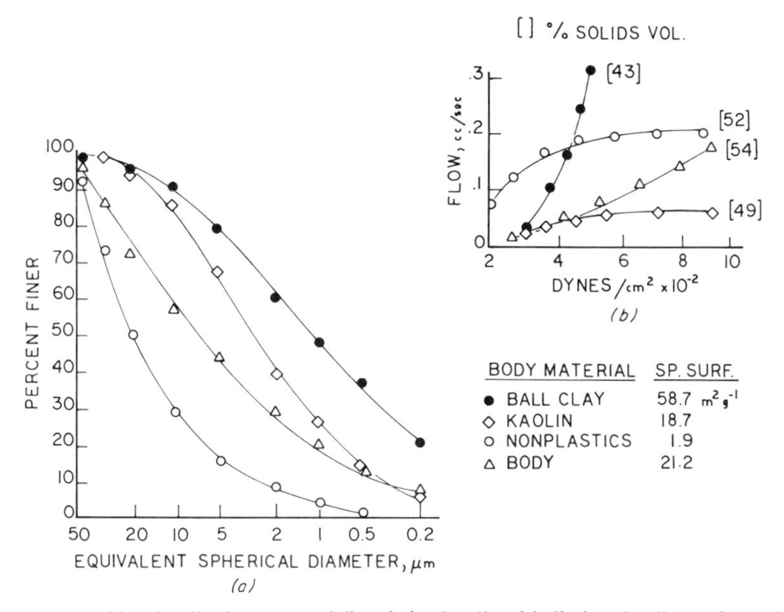

Figure 17.3. Size distributions (*a*) and flow behavior (*b*) of ball clay, kaolin, and nonplastic components of a sanitary casting body. The solid volumes represent pourable loading limits of deflocculated aqueous slips. Specific surface areas are: (*1*) ball clay, 58.7 m^2/g; (*2*) kaolin, 18.7 m^2/g; (*3*) nonplastics, 1.9 m^2/g; and (*4*) body, 21.2 m^2/g. Numbers in brackets are the percent solids by volume.

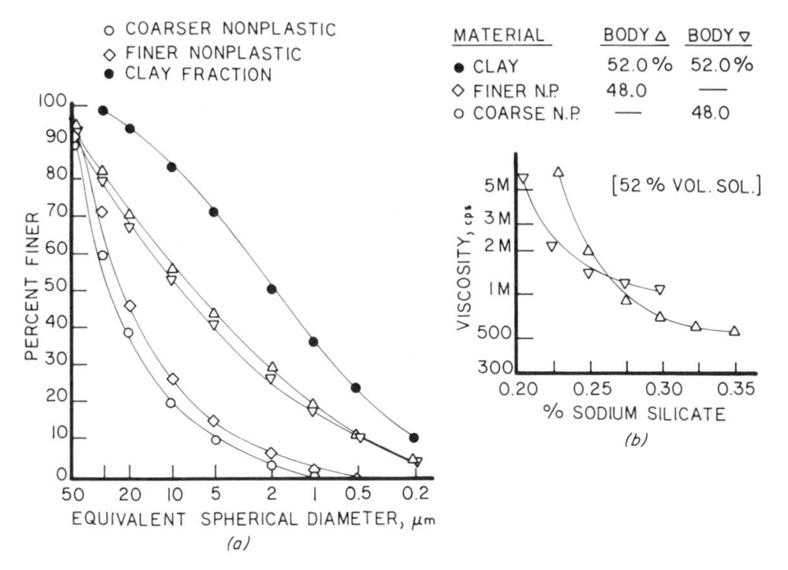

Figure 17.4. Sanitaryware casting body coinciding with moderate variation of intermediate-range particle-size distribution caused by a change in nonplastics component size distribution: (*a*) size distributions and (*b*) flow behavior with sodium silicate added.

the deflocculation curves for the respective bodies (Figure 17.4*b*) indicates a decided increase in initial-flow deflocculant requirement and much greater fully deflocculated fluidity for the body with separately ground nonplastics. Analogous rheological changes also occur where coarse china clays are interchanged in high-solids whiteware casting bodies.

FLOC STRUCTURE IN COMMERCIAL CASTING SLIPS

Commercial casting slips are usually maintained in a state of deflocculation that allows moderate thickening with time. Control laboratories in commercial plants use various flow-tube and rotational viscometric devices for consistency and gel-rate measurements. In Figure 17.5 are given deflocculation-curve and thickening-rate data for a commercial sanitaryware formula, obtained with the commonly used Gallenkamp[26] and Brookfield [27] rotational viscometers. Gel-rate data were determined with the Gallenkamp by rotating its flywheel through 360°, releasing it, recording degrees of overswing past the zero point for fresh slip, and then allowing the spindle to stand for 6 minutes in the cocked position before releasing again. The dif-

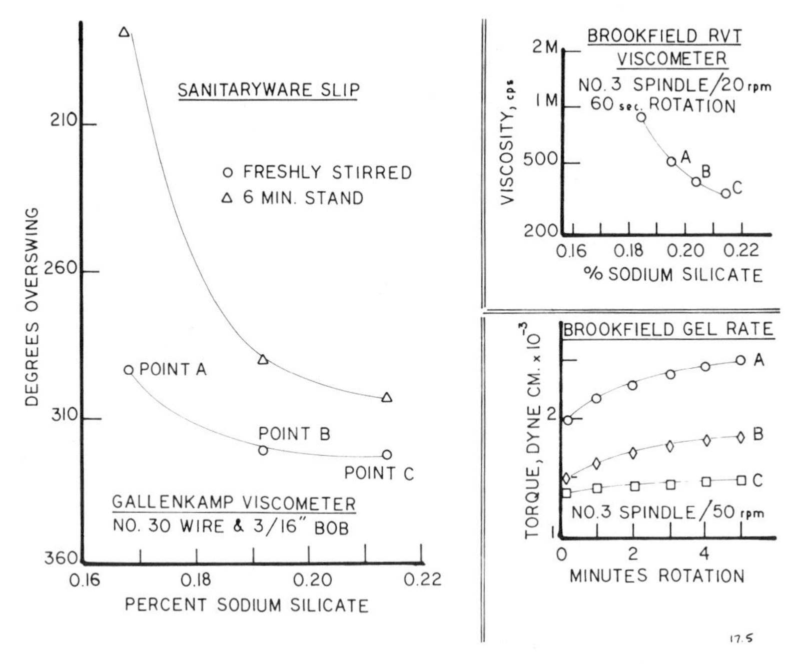

Figure 17.5. Rheological data for a commercial lot of sanitaryware slip using the Gallenkamp and Brookfield RVT rotational viscometer.

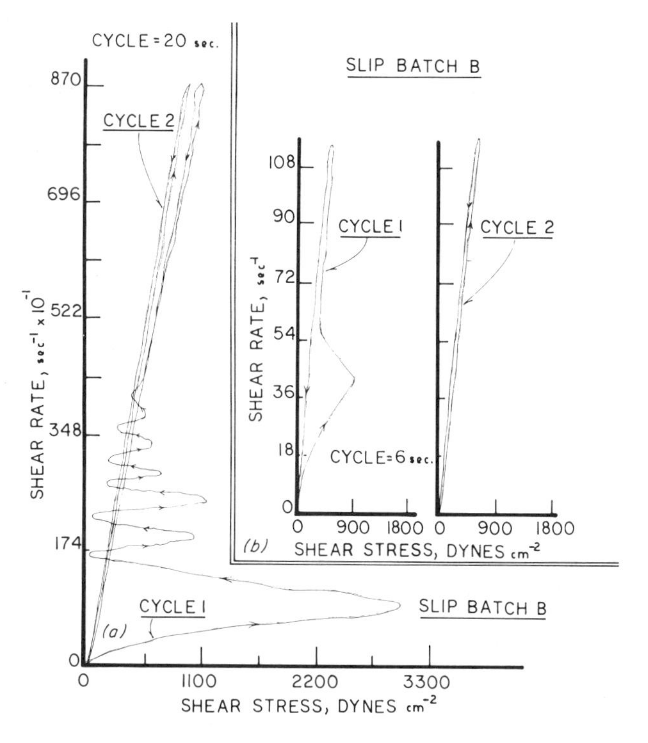

Figure 17.6. Shear rate–shear stress curves for a commercial sanitaryware slip using a Ferranti-Shirley cone-plate viscometer under two different conditions: (*a*) high shear and slow cycle of 20 seconds, (*b*) low shear and rapid cycle of 6 seconds.

ference between fresh and aged slip overswings was reported as "thixotropy." Brookfield time–thickening data were obtained by allowing the spindle to rotate at constant speed over a period of time and observing the increase in torque.

A research viscometer, the Ferranti-Shirley coneplate instrument,[28] was used in obtaining the rate of shear–shear stress curves on slip B shown in Figure 17.6. Two sets of runs were made, one at high shear rate (0/500/0 rpm in a 20 second cycle) and another at low shear rate (0/30/0 rpm in 6 seconds). In each case the up-and-down cycle was repeated immediately following the initial cycle. The marked differences in curve configuration between first and second cycle traces of Figure 17.6 prompted an experiment in which a previously cycled (0/500/0 rpm) sample was left on the plate while cone and plate were separated and then repositioned. The sample was protected with a damp cloth to prevent moisture loss and allowed to stand 15 minutes, and the two-cycle sequence was repeated. The curves obtained were similar to the original first and second high-shear-rate cycle traces. Beazley[29] has proposed that data of this nature show the slip to be in a state of aggregated, liquid-holding flocs (consisting of particles held together by a combination of coulombic and van der Waal's forces), forming a network. The initial impact of shearing shows strong stressing that breaks the network into individual flocs, each of which contains considerable liquid. Breaking of the flocs frees some liquid and extends the distribution of particles so that a sudden reduction in stress occurs. Subsequent oscillations, as rate of shear increases, are joint consequences of increasing stress and alternating shear thickening from Brownian-motion-induced particle contacts and continuing shear breakdown. The more gradual application of shear diagramed in the insert graph of Figure 17.6 indicates, according to Moore,[30] an approach to equilibrium between breakdown and restructuring. Once the structure has been completely disrupted, there must be a period of time before it reforms, thus accounting for the disappearance of the bulge on second cycle traces. The reappearance of the bulge on "aged," previously cycled samples is evidence of time thickening.

CASTING RATE VERSUS PARTICLE SIZE AND SURFACE AREA

Casting rate, consistent with good drain, release, and a state of plastic firmness of the cast, is of primary concern to the manufacturing plant.[13] Particle size has long been recognized as of prime importance in controlling rate of cast formation. One observer[15] noted that the percentage of particles finer

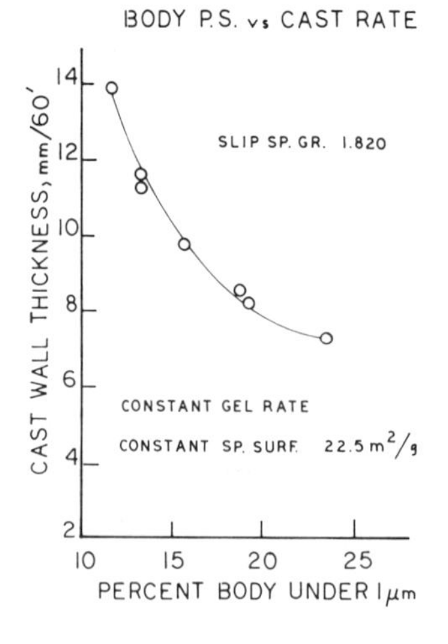

Figure 17.7. Effect of intermediate-range particle-size variation on rate of cast of organic-free sanitary-ware casting body at constant slip weight, gel rate, and specific surface.

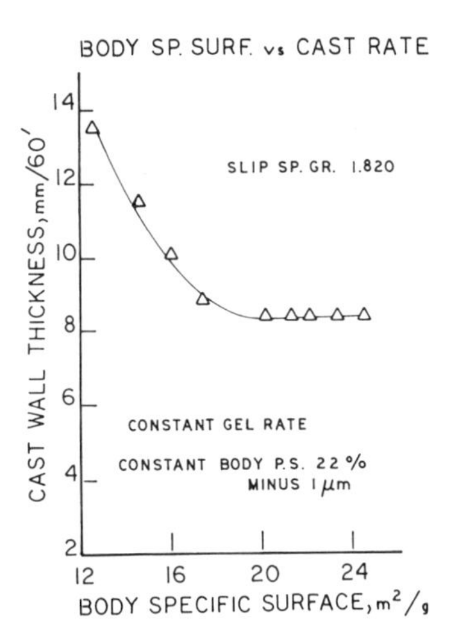

Figure 17.8. Effect of changes in casting rate of sanitary-ware (Figure 17.7) from variation in clay colloid content at constant slip weight, gel rate, and intermediate particle-size distribution.

than 1 μ in a whiteware slip constituted an index of rate of cast. The data of Figure 17.7 show the changes in rate of cast with changes of minus 1 μ variation of the clays component of a sanitary-ware-type organic-free, salts-free body at constant gel rate (Gallenkamp), constant slip weight, and constant specific surface. Although the trend is not linear, casting rate is definitely sensitive to apparently modest -1 μ percentage changes. Figure 17.8 shows that for a constant 1 μ the specific surface can be varied over a wide range with reasonably constant rate of cast. However, once the specific surface (i.e. colloid fineness) is reduced below some critical point, rate of cast will rise sharply.

SUMMARY AND DISCUSSION

The rheological curves of Figure 17.1 show the profound effect of quite small weight percentages of fine (i.e., colloidal and near colloidal) particles on the fluidity of deflocculated, crowded suspensions. The more extended a distribution (see Figures 17.1 and 17.3), the more fluid is the slip at high-volume solids. These data for casting slips parallel findings of Freundlich and Jones[31] and of Pryce-Jones[32] obtained with suspensions of narrow distributions of quartz particles in the range of 1 to 25 μ. Clarke[33] observed sharp reductions in apparent viscosity of 50 vol % solids 30 to 175 μ suspensions of quartz and glass particles upon addition of colloid-size polyanions. Eveson et al[34] reported puzzlement at the anomalously low relative viscosities obtained for mixtures of suspensions of monosize spheres in water. These observations appear to be related to theoretical considerations of

Furnas[18] and of Karlsson and Spring,[35] which point out that voids volume in packed polydisperse powders is a function of (*a*) the percentages of constituent monosize particles and (b) the ratio of the smallest to largest size particle. The smaller the S/L (ratio smallest size S to largest size L) and the more continuous the distribution, the lower the voids volume of the system. Furnas' argument is demonstrated by the graph of Figure 17.9. The voids volume for a monosize powder is assumed to be 40%. The addition of a second, smaller monosize powder to the larger-size powder has the effect of filling the voids of the predominant larger-size mass, thus reducing the voids volume of the combination. As the particles of the finer size are made still finer (i.e., more particles per unit volume), the combined powder voids volume drops to about 17% at $S/L = 0.01$ and remains constant regardless of the value of S. By using progressively smaller multiple monosize powders in combinations approaching a continuous distribution, the reduction in voids as S/L diminishes is less abrupt but approaches 5% at very low S/L (10^{-4} to 10^{-5}). Thus for a fixed liquid volume, that suspension whose particle-size distribution is most nearly continuous and whose S/L is low, would theoretically make maximum use of the liquid in flow. Of course, this neglects any interfacial effects between liquid and solid that might immobilize a portion of the liquid. Such an argument might account for the large

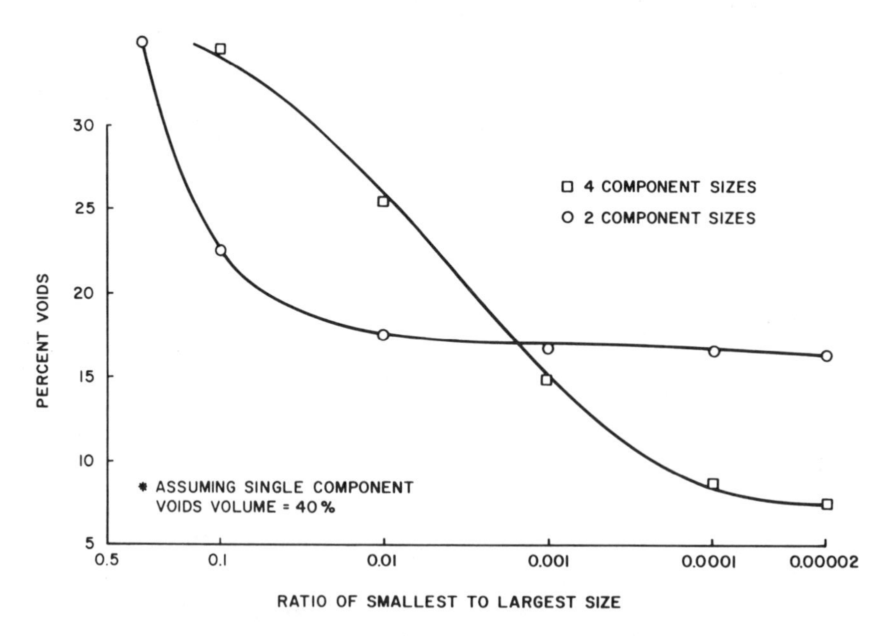

Figure 17.9. Data of Furnas[18] drawn to show the effect of porosity of mixing a small amount of finer monosize spheres with a larger amount of large single-size spheres.

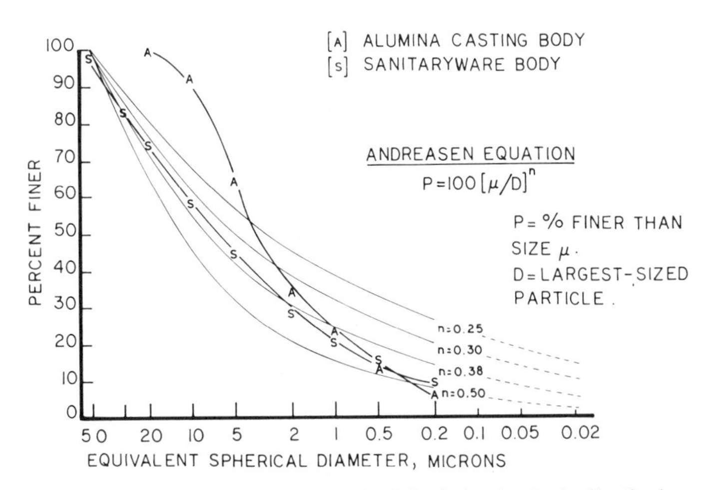

Figure 17.10. Plots of sanitary-ware-body and calcined-alumina-body distributions superimposed on a plot of Andreasen limit curves[19] calculated for a maximum particle-size $D = 50\ \mu$ over a range of 0.25 to 0.50. Note that the sanitary-ware distribution falls within the 0.33 to 0.50 limits established by Andreasen for optimum distributions.

increases in fluidity obtained by combining narrow distributions to make a more extended distribution or by adding small volumes of fines to a coarser distribution.

Andreasen[19] studied continuous distributions of quartz grains and concluded that optimum packing for a distribution whose largest size D appeared in the relation

$$P = 100 \left(\frac{\mu}{D}\right)^{\eta}$$

where P is the percentage of particles finer than size μ, and η a fractional constant. Optimum packing was reported as lying between $\eta = 0.33$ and 0.50. Andreasen limits calculated for $D = 50\ \mu$ over a range of η values are shown by Figure 17.10. Plots for a sanitary-ware body (from Figure 17.3) and an alumina casting body (from Figure 17.1) superimposed on the graph show that the sanitary-ware distribution more nearly approaches the Andreasen theoretical limits for optimum voids. An extrapolation of the A and S curves into the range of colloidal dimensions shows the sanitary-ware body to have a markedly smaller S/L value, hence lower theoretical voids. A similar calculation and plot for a refractory grain casting body is shown by Figure 17.11 and indicates that the distribution conforms to the Andreasen limits only in the lower size range.

The very sharp changes in fluidity and deflocculation response with small changes in intermediate particle size shown by Figure 17.4 and the marked

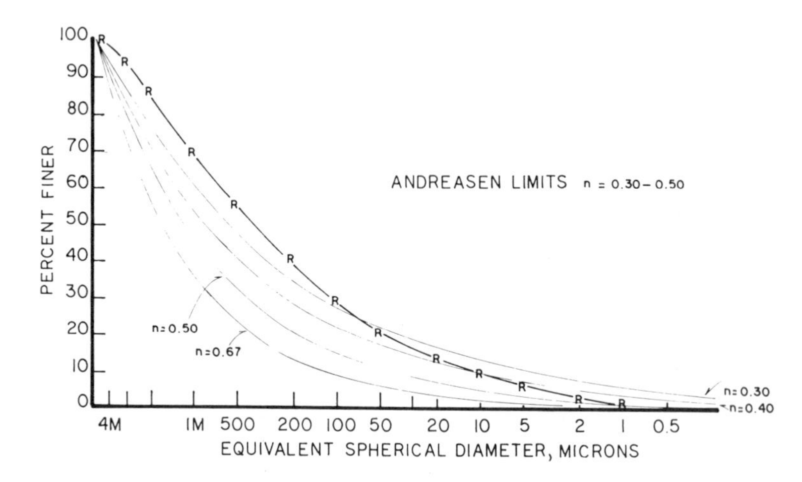

Figure 17.11. Andreasen limits calculated for $D = 4750\ \mu$. A plot of a refractory casting body [R] lies outside the 0.33 to 0.50 limits over most of its range.

changes in rate of cast with variation of the 1 μ particle content at constant colloid shown by Figure 17.7 may possibly be due to changes in void volume (porosity) induced by small shifts in the intermediate range of an extended particle distribution. Figure 17.12 shows plots of data taken from a study by Bo et. al[36] of the role of particle-size distributions on the porosity and air permeability of continuous distributions of combinations of very-narrow-range distributions of glass spheres. Curves *B* and *E* represent optimum distributions of two different ranges. Curve *B* had a more extended range ($S/L = 0.05$) than curve *E* ($S/L = 0.27$). Where there is a deficit of intermediate-size particles (as in curves *A* and *D*), the curve becomes more concave, and porosity and permeability both increase. When there is an excess of intermediate-size particles (curves *C* and *F*), the curve configurations lie above the optimum traces *B* and *E*, and permeability diminishes, with porosity again increasing. The porosity and permeability changes coinciding with intermediate-particle-size shifts are proportionately very much larger for $S/L = 0.05$ than for $S/L = 0.27$. If this observation could be applied to the sanitary-ware body distribution where $S/L = 10^{-3}$, one might expect very small intermediate-size deficits to produce relatively quite large increases in porosity (higher liquid entrapment = higher viscosity) and permeability (higher casting rate). Moves in the direction of excess intermediate-size material would be expected to sharply reduce rate of cast (decreased permeability) and raise viscosity (increased porosity).

In recent years a number of rheological models have been proposed to describe the flow properties of colloidal suspensions. One model based on

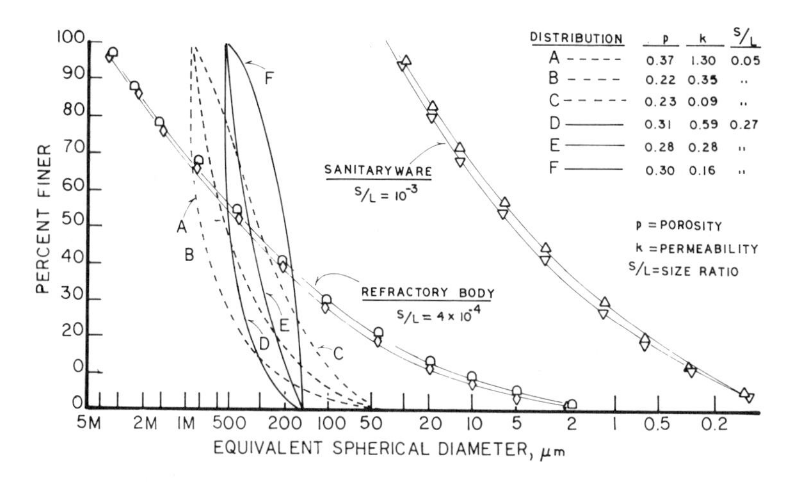

Figure 17.12. Graphical presentation of data of Bo et al.[36] showing the effect of particle distribution on porosity and permeability of packed distributions of glass spheres.

elementary reaction rate theory[37,38] treats shear-thinning behavior in terms of a few empirically determined constants but does not take into account dependence of these constants on colloidal and surface properties of the dispersed particles. Although these models do appear to successfully picture suspension rheology over a fairly wide range of shear rate,[39] their basic assumptions were questioned[40] and a modified model was proposed in which the hydrodynamic–structural theory was extended to account for a distribution of relaxation times. The effect would be to widen the range over which shear thinning can occur. Another model based on particle collision and aggregation processes[41] accounts for increased viscosity at both low and high rates of shear. A recent comparison of these two theories[42] for shear thinning found the latter to more nearly agree with observed shear rate–shear stress data.

In light of these observations it seems possible to consider investigating deflocculated, high-solids ceramic suspensions within the context of packed powders. The particles of colloidal dimensions constitute a complicating factor because of their tendency to agglomerate and bind the system into a unit that breaks up into a distribution whose extension depends on the rate of shearing.

REFERENCES

1. A. L. Johnson and F. H. Norton, "Fundamental Study of Clay. II. Mechanism of Deflocculation in the Clay–Water System," *J. Amer. Ceram. Soc.*, **24,** 189–203 (1941).

2. R. K. Schofield and H. R. Samson, "Flocculation of Kaolinite Due to the Attraction of Oppositely Charged Crystal Faces," *Discuss. Faraday Soc.*, **18,** 134–145 (1954).
3. G. W. Phelps, "The Organic Colloid Fraction in Clay Slip Casting," *Proc. Fall Meetings ME/WW Div, Amer. Ceram. Soc.*, **1970,** 9–15.
4. F. J. Stevenson and J. H. A. Butler, "Chemistry of Humic Acids and Related Pigments," *Organic Geochemistry,* G. Eglinton and M. T. J. Murphy, eds., Springer-Verlag, Berlin–Heidelberg, 1969, Chap. 22.
5. H. J. Modi and D. W. Fuerstenau, "Streaming Potential Studies on Corundum in Aqueous Solutions of Inorganic Electrolytes," *J. Phys. Colloid Chem.*, **61,** 640–643 (1957).
6. G. A. Parks and P. L. De Bruyn, "The Zero Point Charge of Oxides," *J. Colloid Sci.*, **63,** 967–973 (1962).
7. G. Y. Onoda and P. L. De Bruyn, "Proton Adsorption at the Liquid-Solid Interface. Pt. I. A Kinetic Study of Adsorption," *Surf. Sci.*, **4,** 48–63 (1966).
8. (a) S. M. Ahmed, "Studies of the Dissociation of Oxide Surfaces at the Liquid–Solid Interface," *Can. J. Chem.*, **44,** 1663–1670 (1966); (b) S. M. Ahmed and D. Marsimov, "Studies of the Oxide Surfaces at the Liquid–Solid Interface. Pt. II. Fe Oxides," *Can. J. Chem.*, **46,** 3841–3846 (1968).
9. D. W. Fuerstenau, "Streaming Potential Studies on Quartz in Solutions of Aminium Acetates in Relation to the Formation of Hemimicelles at the Quartz–Solution Interface," *J. Phys. Chem.*, **60,** 981–985 (1956).
10. P. Somasundaran, T. W. Healy, and D. W. Fuerstenau, "Surfactant Adsorption at the Solid–Liquid Interface–Dependence of Mechanism on Chain Length," *J. Phys. Chem.*, **68,** 3562–3566 (1964).
11. P. Somasundaran and D. W. Fuerstenau, "Mechanisms of Alkyl Sulfonate Adsorption at the Alumina–Water Interface," *J. Phys. Chem.*, **70,** 90–96 (1966).
12. P. Somasundaran, T. W. Healy, and D. W. Fuerstenau, "The Aggregation of Colloidal Alumina Dispersions by Adsorbed Surfactant Ions," *J. Colloid Interface Sci.*, **22,** 599–605 (1966).
13. S. R. Hind, "Study of Factors Involved in Slip Casting," *Trans. Ceram. Soc.*, **22,** 90–104 (1923).
14. F. P. Hall, "The Casting of Clay Ware—A Resume," *J. Amer. Ceram. Soc.*, **13,** 751–766 (1930).
15. G. A. Loomis, "Properties of Clay Casting Slips," *J. Amer. Ceram. Soc.*, **23,** 159–162 (1940).
16. A. Dietzel and H. Mostetzky, "Mechanism of Dewatering of a Ceramic Slip by The Plaster Mold: I. Experimental Investigation of the Diffusion Theory of the Slip-Casting Process," *Ber. Deut. Keram. Ges.*, **33,** 7–18 (1956).
17. W. E. Worrall and W. Ryan, "The Mechanism of Slip Casting," *J. Br. Ceram. Soc.*, **1,** 270–271 (1964).
18. (a) C. C. Furnas, "The Relations Between Specific Volume, Voids, and Size Composition in Systems of Broken Solids of Mixed Sizes," U.S. Bur. Mines Report Invest. No. 2894 Oct. 1928. (b) C. C. Furnas, "Grading Aggregates, I. Mathematical Relations for Beds of Broken Solids of Maximum Density," *Ind. Eng. Che.*, **23,** 1052–1058 (1931).
19. G. Herden, *Small Particle Statistics,* 2nd Ed., Butterworths, London, 1960, pp. 188–189.
20. M. Reiner, *Deformation, Strain and Flow,* Interscience, New York, 1960, Chap. XXI.

21. A. B. Metzner and M. Whitlock, "Flow Behavior of Concentrated (Dilatant) Suspensions," *Trans. Soc. Rheol.*, **II,** 238–254 (1958).

22. F. R. Eirich, Ed., *Rheology: Theory and Applications,* Vol. *4,* Academic, New York, 1967, Chap. 8.

23. R. D. Cooke, "The Plastic Properties of Enamel Slip," *J. Amer. Chem. Soc.*, **7,** 651–655 (1924).

24. W. N. Harrison, "Vitrenous Enamel Slips and Their Control," *J. Amer. Ceram. Soc.*, **10,** 970–994 (1927).

25. G. W. Phelps and A. Silwanowicz, "The Role of Nonclay Particle Size in Whiteware Slip Casting," *J. Can. Ceram. Soc.*, **39,** 17–19 (1970).

26. A. Clark and H. J. Hodsman, "Viscometer," *J. Soc. Chem. Ind.*, **56,** 67t (1937).

27. J. R. van Wazer, J. W. Lyons, K. Y. Kim, and R. E. Colwell, *Viscosity and Flow Measurement,* Interscience, New York, 1963, pp. 139–150.

28. R. McKennell, "Cone-Plate Viscometer Comparison with Coaxial Cyliner Viscometer," *Anal. Chem.*, **28,** 1710–1714 (1956).

29. K. M. Beazley, "Breakdown and Buildup in China Clay Suspensions," *Trans. Br. Ceram. Soc.*, **63,** 451–471 (1964).

30. F. Moore, "The Rheology of Ceramic Slips and Bodies," *Trans. Br. Ceram. Soc.*, **58,** 470–494 (1959).

31. H. Freundlich and A. D. Jones, "Sedimentation Volume, Dilatancy and Plastic Properties of Concentrated Suspensions," *J. Phys. Chem.*, **40,** 1217–1236 (1936).

32. J. Pryce-Jones, "Flow of Suspensions, Thixotrophy and Dilatancy," *Durham Philoso. Soc.*, **10,** 427–467 (1948).

33. B. Clarke, "Rheology of Coarse Settling Suspensions," *Trans. Inst. Chem. Eng. (London)* **45,** T251–T256 (1967).

34. G. F. Eveson, S. G. Ward, and R. L. Whitmore, "Theory of Size Distribution in Paints, Coals, Greases, etc. Anomalous Viscosity in Model Suspensions," *Discuss. Faraday Soc.*, (**11**), 11–14 (1951).

35. K. Karlsson and L. Spring, "Packing of Irregular Particles," *J. Mater. Sci.*, **5,** 340–344 (1970).

36. M. K. Bo, D. C. Freshwater, and B. Scarlett, "The Effect of Particle Size Distribution on the Permeability of Filter Cakes," *Trans. Inst. Chem. Eng.*, **43,** T228–T232 (1965).

37. D. A. Denny and R. S. Brodkey, "Kinetic Interpretation of Non-Newtonian Flow," *J. Appl. Phys.*, **33,** 2269–2274 (1962).

38. M. M. Cross, "Rheology of Non-Newtonian Flow: A New Flow Equation for Pseudoplastic Systems," *J. Colloid Sci.*, **20,** 417–439 (1965).

39. M. M. Cross, "Kinetic Interpretation of Non-Newtonian Flow," *J. Colloid Interface Sci.*, **33,** 30–35 (1970).

40. T. Gillespie, "Application of the Hydrodynamic-Structural Theory of Non-Newtonian Flow to Suspensions which Exhibit Moderate Shear Thickening with Particular Reference to 'Dilatant Plastisols'," *J. Colloid Interface Sci.*, **22,** 554–562 (1966).

41. A. S. Michaels and J. C. Bolger, "The Plastic Behavior of Flocculated Kaolin Suspensions," *Ind. Eng. Chem. Fund,* **1,** 153–164 (1962).

42. R. J. Hunter and S. K. Nicol, "The Dependence of Plastic Flow Behavior of Clay Suspensions on Surface Properties," *J. Colloid Interface Sci.*, **28,** 250–259 (1968).

18

Viscosity of Concentrated Newtonian Suspensions

K. Sommer

The viscosity η of a dilute suspension of rigid spheres in a liquid was deduced by Einstein in 1905 to be given by

$$\eta = \eta_{fl}(1 + 2.5c_v)$$

where η_{fl} is the viscosity of the liquid and c_v is the ratio of the volume of the suspended matter to the total volume.

Subsequently, many new formulas have been developed for highly concentrated suspensions and nonspherical particles; however, these formulas are semiempirical and apply only to the particular suspension of the investigation. The differences in the various suspensions can be explained in terms of the differences in the materials and the interactions between their particles. New theoretical and experimental approaches are needed to separate the different effects in concentrated suspensions.

THEORY

The viscosity of a concentrated suspension is dependent on the rheology of the liquid phase and on the direct interactions between particles. These two effects can be separated theoretically as shown below.

For a unit volume of suspension, the rate of energy dissipation $\dot{E}_v$ due to a shear flow is given by

$$\dot{E}_v = \tau\dot{\gamma} = \eta\dot{\gamma}^2 \tag{1}$$

where τ = external shear stress rate
$\dot{\gamma}$ = shear strain rate
η = apparent viscosity of the suspension

For a Newtonian liquid, η is independent of the shear strain rate. For other cases η depends on $\dot{\gamma}$.

The total dissipated energy is the sum of the energy E_{fl} dissipated by the liquid phase and the energy $\dot{E}_s$ dissipated by the direct particle–particle interactions (or solid friction energy), that is,

$$\dot{E} = \dot{E}_{fl} + \dot{E}_s \tag{2}$$

Dividing Equation 2 by the volume V of the suspension and dividing the first term on the right side of the equation by the volume V_{fl} of the liquid yields

$$\dot{E}_v = \frac{V_{fl}}{V}\frac{\dot{E}_{fl}}{V_{fl}} + \frac{\dot{E}_s}{V} \tag{3}$$

Letting C_v equal the volume concentration of the solid,

$$\dot{E}_v = (1 - C_v)\dot{E}_{v_{fl}} + \frac{\dot{E}_s}{V} \tag{4}$$

where $\dot{E}_{v_{fl}}$ is the dissipated energy per unit volume of liquid.

A parameter $\dot{\gamma}^*$ can be defined by

$$\dot{E}_{v_{fl}} = \eta_{fl}\cdot\dot{\gamma}^{*2} \tag{5}$$

where $\dot{\gamma}^*$ is a mean shear rate for the liquid phase and η_{fl} is the apparent viscosity of the liquid.

Combining Equations 1, 4, and 5 gives

$$\tau\cdot\dot{\gamma} = (1 - C_v)\eta_{fl}\dot{\gamma}^{*2} + \frac{\dot{E}_s}{V} \tag{6}$$

All the quantities in Equation 6 are measurable except $\dot{\gamma}^*$ and $\dot{E}_s$.

The solid friction energy $\dot{E}_s$ can be determined if two suspensions are selected that allow certain simplifications to occur:

1. τ is measured for both suspensions at the same value of γ (thereby giving τ_1 and τ_2 for the first and second suspension).

2. Both suspensions have the same solid at the same C_v.
3. γ^* is assumed to be the same for both because $\dot{\gamma}$ and C_v are the same for both suspensions.
4. The two liquids have similar physical and chemical properties so that $\dot{E}_{s_1}$ is the same as $\dot{E}_{s_2}$.
5. The only differences in the two liquids are their viscosities (η_{fl_1} and η_{fl_2}).

Writing Equation 6 for both suspensions, and making the above assumptions, we can combine the two equations to yield

$$\dot{\gamma}^* = \left[\frac{\dot{\gamma}(\tau_1 - \tau_2)}{(1 - C_v)(\eta_{fl_1}\eta_{fl_2})} \right]^{1/2} \tag{7}$$

and

$$\frac{E_s}{E} = 1 - \frac{(\tau_1/\tau_2) - 1}{(\eta_1/\eta_2) - 1} \tag{8}$$

EXPERIMENT

Figure 18.1 shows two flow curves of closely sized, glass spheres (mean size of 23 μm) suspended in two oils with similar chemistry but different vis-

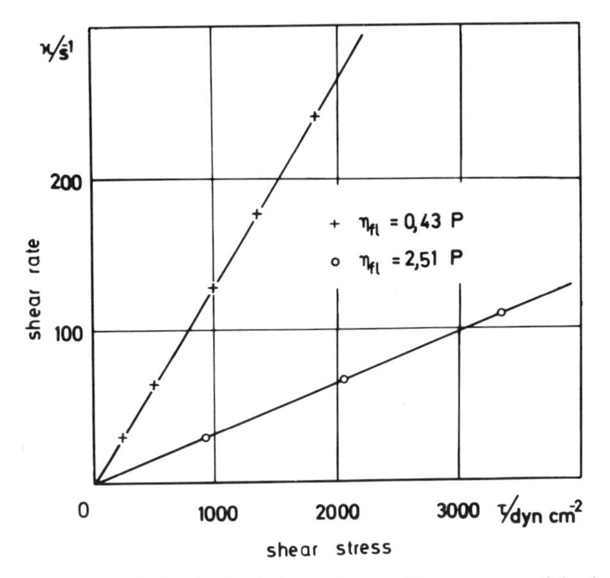

Figure 18.1. Flow curves of closely sized glass spheres with a mean particle size of 23 μm in two different oils with viscosities η_{fl} = 0.43 and 2.51 P.

cosities (0.43 and 2.51 P). For both suspensions $C_v = 0.4$. Newtonian behavior is found in both cases.

From the two sets of data, the value of $\dot{\gamma}^*$ is calculated with Equation 7 and plotted versus $\dot{\gamma}$ in Figure 18.2. A linear relation is found between $\dot{\gamma}^*$ and $\dot{\gamma}$.

Also shown in Figure 18.2 are similar results for cases where C_v is 0.20 and 0.35. These are also linear. The mean shear rate for a given external shear rate decreases as C_v decreases.

Experiments using different solid particles but the same liquid have been carried out. Two suspensions with the same liquid (and same liquid viscosity) are formed using glass spheres in one case and Plexiglas spheres of the same size in the second case. For both suspensions C_v was 0.4. The flow curves for the two cases are given in Figure 18.3. Both are Newtonian, but the suspension with glass spheres had a significantly larger apparent viscosity. However, the mean shear rates in the liquids were the same

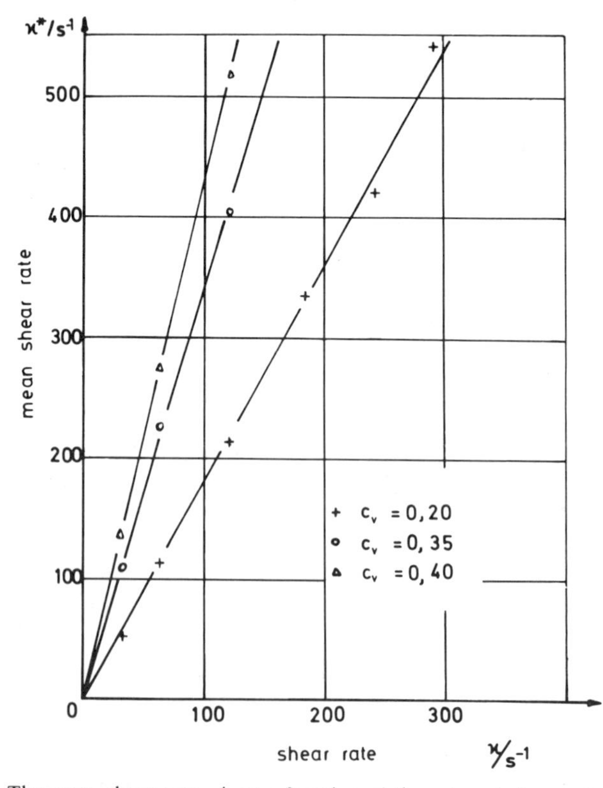

Figure 18.2. The mean shear rate γ^* as a function of the external shear rate γ by a glass sphere–oil suspension and different concentrations c_v.

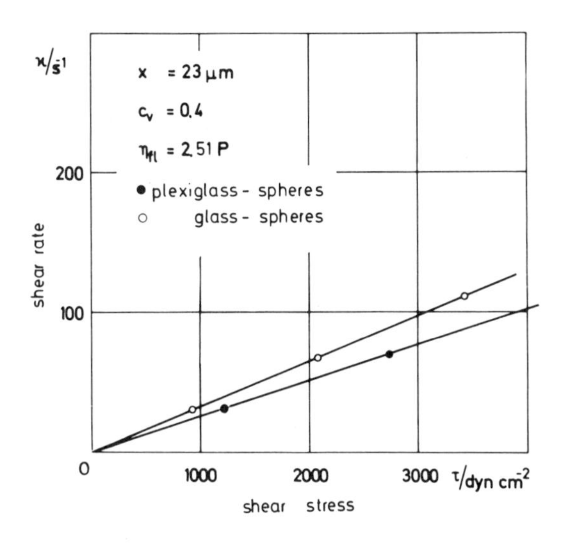

Figure 18.3. Flow curves of plexiglas spheres and glass spheres suspended in the same oil (η_{fl} = 2.51 P).

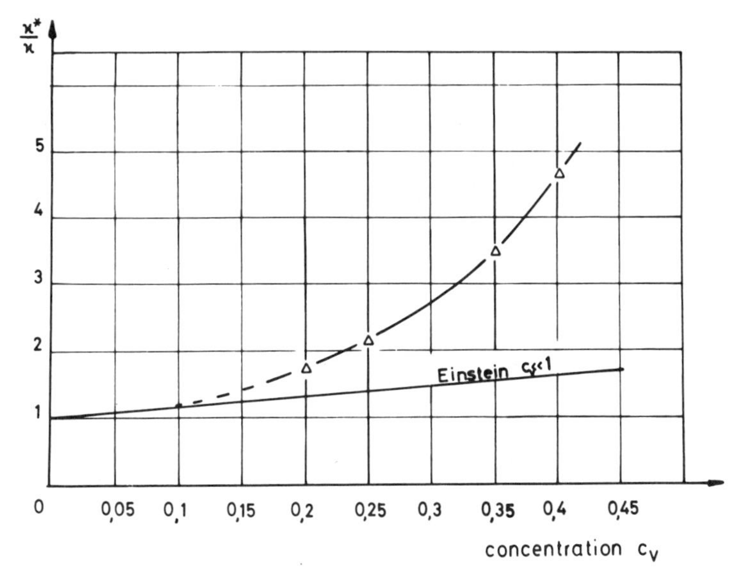

Figure 18.4. The ratio γ^*/γ as a function of concentration C_v of the glass spheres–oil suspensions (x = 23 μm).

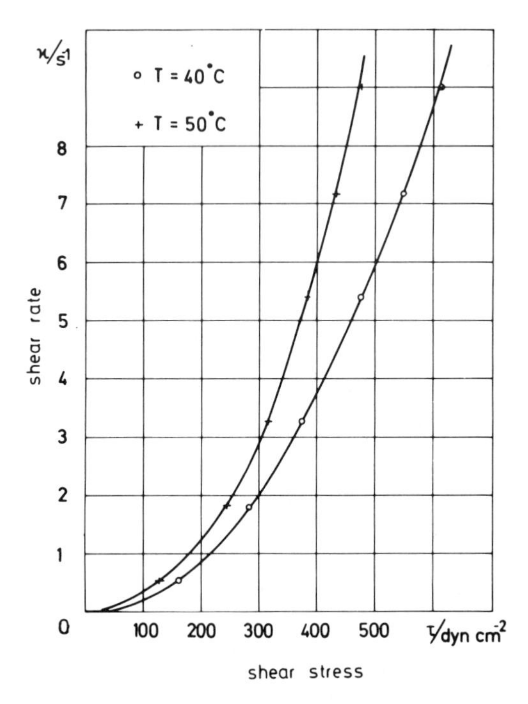

Figure 18.5. Flow curves of chocolate at two temperatures ($T_1 = 40°C$ and $T_2 = 50°C$).

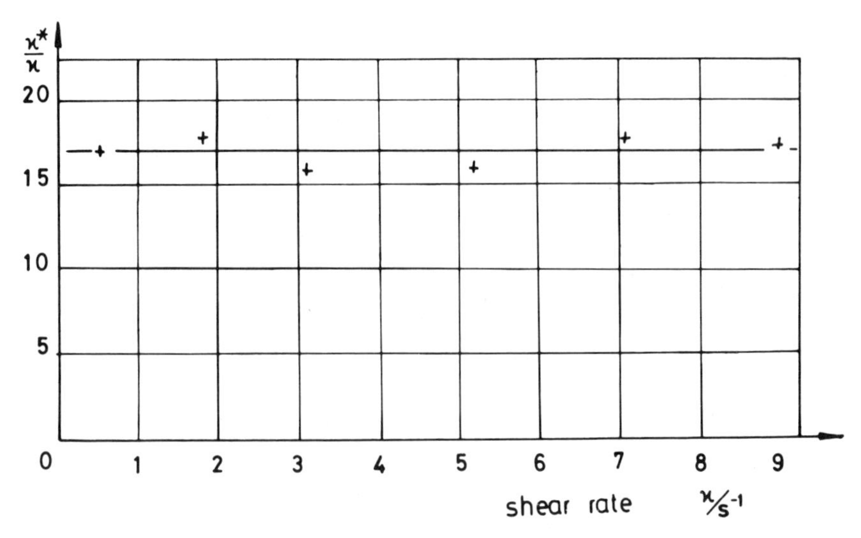

Figure 18.6. The ratio γ^*/γ as a function of the external shear γ for chocolate.

throughout the whole range of the external shear rate examined. In both cases $\dot{\gamma}^*/\dot{\gamma}$ is equal to 4.45. Thus the flow properties of the liquid phase did not change. The differences in viscosity were due to the different particle–particle interactions.

The ratio $\dot{\gamma}^*/\dot{\gamma}$ was measured as a function of C_v for the glass-spheres suspension. The results, presented in Figure 18.4, show that a straight line is approached at dilute concentrations (at C_v values below 0.1). This straight line shows that Einstein's relationship is being approached at low concentrations. At higher concentrations the shear rate within the liquid phase increases more rapidly than expected by Einstein's theory because of the hydrodynamic interactions between particles.

The fraction $E_s/\dot{E}$ calculated from Equation 8 is very small at dilute concentrations. Above $C_v = 0.2$, $\dot{E}_s$ becomes a significant fraction of $\dot{E}$ because of the influence of the direct particle–particle interactions.

The above results demonstrate that the relative contributions of particle–particle interactions to the viscosity of suspensions can be determined experimentally. It was therefore of interest to examine chocolate, which exhibits shear thinning behavior that has been attributed often to the changing of the structure of the suspension. The viscosity of the liquid in the chocolate is altered by changing temperature. The overall flow behaviors at 40 C and 50°C are compared in Figure 18.5. Calculation of $\dot{\gamma}^*/\dot{\gamma}$ reveals that the quantity is independent of $\dot{\gamma}$, as shown in Figure 18.6. This indicates that the thinning behavior of chocolate cannot be due to changes in the flow properties but rather to a decrease of the direct particle–particle interaction; that is, the solid friction has changed.

SUMMARY

By defining a characteristic mean shear rate $\dot{\gamma}^*$, it is possible to determine the changing "structure" and flow properties of concentrated suspensions. A method of measurement has been given that permits the separation of the influence of a viscous liquid phase from that of the direct interactions between particles on the viscosity of the solution.

19

The Rheology of Organic Binder Solutions

G. Y. Onoda, Jr.

Organic binders are essential additives for the processing of many commercial ceramics. The binders provide green strength and plasticity to clay-free ceramics so that bodies can be molded and retained in the desired shapes before firing. Ceramic processes that utilize organic binders include dry pressing, extrusion, tape casting, roll forming, thick-film printing, slip casting, injection molding, and compression molding.

A large number of organic substances have been utilized or advocated as binders for ceramics during the past 50 years. Several useful compilations and descriptions of various organic binders have appeared in the literature.[1-9] Certain classification schemes have been proposed for binders, based on origin,[9] chemical composition, physical characteristics, molecular weight, polarity, affinity for water, and dominant adhesion mechanisms.[10] However, actual comparisons of organic binders according to the proposed classification schemes were not carried out in most cases.

Of the ceramic processes listed above, all except injection molding and compression molding involve the use of binders that are dissolved or dispersed in a liquid. The binders are dissolved molecularly in water or an organic solvent or are dispersed in liquid as an "emulsion." The liquid phase is important for uniformly dispersing the binder throughout the particulates. Also, the liquid is necessary for providing fluidity for slips or plasticity for extrusion or semidry pressing. Dry green strength in the body is developed by the evaporation of the liquid. The binder is retained in the body and provides organic bridges between the ceramic particles.

Organic binders strongly affect the rheology of the liquid phase. They increase the viscosity and change the flow characteristics from Newtonian (for pure water) to pseudoplastic in most cases. Also, with some binders gels can develop under various conditions. The rheology of the solution directly affects the behavior of suspensions and pastes formed by adding particulates to the solution.

The applicability of a given organic binder for a specific ceramic process is very much dependent on the rheological characteristics of the binder solution. In this chapter relationships between rheological variables and ceramic processing phenomena are discussed. A wide variety of organic binders are compared on the basis of rheology and in relation to processing.

CHEMICAL STRUCTURE

Most soluble organic binders are long-chain polymer molecules. The backbone of the molecule consists of covalently bonded atoms such as carbon, oxygen, and nitrogen. Attached to the backbone are side groups located at frequent intervals along the length of the molecule. The chemical nature of the side groups determines in part what liquids will dissolve the binder. If the side groups are highly polar, solubility in water is promoted. Binders soluble in polar organic solvents have side groups of intermediate polarity. Solubility in nonpolar liquids is promoted by nonpolar side groups.

The chemical polarity and resulting affinity for a liquid is a necessary but not a sufficient condition for effecting solubility in the liquid. The side groups of a molecule bond strongly to the side groups of adjacent molecules, thereby giving the binder a high cohesive strength. If the solvent molecule cannot disrupt this intermolecular bonding, solvation does not take place. This condition occurs, for example, with pure cellulose and pure poly(vinyl alcohol) binders. Both of these polymers must be modified chemically at their side groups to effect solubility. This is accomplished by making the side groups bulkier, thereby reducing the amount of interaction that can take place between molecules. In the case of poly(vinyl alcohol), approximately 12 to 20% of the OH side groups must be replaced by acetate groups to convert the polymer to the cold-water-soluble form. Cellulose is modified to water-soluble form by substituting hydroxyethyl, methyl, carboxymethyl, or hydroxypropyl side groups. These groups are not only hydrophilic but also provide enough bulk so that solute molecules can penetrate between polymer molecules.

The long-chain molecules of binders consist of smaller units, called monomers, that are linked together. The number of monomers in a polymer is called the "degree of polymeration." The monomer is the smallest unit that characterizes the chemical structure of the molecule.

The carbohydrate-derived binders are characterized by a ring-type monomer, the ring being a modified α-glucose structure. The ring has five carbon atoms, as shown in Figure 19.1. Each carbon atom can be identified by a number, such as C-1 or C-2, according to its position indicated by the numbers in parenthesis in Figure 19.1. The linkages between rings occur by an ether bond (—O—) between C-1 of one monomer and C-4 of another.

The various modifications occur by side groups (R) attached to C-5, C-2, or C-3. The degree of substitution (DS) is the number of sites on which modifications are made in a monomer. When DS is one, the C-5 site possesses the side group. As the DS values increase above one (maximum of 3), the C-2 is the next favored position for substitutions (after C-5), and C-3 is the least favored. When not possessed by a special side group the R position on C-2 and C-3 is filled by an OH group.

The monomer formulas for a variety of water-soluble, carbohydrate-derived binders are given in Figure 19.2. The typical DS values are also given.

The monomer formulas for some important binders that are not of vegetable origin are given in Figure 19.3. These include the vinyls, acrylics, and polyethylene oxides. The vinyls are characterized by a linear backbone consisting of carbon–carbon linkages, with a side group attached to every other carbon atom. The acrylics are similar to vinyls in that they have the same backbone structure. However, some acrylics have two side groups attached to the carbon atom.

Both vinyls and acrylics have flexible backbones because of the rotatable carbon–carbon bonds and these can lead to a molecule with a coiled and curving configuration. Thus the spatial length of a molecule may be much

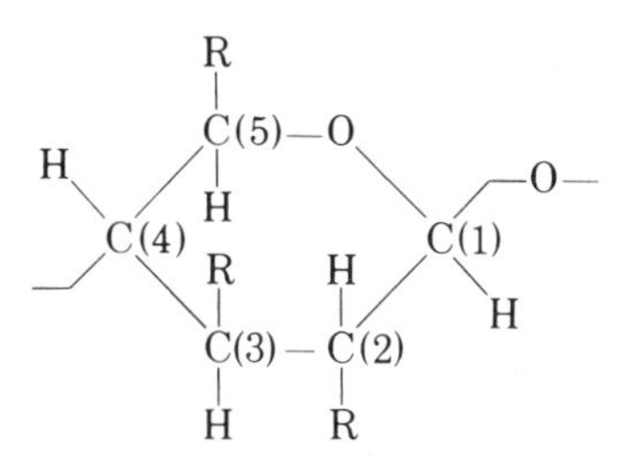

Figure 19.1. Modified α-glucose monomer.

Binder	R Groups	DS
Methylcellulose	$-CH_2-O-CH_3$	2
Hydroxypropylmethylcellulose	$-CH_2-O-CH_2-CH(OH)-CH_3$	2
Hydroxyethylcellulose	$-CH_2-O-C_2H_4-O-C_2H_4-OH$ $-CH_2-O-C_2H_4-OH$	0.9–1.0
Sodium carboxymethylcellulose	$-CH_2-O-CH_2-C(=O)ONa$	
Starches and dextrins	$-CH_2-OH$	
Sodium alginate	$-C(=O)ONa$	
Ammonium alginate	$-C(=O)ONH_3$	

Figure 19.2. Formula for some water-soluble, carbohydrate-derived binders.

smaller than the actual length of the molecule backbone. This has an important effect on the rheology of binder solutions, as is discussed later. The backbone of the carbohydrate-type binders is rather inflexible compared with those of the vinyls and acrylics. These molecules tend to remain extended, thereby having a larger spatial effect than those molecules that can coil and ball up into a compact configuration.

The monomer formulas for some binders that are soluble in organic solvents are given in Figure 19.4.

RHEOLOGY AND MOLECULAR STRUCTURE

The increase in viscosity of a liquid as large molecules (and colloids) are added is strongly influenced by their shape. The effectiveness of a molecule or colloid for increasing viscosity is dependent on its "sphere of influence" in the liquid. Molecules and colloids are constantly in motion (Brownian) because of thermal vibrations and impact by the liquid molecules. The solute molecules or colloids are tumbling and rotating, and the time-average volume swept out by molecules or colloids is defined as their "sphere of influence." For a given molecular weight, a molecule has the largest sphere of influence when it has the largest length in one direction. Thus a linear molecule has a greater sphere of influence than a branched molecule of equivalent molecular weight. A fully extended linear molecule has a greater sphere of influence than the same molecule when it is coiled. The often dramatic increases in solution viscosity that are found with binders is a result of their long, linear chemical structure.

To a first approximation, the viscosity η of a solution is related to the concentration c of binder by an equation of the form

$$\log \eta = kc \tag{1}$$

where η is in centipoise and c is in weight percent. Although some significant departures from this equation exist, usually the viscosity changes are more easily represented on plots of $\log \eta$ versus c, and this has become common practice. Another limitation of representing rheology data by these plots is associated with the phenomenon of "pseudoplasticity," as is discussed below.

Pseudoplasticity is another important rheological characteristic that develops as binders are added to a liquid. The time-average volume of the binder molecule can be distorted by shear forces in the liquid. The molecules tend to line up in a manner than reduces the resistance to flow.

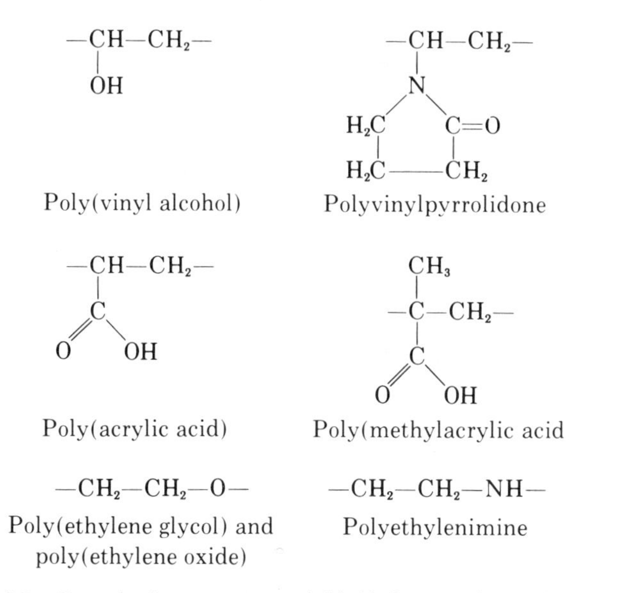

Figure 19.3. Formulas for some water-soluble binders not of vegetable origin.

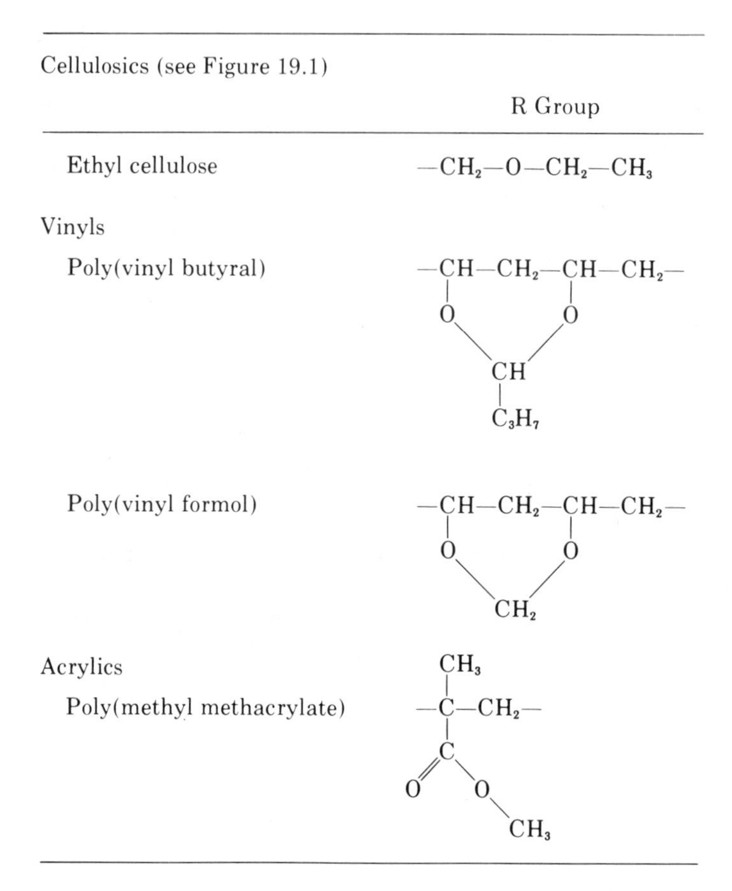

Figure 19.4. Formulas for some binders soluble in organic solvents.

This alignment has the effect of reducing the instantaneous viscosity with increasing shear strain rate. When the shear stress is removed, the alignment of molecules may disappear immediately. In this case the shear stress–strain rate dependency is reversible. If the alignment dissipates gradually, the lower-viscosity state is retained for some time after the higher shear stresses are removed. This time-dependent rheological behavior is known as thixotrophy.

Because of the common pseudoplastic behavior of binder solution, the viscosity of binder solutions must be defined in terms of the specific measuring conditions (e.g., method of measurement and flow rate). Data available on viscosity for different binders are often obtained under different conditions, so comparing the viscosity of different binders must be made with appropriate precautions.

"Gelation" is a third rheological condition of importance in binder solutions. Gels can be obtained by (1) cooling a warm solution, (2) heating a solution (thermal gelation), and (3) action of a chemical gelling agent (chemical gelation). Gels have elasticity up to a certain yield stress, beyond which tearing and fracture may occur. Therefore, gelled liquids do not flow and their behavior cannot be represented in terms of flow response to various shear strain rates.

Bonding between polymer molecules is responsible for gel formation. Gels formed by heating or cooling are usually reversible in that a liquid state can be reobtained by reversing the thermal conditions or by diluting the gel with the solvent. Formation of gels by a chemical gelling agent tends to be irreversible.

COMPARISON OF BINDER VISCOSITIES

Organic binders are often arbitrarily designated according to how effectively they increase the viscosity of a solution. Designations such as low, medium, and high viscosity grades of binders are commonly employed for specific binders that are available in different molecular weights. Such definitions have not been established for comparing different binders, however. For purposes of comparison, it is convenient to establish definitions for viscosity grades.

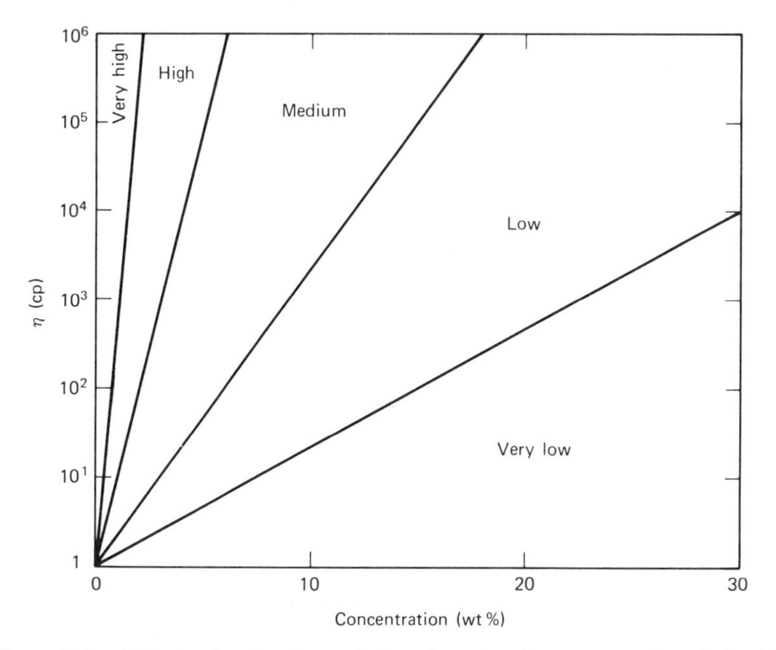

Figure 19.5. Criterion for viscosity grades based on viscosity–concentration relationship.

The graph in Figure 19.5 defines the criterion for categorizing viscosity grades that are used in this chapter. While the borderlines between grades are arbitrarily selected, they are convenient for the present purposes. The viscosity grade of a binder is determined by superimposing the viscosity-versus-concentration curve on the graph of Figure 19.4. The field within which the curve falls defines the viscosity grade (very low, low, medium, high, or very high).

From data given by binder suppliers and available in the literature, the viscosity grades of a wide variety of water-soluble binders have been

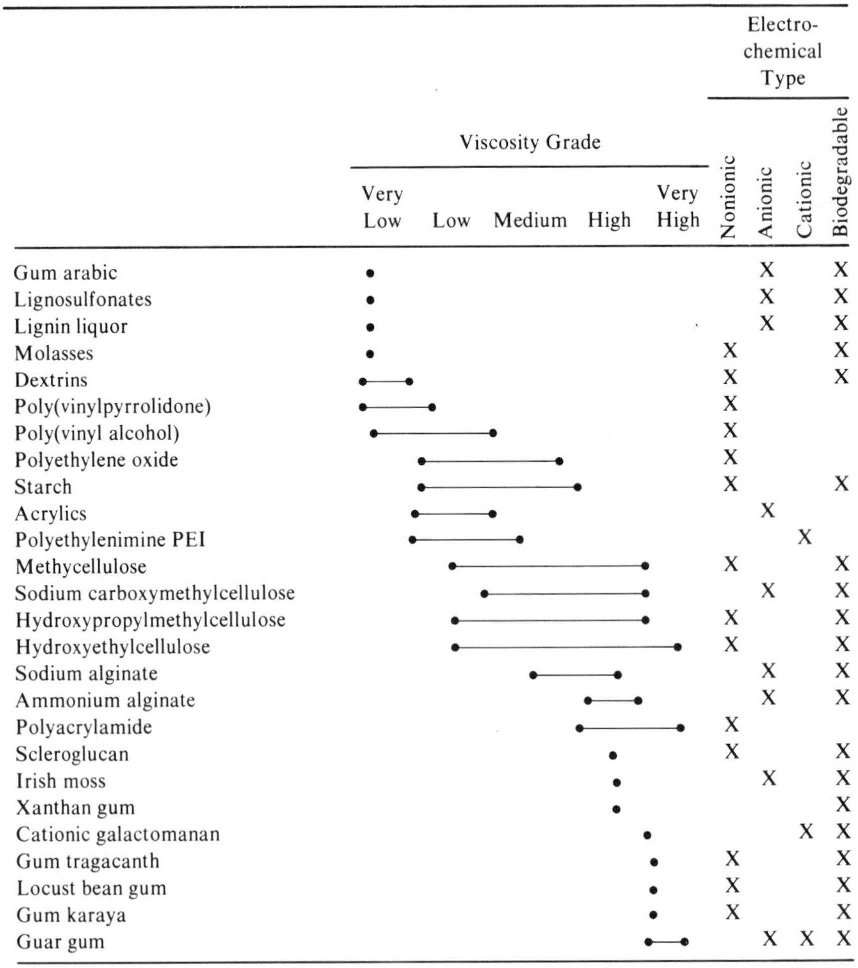

	Viscosity Grade					Electrochemical Type			
	Very Low	Low	Medium	High	Very High	Nonionic	Anionic	Cationic	Biodegradable
Gum arabic	•						X		X
Lignosulfonates	•						X		X
Lignin liquor	•						X		X
Molasses	•					X			X
Dextrins	•—	•				X			X
Poly(vinylpyrrolidone)	•—	•				X			
Poly(vinyl alcohol)	•	—	•			X			
Polyethylene oxide		•	—	•		X			
Starch		•	—	•		X			X
Acrylics		•	•				X		
Polyethylenimine PEI		•	—•					X	
Methycellulose		•	—	—	•	X			X
Sodium carboxymethylcellulose		•	—	—	•		X		X
Hydroxypropylmethylcellulose		•	—	—	•	X			X
Hydroxyethylcellulose		•	—	—	•	X			X
Sodium alginate			•	•			X		X
Ammonium alginate				•—•			X		X
Polyacrylamide				•	•	X			
Scleroglucan				•		X			X
Irish moss				•			X		X
Xanthan gum				•					X
Cationic galactomanan					•			X	X
Gum tragacanth					•	X			X
Locust bean gum					•	X			X
Gum karaya					•	X			X
Guar gum					•—•		X	X	X

Figure 19.6. Viscosity grades for some water-soluble binders.

determined and are summarized in Figure 19.6. For many binders a range of viscosity grades are available by variation in molecular weight.

Examination of Figure 19.6 shows the trends in the viscosities for vinyl-, acrylic-, and carbohydrate-derived binders. The vinyls and acrylics have very low or low viscosity. The cellulose derivatives are predominantly in the medium-and high-viscosity ranges. Also, the alginates and most of the natural gums (which are carbohydrates) have high or very high viscosities.

EMULSIONS

A class of binders, called "emulsions," can also be considered as very-low viscosity binders. Emulsions are organic substances that are finely divided and dispersed in a liquid (e.g., water) with the aid of emulsifying agents (certain surfactants).

Wax emulsions were among the first important binders of the emulsion type. The properties of waxes have long been recognized as useful for binder applications. They provide a reasonable level of green strength for many applications. At the same time they are soft enough so that granules are readily crushed at relatively low pressing pressures. However, the use of waxes in processing was hindered for a time by the fact that they were available only in the solid (bulk) state. It was necessary to melt or dissolve them in organic solvents when adding them to a ceramic batch. When wax emulsions were developed, waxes could readily be incorporated into ceramic bodies with water.

During the past several decades, a variety of wax or waxlike binders in emulsion form have been developed, including paraffin wax, modified paraffin wax, microcrystalline wax, mineral wax, and synthetic waxes such as stearic acid derivatives.

SIGNIFICANCE OF VISCOSITY IN CERAMIC PROCESSING

The viscosity imparted by binders is one of the primary considerations in the selection of a binder for a specific process. While binder is added to provide the necessary green strength, it must also impart the appropriate viscosity to the liquid in the batch material. For slip casting, doctor blading, and spray drying, the slip must have a low enough viscosity to carry out the process. This requires that the solution in the batch have low enough viscosity. In contrast, the liquid present for extrusion processes must have high viscosity[11] (10^4 to 10^6 cP). If the liquid is too fluid, it would be too easily squeezed out and separated from the ceramic mass.

An equation useful for estimating what viscosity grade of binder is necessary for a ceramic application is derived below. In Equation 1 the parameter k is a measure of the viscosity grade of a binder. Using the definitions of Figure 19.4, the range of k for each viscosity grade is as follows:

viscosity grade	k value range
very low	0–0.133
low	0.133–0.333
medium	0.333–1.00
high	1.0–3.0
very high	3.0–∞

The value of k required for a given ceramic process depends on the amount of liquid, the amount of binder, and the viscosity needed for the process.

The amount of liquid can be estimated from space-filling considerations. A dry powder packs to a certain packing density. If V is the total volume of the pack, this volume is the sum of the actual volume V_s occupied by the solid and the volume V_v of the void spaces. The packing density p expressed as percent of theoretical density, is given by

$$p = \frac{100 V_s}{V_s + V_v} \tag{2}$$

When the volume V_L of liquid added to the powder exceeds V_v, the mixture is fluid. A plastic state arises when $V_L \approx V_v$ (assuming the liquid has a high viscosity). A semidry pressing condition exists when $V_L \approx 0.3\ V_v$. It is useful to define a parameter y given by

$$y = \frac{V_L}{V_v} \tag{3}$$

which represents the fraction of void space actually occupied by the liquid.

The concentration of binder c, expressed as percent of liquid weight, is given by

$$c = \frac{100 \rho_B V_B}{\rho_L V_L} \tag{4}$$

where ρ_B and ρ_L are the densities of the dry binder and liquid, respectively.

Combining Equations 1 through 4 by eliminating c, V_v, and V_L, we arrive at

$$k = \left(\frac{100 - \rho}{\rho}\right)\left(\frac{\rho_L}{\rho_B}\right) \frac{y \log \eta}{100\,(V_B/V_s)} \tag{5}$$

Thus k can be calculated by estimating the percent of theoretical packing density of the powder (p), the viscosity η, and the amount of binder (V_B/V_s) needed for the process.

To illustrate the use of the relation given by Equation 5, some typical cases for extrusion, slips, and semidry pressing are considered.

Typically, ceramic powders have a packing density of around 55% of the theoretical value. Binders have a density of around 1.25 g/cc in many cases. If the liquid is water, its density is 1.0 g/cc. Thus

$$p = 55$$
$$\rho_B = 1.25$$
$$\rho_L = 1.0$$

For *extrusion* y is approximately 1 and η is around 10^5 cP; substituting these numbers and the typical values for p, ρ_B, and ρ_L, we can calculate k as a function of V_B/V_s from Equation 5. From the k values we can determine the viscosity grades corresponding to the specific conditions. In Figure 19.7, the grade binders required for different binder contents are shown based on the typical data of this example.

Similar results are shown in Figure 19.7 for slips and semidry pressed bodies. For slips y may range from 1.5 to 2.0 and so these two extreme cases are illustrated. The viscosity for slip solutions is assumed to be 10 cP. For the semidry batch material, y is around 0.3 and viscosity is around 10^3 cP.

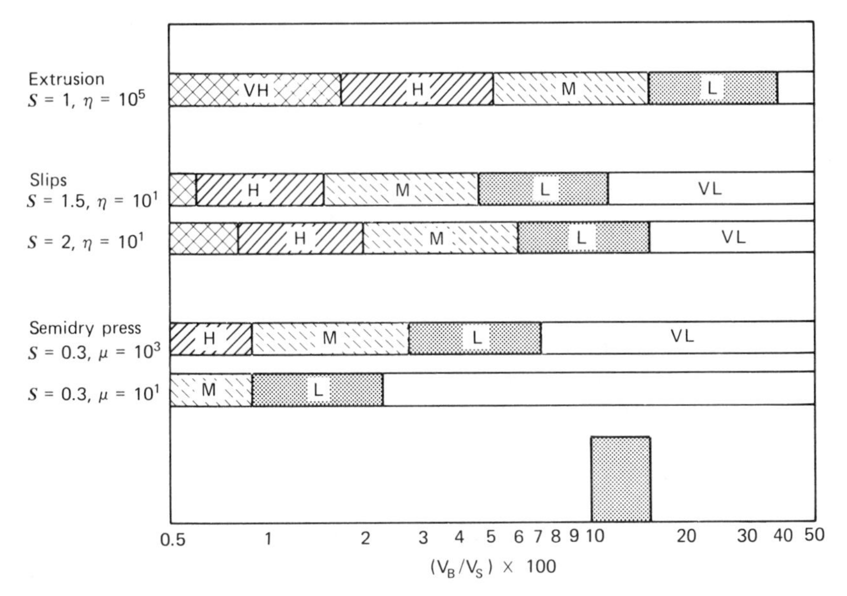

Figure 19.7. Binder grades calculated for different ceramic processes.

It is common in practice to find that the amount of binder used corresponds to V_B/V_s values between 0.08 to 0.15. This quantity is approximately the maximum amount that could be added without closing up the pore spaces in the body (which could be detrimental to burnout) The binder content range is designated by the width of the marker in Figure 19.7. We see that specific binder viscosity grades are identified for the three types of processes that fall into this V_B/V_s range. For extrusion processes medium-viscosity grade binders are needed. Low- or very-low-viscosity binders are needed for slips. Simidry pressing clearly requires very-low-viscosity binders.

The above conclusions explain why certain types of binders are used for certain processes. Starch, methylcellulose, and hydroxyethylcellulose are common binders for extrusion, for example, because they are of medium viscosity grade. Dry pressing preceded by spray drying utilizes gum arabic, poly(vinyl alcohol), and low-viscosity-grade starches. Semidry pressing utilizes very-low-viscosity binders of lower cost, including lignin liquor, lignosulfonates, dextrine, molasses, and poly(vinyl alcohol).

It should be remembered that Figure 19.7 was constructed on the basis of the "typical" conditions described in the text. For any specific application, the variables may be different and the necessary viscosity grades should be calculated from these variables.

Another reason for worrying about binder solution viscosities is that it is the high viscosities that limit how much binder can be dissolved in the liquid. Generally, once the viscosity reaches around 10^6 cP, the binder cannot dissolve any further because of kinetic reasons. If too much binder is added, undissolved binder would exist in the body. This can result in large pores after firing and in springback problems because of the low modulus of elasticity of the binder.

PSEUDOPLASTICITY AND ITS SIGNIFICANCE

The pseudoplasticity of solutions is important in many technologies, including ceramics. This property is very important in paints for the identical reasons that it is important in ceramics.

A suspension of solid particles tends to settle out in simple liquids such as water if the particles are larger than colloidal ($> 10^{-5}$ cm). The slips utilized in ceramics would not remain homogenous if settling occurred. One approach to slow down the settling is to increase the viscosity of the liquid. However, slips must be fluid enough to be sprayed or painted. To solve this problem, a pseudoplastic solution is utilized. The settling of a particle involves very small shear forces and shear strain rates. Under these conditions a pseudoplastic solution may have a very high viscosity. At high shear forces, as in pouring, spraying, or painting, the viscosity of the solution may be several orders of magnitude lower.

Figure 19.8 shows a solution viscosity–shear rate relationship for a 2% solution of hydroxyethylcellulose, a typical, strongly pseudoplastic binder. The ranges of shear rate corresponding to particle suspensions, leveling of paints, pouring of solutions, mixing and stirring, spraying, brushing, and rolling are indicated. For this particular binder, the viscosity resisting the settling of particles is more than three orders of magnitude greater than the viscosity under spraying or painting conditions.

In thick-film printing, pseudoplasticity is important for having low viscosity during screening but high viscosity once deposited so that the slip

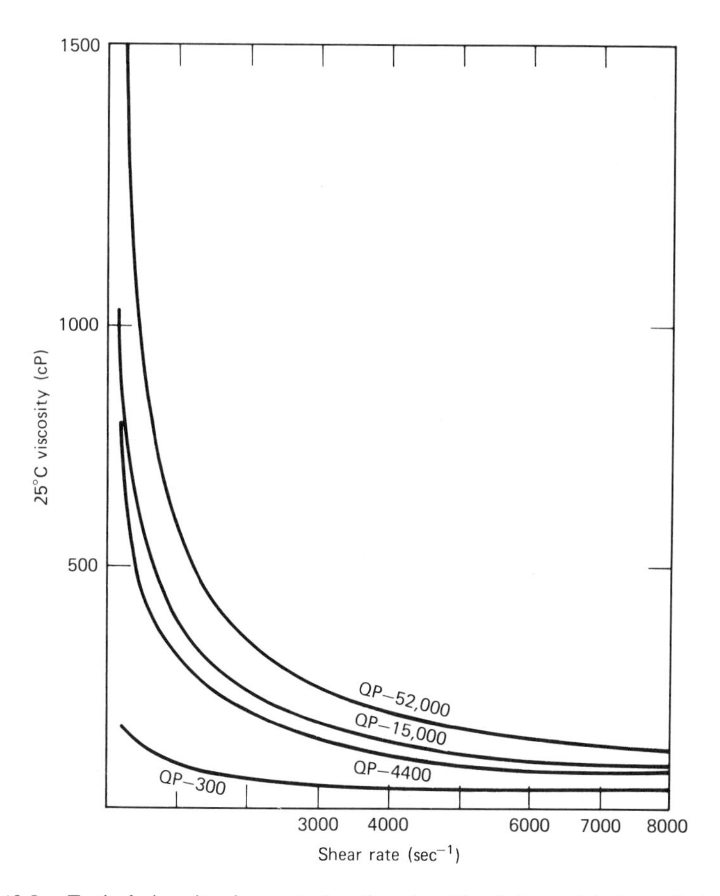

Figure 19.8. Typical viscosity-shear rate functions for 2% solutions of hydroxyethylcellulose. Different curves represent different HEC grades of varying molecular weight.

does not run and level out throughout the surface area. The time available before the slip is dry is only sufficient for a localized leveling between adjacent screen hole patches.

Thixotropy often is not a desirable feature when the pseudoplastic behavior is important. If the slip remains of low viscosity after being acted upon by the process step, the advantage of high viscosity at low shear rates become lost. A fast recovery of high viscosity when shear is reduced is essential.

Pseudoplasticity may also be an important factor in extrusion processes, although this has not yet been proven. Without external forces, the solution would have only small shear forces acting on it and may make the body more rigid by being at high viscosity. Under extrusion forces, the solution viscosity may be lower, thereby aiding the deformation process.

Most organic binders are pseudoplastic to some degree. While the effect is small in some, appreciable pseudoplasticity is exhibited by binders such as hydroxyethylcellulose, scleroglucan (Polytran FS), sodium carboxylethylcellulose, poly(ethylene oxide), poly(vinyl alcohol), and alginates. The pseudoplasticity increases with increasing binder concentrations.

MISCELLANEOUS PROPERTIES

Gelation

Under certain conditions, binder solutions can form gels. For example, poly(vinyl alcohol) (fully hydrolyzed grades) can be gelled by the addition of certain dyes (e.g., Congo red) and borax. Sodium alginates can be gelled with alkaline earth and heavy metal ions. Locust bean gum is gelled by borax. Gum karaya, irish moss, and agar gel by the gradual swelling of the binder. Heating increases the fluidity, and cooling brings about gel formation. Under certain conditions starches and starch derivatives can also be gelled.

Thermal gelation is an unusual type of gel formation exhibited by methylcellulose and hydroxypropylmethylcellulose. They gel when the solution is heated to temperatures between 50 and 90°C, depending on the specific binder type.

Gel formation has not been stressed as useful in the ceramic literature. Bodies having gelled liquids dry much more slowly because the liquid does not flow to the surface during drying. Water must leave the body by diffusion within the gel structure. However, one advantage of a gelled structure is that the binder does not migrate to the drying surface,[12] as it would if it were carried there by the flowing liquid as it moves to the surface. This advantage can be utilized with thermogelling solutions, since the gel is stable at higher temperatures. With higher temperatures the diffusion rate of water out of the gel is high and drying can be very rapid. Normal gels, which become liquid with heating, can only be dried at lower temperatures (if the gel is to be retained) where drying would be very slow.

Compatability with Electrolytes

The rheological behavior of binder solutions can be affected by the presence of electrolytes in solution. The electrolytes may increase or decrease the viscosity, or cause gelation or precipitation of the binder.

In general, nonionic binders have higher tolerances for most electrolytes than ionic binders. For example, the nonionic binders of methylcellulose, hydroxethylcellulose, and poly(vinyl alcohol) have stable viscosities over a wide pH range and can tolerate monovalent cations and multivalent metallic ions to fairly high concentrations with only a few exceptions. Ionic binders, such as carboxymethylcellulose, gum arabic, guar gum, and alginates, are particularly susceptible to alkaline earth ions, heavy metal ions, and trivalent cations. Stability with respect to pH changes exists over a smaller pH range.

When electrolytes are a problem in binder stability, one approach to remedy the problem, other than changing binders, by adding chelating agents that will tie up the detrimental cations.

Biodegradation

All carbohydrate-derivative binders are susceptible to degradation by enzymatic attack as a result of bacteria or fungi in the solution. Natural gums are particularly susceptible. Hydroxyethylcellulose is somewhat more resistant, while methylcellulose has rather high resistance to attack. In contrast, binders such as poly(vinyl alcohol), acrylics, polyethylene oxides, which are not derived from carbohydrates, are not susceptible to biodegradation.

Biodegradation reduces the molecular weight of the binders and consequently reduces the viscosity of the solutions. When this problem is encountered, a preservative or microbicide should be used to inhibit the degradation. The suppliers of the specific binder usually can recommend appropriate preservatives or microbicides. Care must be exercised that these agents do not introduce metallics that would be undesirable to the finished product.

Mixed-Binder Systems

Many binders are compatible with other binders. That is, a combination of two binders does not cause precipitation, unexpected gelation, or other detrimental effects. In some cases the use of two binders has certain merit. For example, poly(vinyl alcohol) has good bonding power and may be desired for green strength. However, this binder may not have enough thickening power for the desired application. It is possible to add hydroxyethylcellulose, for example, to obtain high viscosities. In another case the dispersion action of gum arabic may be very useful, but the strength and plasticizing capability of poly(vinyl alcohol) may be needed for dry pressing. Mixtures of poly(vinyl alcohol) and gum arabic are possible because of their compatability. The compatability of different binder combinations can often be ascertained by consulting the suppliers of the binders.

Plasticizers

When dry, binders often became hard and stiff. Granules for pressing often must be sufficiently soft and deformable to break down in the die under pressure. Plasticizers are additives that soften the binder in the dry or near-dry state. They are lower-molecular-weight organic species that dissolve in the same liquid as the binder. After drying, the binder and plasticizer are intimately mixed as a single material. The plasticizer disrupts the close aligning and bonding of the binder molecules, thereby increasing the flexibility of the material. While softening the binder, the plasticizer tends to reduce the strength.

Water-soluble plasticizers include substances such as glycerine, poly(ethylene glycol), poly(propylene glycol), and propylone glycol. These are effective for poly(vinyl alcohol), methylcellulose, hydroxethylcellulose, poly(ethylene oxide), polyacrylamides, and several other binders.

Binder Burnout

The amount of foreign contaminants due to the binder left in the fired body depends largely on the chemistry of the binder. Some binders are anionic polyelectrolytes. These include lignosulfonates, carboxymethylcellulose, and alginates. They are often effective as dispersants for clay-based systems. However, these polymers are salts and contain cations such as sodium, potassium, calcium, and ammonium. The metals can be deleterious to many ceramics requiring high purity. The use of these polyelectrolytes as binders is usually with clay-based systems and certain refractories in which the introduced cations can be tolerated. Those with ammonium cations are employed when higher purity is required.

The nonionic binders do not contain metals except as impurities. However, some contain nitrogen, as with PVP, polyacrylamide, polyethylenimine, and PEI. The use of these nitrogen-containing binders appears to be very limited in ceramic processing, possibly because the nitrogen can be deleterious in firing or because the cost of these binders tends to be high. The use of protein binders during the 1940s suggests that nitrogen may not be harmful in many cases, however.

For those binders having only carbon, hydrogen, and oxygen as primary constituents, burnout problems can still be encountered. The ash content is a measure of the level of metallic contaminants. Most of the purer binders have ash contents in the 0.5 to 2% range. This can introduce metallic contaminants on the order of 100 to 500 ppm to a ceramic. Also, significant levels of carbon can remain if the binder does not undergo complete thermal degradation. Particularly in reducing atmospheres, the binders with pure carbon–carbon linkage in their backbone and with smaller side groups tend to carburize. Oxygen is required to convert the carbon to CO or CO_2.

REFERENCES

1. J. W. Whittemore, "Industrial Use of Plasticizers, Binders, and Other Auxiliary Agents," *Bull. Amer. Ceram. Soc.*, **23** (11), 427–432 (1944).
2. E. P. McNamara and J. E. Comefora, "Classification of Natural Organic Binders," *J. Amer. Ceram. Soc.*, **28** (1), 25–31 (1945).
3. C. C. Treischel and E. W. Emrich, "Study of Several Groups on Organic Binders Under Low Pressure Extrusion," *J. Amer. Ceram. Soc.*, **29** (5), 129–132 (1946).
4. A. Wild, "Review of Organic Binders for Use in Structural Clay Products," *Amer. Ceram. Soc. Bull.*, **33** (12), 368–370 (1954).
5. T. Knapp, "Glaze Binders," *Bull. Amer. Ceram. Soc.*, **33** (1), 11 (1954).
6. H. Thurnauer, "Controls Required and Problems Encountered in Production Dry Pressing," *Ceramic Fabrication Processes*, W. D. Kingery, ed., Massachusetts Institute of Technology and Wiley, 1958.
7. S. Levine, "Organic (Temporary) Binders for Ceramic Systems," Ceram. Age, **75** (1), 39–42 (1960); **75** (2), 25–36 (1960).
8. T. A. Smith, "Organic Binders and Other Additives for Glazes and Engobes," *Trans. Br. Soc.*, **61** (9), 523–549 (1962).
9. A. R. Teter, "Binders for Machinable Ceramics," *Ceram. Age*, **82** (8), 30–32 (1966).
10. A. Pincus and L. Shipley, "The Role of Organic Binders in Ceramic Processing," *Ceramic Ind.*, **92** (4), 106 (1969).
11. J. F. White and A. L. Clavel, "Extrusion Properties of Non-Clay Oxides," *Bull. Amer. Ceram. Soc.*, **42** (11), 698–702 (1963).
12. J. E. Comeforo, "Migration Characteristics of Organic Binders," *Ceram. Age*, 132 (April 1945).

20

Mixedness of Suspensions

K. Sommer

H. Rumpf

Obtaining homogeneous suspensions is one of the goals of processing slips and pastes for a variety of operations. To assess how well a slip or paste is mixed requires defining some parameter that serves as a measure of the degree of mixedness. The use of the variance, a statistical parameter, for various types of suspensions is described in this chapter.

CONCEPT OF VARIANCE

For a simple two-component system, the volume concentration X_1 for component 1 is given by

$$X_1 = \frac{V_1}{V_1 + V_2}$$

where V_1 and V_2 are the volumes of components 1 and 2. The system as a whole has an average value of X_1. If small samples are taken that all have the same values of X_1 as the overall sample, the mixture is called ideally homogeneous.

Ideal homogeneity cannot be obtained in practice; that is, the concentration of a sample in a real mixture is generally larger or smaller than that for the ideally homogeneous value. Obviously, the more the concentration varies from the ideal, the poorer the degree of mixing.

The well-known statistical parameter variance is a useful measure of mixedness. The variance σ^2 is the mean square deviation of concentration from the average:

$$\sigma^2(X_1) = \lim_{K \to \infty} \frac{1}{K} \sum_i (X_{1_i} - P_1)^2$$

where P_1 = average, or ideal, value for the volume concentration of component 1

K = number of samples taken.

For a finite number of samples

$$\sigma^2(X_1) = \frac{1}{K} \sum_i (X_{1_i} - P_1)^2$$

In the remaining text, the variance is also referred to as the mixedness.

If the two constituents of the mixing are entirely separated (e.g., at the beginning of mixing), every sample will contain either only constituent 1 or constituent 2. The variance for this system is

$$\sigma_0^2(X_1) = P_1(1 - P_1)$$

and is independent of the size of the sample and of the size of individual particles.

For an ideally homogeneous mixture,

$$\sigma_{\text{ideal}}^2 (X_1) = 0$$

STOCHASTIC HOMOGENEITY

In contrast to the ideal homogeneity there is another final state of mixing that is characterized by the fact that it is not the composition of the samples that is the same in any part of the mixed material, but only the probability that their composition is the same. We call this a uniform random mix and its state is called stochastic homogeneity. What is often called a random mix is really a uniform random mix. A random mix can exist when the distribution of probabilities is different. A uniform random mix is the most homogeneous state of mixing that can be obtained by motion if there are no selective forces present.

When samples with a constant number of particles n are withdrawn from a composition, a generally valid variance of the numerical concentration for stochastic homogeneity can be derived and is given as:

$$\sigma_z^2(x) = \frac{p(1 - p)}{n}$$

Pawlowski[1] gave the following solution for the stochastic homogeneity of low-concentration suspensions, always assuming a constant total sample volume V:

$$\sigma_z^2(X_1) = P_1 \cdot \frac{v_1}{V}; \qquad V = \text{const.} \qquad P_1 \ll 1$$

where v_1 is the volume of a single particle.

Stange[2] dealt with a two-component system whose particle sizes differ only slightly. He assumed that for samples with a constant volume of all particles V_p, the mean number of particles per sample can be regarded as approximately constant, but in practice this is not so. Even when the ratio of the diameters of the two particle fractions to be mixed is $d_1/d_2 = 1.2$ there is, for $n = 150$, a range of variation between $n = 100$ and $n = 200$. This diameter ratio is still regarded as a monosize fraction for most practical chemical engineering purposes. Stange's equation

$$\sigma_z^2(X_1) = \frac{P_1(1 - P_1)}{V_p}(P_1 \cdot v_2 + (1 - P_1) \cdot v_1)$$

therefore only holds when $v_2 \approx v_1$.

Since the conditions of Stange and Pawlowski are sometimes not fullfilled, the present authors derived a general solution for samples having particle sizes that vary.[3] For sampling at a constant particle volume V_p, the variance is given by:

$$\sigma_z^2(X_1) = P_1(1 - P_1) \cdot \frac{j \cdot v_1}{V_p}$$

where j is a number defined by the simple fraction:

$$\frac{v_1}{v_2} = \frac{i}{j} \qquad i \cdot j = 1, 2, 3 \cdots$$

For the special case when the particle volumes v_1 of one component are an integral multiple of the particle volume v_2 of the other, the variance of stochastic homogeneity reduces to:

$$\sigma^2(X_1) = P_1(1 - P_1) \cdot \frac{v_1}{V_p} \tag{1}$$

SUSPENSION

When a monosize-particle fraction is mixed into a suspension, the liquid molecules v_2 are so small that it is perfectly fair to claim that the particle volume v_1 is an integral multiple of v_2. Equation 1 therefore applies to this case, too. For a suspension the "sample particle volume" V_p consisting of the solid component and the liquid is identical to the total sample volume V. A comparison with Pawlowski's equations shows that in the limiting case $P_1 \rightarrow 0$ Equation 1 converts to that of Pawlowski. It differs from Stange's equation by the factor $(1 - P_1)$.

The same analysis as above can also be applied to random packing. In this case we do not regard liquid molecules but rather vanishingly small space units as the monosize-particle volume of the second component.

MIXING OF SEVERAL MONOSIZED-PARTICLE FRACTIONS OF UNEQUAL PARTICLE SIZES

For the mixing of several monosized-particle fractions it is assumed that the single-particle volumes v_i are always an integral multiple of the next smaller particle volumes $_{i+1}$:

$$v_1 = \mu_2 \cdot v_2 = \mu_2 \cdot \mu_3 \cdot v_3 = \dots \qquad \mu_i = 0, 1, 2 \dots \tag{2}$$

v_1 is the greatest single-particle volume and v_n is the smallest. From this we determine that the smallest ratio of different single-particle volumes v_i/v_{i+1} is 2, that is, the ratio of the equivalent particle sizes d_i/d_{i+1} is 1.26, where the value 1.2 is often regarded as a monosized fraction. Therefore Equation 2 is a practical simplification. With these assumptions an equation for the variance of the stochastic homogeneity can be formulated[4]:

$$\sigma_z^2(X_k) = P_k(1 - P_k')\cdot\frac{v_k}{V_p} + P_k'^2\cdot\sigma^2\left(\sum_1^{k-1} X_i\right) \tag{3}$$

$$P_k' = \frac{P_k}{\left(1 - \sum_1^{k-1} X_i\right)}$$

and

$$\sigma_z^2\left(\sum_1^{k} X_i\right) = P_k(1 - P_k')\cdot\frac{v_k}{V_p} + (1 - P_k')^2\cdot\sigma^2\left(\sum_1^{k-1} X_i\right)$$

X_i is the volume concentration in a sample and P_i is the theoretical volume concentration of the component i. The above formulas are relatively complicated, because the concentration of a component i in a sample is correlated to the concentrations of each of the other components.

MIXEDNESS AND TEST ACCURACY

It can be shown that the following equation of the variance of the stochastic homogeneity of a two-component suspension holds quite generally

$$\sigma_z^2(X_i) = P_1(1 - P_1)\cdot\frac{v_1}{V}$$

where v_1 is the particle volume of component 1 and V is the total volume of the sample. It has also become clear that this calculable final state depends only on the composition of the mix, the particle size or particle volume, and the size of the sample. Figure 20.1 shows the two latter correlations.

The calculations are made for a highly concentrated suspension with $P_1 = 0.5$. The total volume of the sample is plotted along the x-axis and the standard deviation σ_z along the y-axis. The particle size d has been taken as the parameter of the straight line. The larger the sample volume and the smaller the particle size, the smaller is σ_z. Just as the final state of stochastic homogeneity is not a characteristic of the mixer, neither is the initial

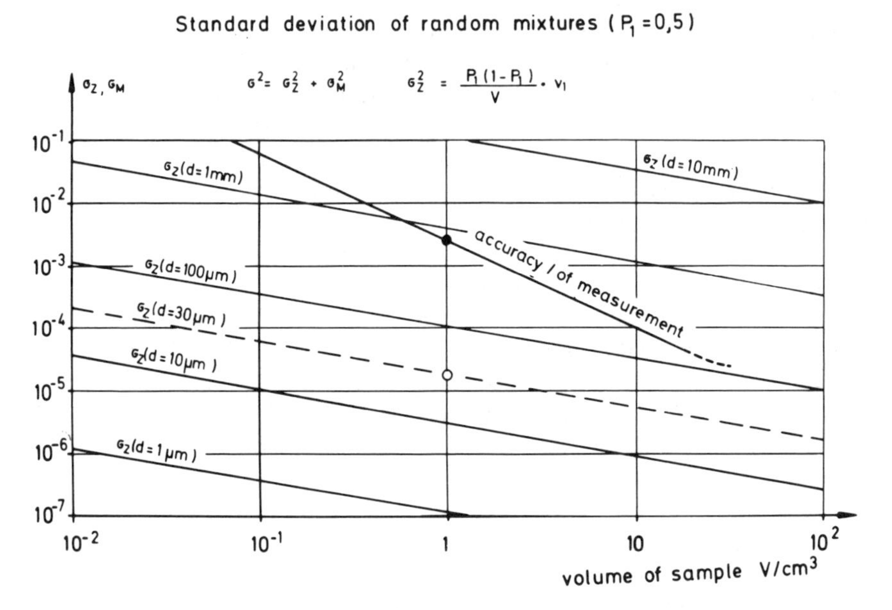

Figure 20.1. Standard deviation of random mixtures as a function of the sample volume V.

state of the completely separated components, for both are only characteristic of the composition of the mix.

For the assessment of mixing, the most important point is the way the variance changes from the initial to the final state. It is therefore necessary to examine which amount of the total variance is a material attribute and which is an attribute of the mixing.

Assuming mixing in which the components at time $t = t_0$ are completely separated the concentration distribution in space will approach a balance in the course of time ($t > t_0$). When repeating this mixing test under identical starting conditions, a perfectly characteristic concentration profile $P(r, t)$ will on the average develop at time t. When we compare the concentrations $X(r_j, t)$ of the various tests in position r_j at time t,, they will vary randomly about the expected value $P(r_j, t)$ (Figure 20.2).

The deviation from the theoretical value is therefore affected by two variation factors: first by the systematic deviation of the mean concentration $P(r_j, t)$ and secondly by random variations of the concentration $X(r_j, t)$.

It can now be shown that the total variance in the mixing is given by:

$$\sigma_{\text{total}}^2 (t) = \sigma_M^2 + \sigma_Z^2 + \left(1 - \frac{V_1}{V}\right) \sigma_{\text{syst}}^2 (t)$$

σ_M^2 and σ_Z^2 are variances that are independent of time. The change with time of the total variance $\sigma_{\text{total}}^2 (t)$ is only determined by the change with time of the systematic deviation, and σ_{syst}^2 is the real criterion for assessing a mixer. Mixing is completed when this systematic variance becomes zero. The total variance in the final state is therefore a constant composed of σ_M^2 and σ_Z^2.

In addition to the computable standard deviation σ_z, the inaccuracy of measurement σ_M determined in tests with chocolate or cocoa butter–sugar mixes is also presented in Figure 20.1. It can be seen that it is generally much greater in the given range than the random variations of stochastic homogeneity. That is why, for a discussion of $\sigma_{\text{total}} (t)$, the inaccuracy of measurement must at the latest be noted at or shortly before the end of mixing, and this is unfortunately often neglected in the literature.

Figure 20.3 shows how the standard deviation s of the sample composition of a mix changes during mixing. The mixing time is plotted along the x-axis and the empirically determined standard deviation s along the y-axis. A cocoa butter–sugar suspension containing 50% sugar by volume was mixed in a Werner-Pfleiderer kneader, which was completely filled.

The test points were obtained during a mixing test. They cannot lie below the accuracy of measurement. After a certain mixing time they are therefore bound to enter the band between σ_{Mo} and σ_{Mu}. During the tests the mixing time was $t_M \approx 12$ minutes.

When σ_M is greater than σ_z it is impossible to ascertain whether

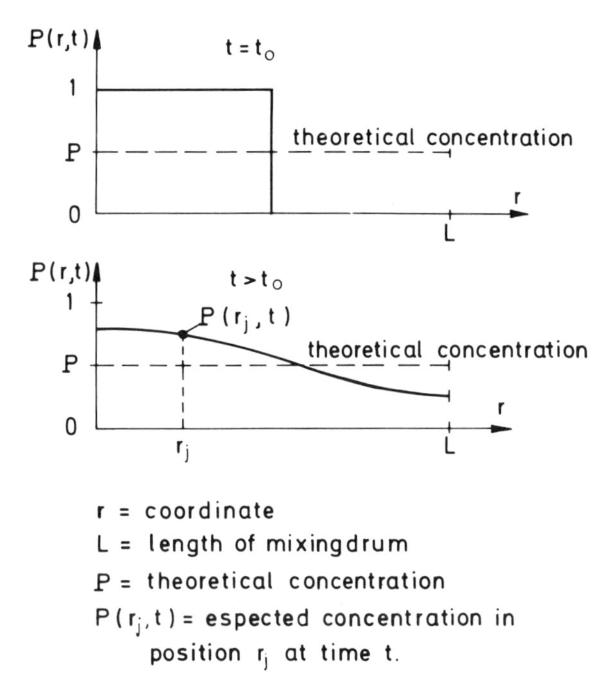

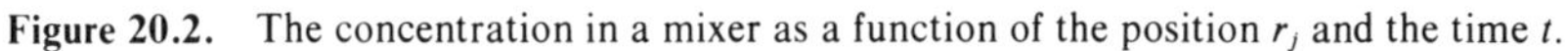

Figure 20.2. The concentration in a mixer as a function of the position r_j and the time t.

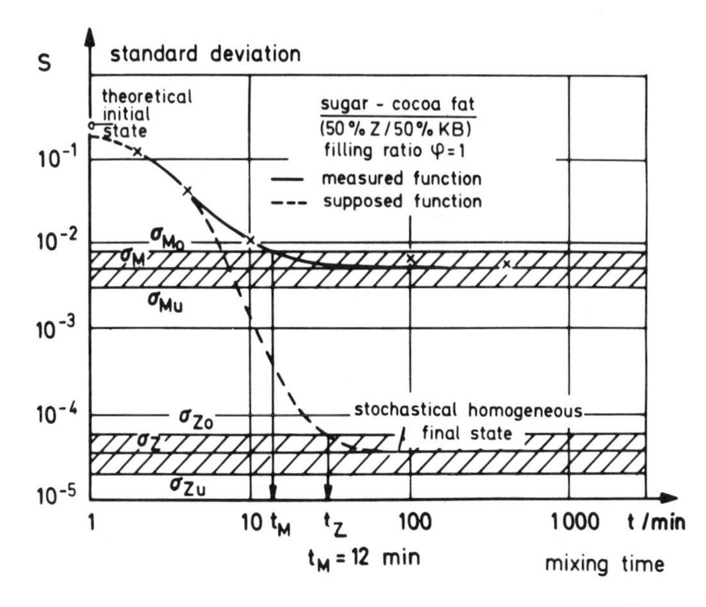

Figure 20.3. Time-dependent mixedness for a sugar–cocoa fat mixture.

stochastic homogeneity has been reached. During actual mixing it is only possible to determine the effective state of mixing when the inaccuracy of the method of measurement is less than σ_z. This would give a curve that enters the confidence limits for σ_z after a mixing time t_z. Stochastic homogeneity is reached at t_z. It is certain that t_z is not less than t_M, but is cannot safely be asserted that t_z is greater than t_M, nor by how much it is greater than t_M.

The question of how homogeneous a mixture has to be will ultimately be decided by technical requirements if, in the application, variations up to σ_M are admittable, the mixing test can be applied, and t_M is a sufficient mixing time. A longer mixing time produces neither a measurable nor a technically required advantage. In many cases the analytical determination of mixedness by measurement of the concentration is worse than the sensory judgement (e.g., homoeopathy, taste, texture). If technical requirements demand a more homogeneous mix, the more accurate measuring method must be used, if indeed this is possible. If not, the demand for greater mixing homogeneity is absurd, because it cannot be checked.

REFERENCES

1. J. Pawlowski, "Zur Statistik der Homogenisierprozesse," *Chem.-Ing.-Techn.*, **36**, (11), 1089/98 (1964).
2. K. Stange, "Die Mischgüte einer Zufallsmischung als Grundlage zur Beurteilung von Mischversuchen," *Chem.-Ing.-Techn.*, **26**, 331/337 (1954).
3. K. Sommer and H. Rumpf, "Varianz der stochastischen Homogenität bei Körnermischungen und Suspensionen," *Chem.-Ing.-Techn.*, **46**, 257 (1974).
4. K. Sommer, "Einfluß des Probevolumens auf den mittleren Fehler der Korngrößenanalyse," *Chem.-Ing.-Techn.*, **47**, 1 (1975).

21

Quantitative Theory of Cracking and Warping During the Drying of Clay Bodies

A. R. Cooper

Cracking and warping of ceramic bodies during drying are a result of the nonuniform shrinkage that takes place throughout the body. From a physical viewpoint, drying behavior has been well described.[1-8] However, no quantitative treatment exists for describing the stresses and strains that arise during drying. To make this understanding quantitative, we have utilized the well-established theory describing the stresses, strains, and deformations that are caused by nonuniform temperature distributions in a material.[9] The same theory can be applied to the problem of drying stresses and deformations caused by nonuniform water contents. Similar considerations have already been utilized for describing the stresses and deformations that develop from nonuniform concentration distributions.[10,11]

The transport of water in soils and particularly in clays is a subject of intense interest in the field of soil science. While several methods[12,13] have evolved for treatment of the process of water transport in soil–water systems, most workers seem to follow the approach of Philip[14] and treat water transport as a diffusion process. Since this approach is directly applicable to "thermal stress theory," and since careful measurements[15] confirm the validity of the diffusion equations over an order of magnitude of time, we choose to follow this scheme. Further, progress[16-18] in techniques of solution of the nonlinear concentration-dependent diffusion equation for cases of interest in soil science and the establishment of a thermodynamic basis[19-21] for the drying process in soils may prove to be of direct utility in ceramics.

In this paper we do not utilize the sophisticated methods of the soil scientists for solution of the nonlinear diffusion equations. Rather, it is shown below that by being concerned about the magnitude of the effects due to variable water diffusivity and by accounting explicitly for approximations that will lead to significant errors, we can utilize a relatively simple theory to gain a quantitative understanding of drying stresses.

PHENOMENOLOGY OF SWELLING

A typical water-expansion curve for clay is shown schematically in Figure 22.1, where V is the total volume of the body, V_0 is the dry volume, and C is the volume fraction of water. If V_w and V_s are the actual volumes of water and solid in the body, C is equal to $V_w/(V_w + V_s)$. An abrupt change in expansion occurs at $C = \theta$. When a body is drying, it shrinks until the clay particles touch. Further drying occurs without shrinkage by replacing the water in the pores by air. The volume fraction θ is that point where the particles touch during drying.

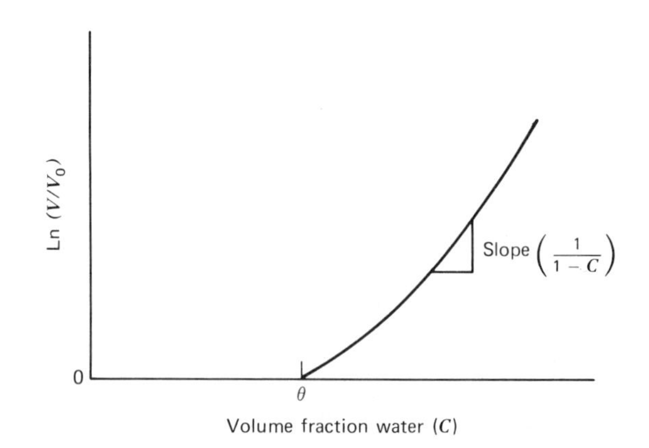

Figure 21.1. Schematic curve of water expansion of clay–water system.

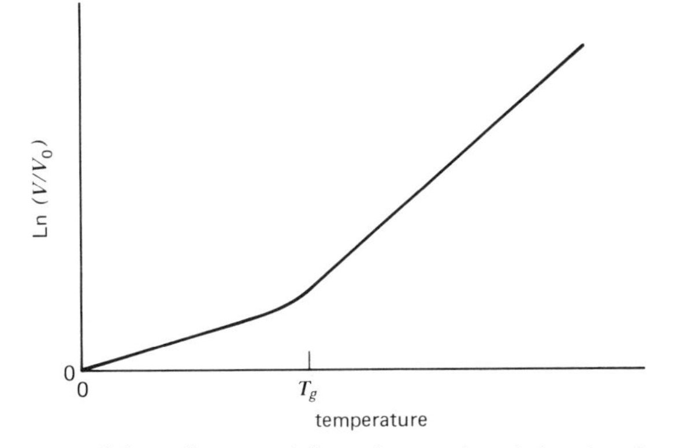

Figure 21.2. Schematic curve of thermal expansion of glass-forming system.

The water expansion can be described by a coefficient α_C as follows:

$$\alpha_C \equiv \frac{1}{V}\frac{\partial V}{\partial C} \equiv \frac{1}{V}\frac{\partial V}{\partial[V_w/(V_w + V_s)]} \tag{1}$$

Note that for $C > \theta, dV = dV_w$ and for $C < \theta, dV = 0$, we see that

$$\begin{aligned} \alpha_C &= 0 & C &< \theta \\ \alpha_C &= \frac{1}{(1-C)} & C &\geq \theta \end{aligned} \tag{2}$$

Since calculation of stresses involves linear strains, we are more concerned with linear expansion than with volume expansion. If L_x is a characteristic length in the x direction, the linear expansion coefficient a_{xC} is defined by $a_{xC} = 1/L_x(\partial L_x/\partial C)$. The linear expansion coefficients in three orthogonal directions must always sum to the volume expansion coefficient, for example, $a_x + a_y + a_z = \alpha$. When the material is isotropic, we can further simplify, since $a_x = a_y = a_z = \frac{1}{3}\alpha$. Although we henceforth assume isotropy for simplicity, we realize that anisotropic water expansion exists[1,6-8] and can be of decisive importance[1] in fracture and warping.

DEFORMATION BEHAVIOR

The water-expansion curve for clay has some similarity to the thermal-expansion curve of a glass-forming system, shown in Figure 21.2. For the glass, V_0 represents the volume at zero temperature. The temperature T is analogous to C for the water-expansion case. The expansion curve for glass also has an abrupt change in slope at the glass transition temperature (T_g). However, the glass still has a finite thermal expansion at $T < T_g$.

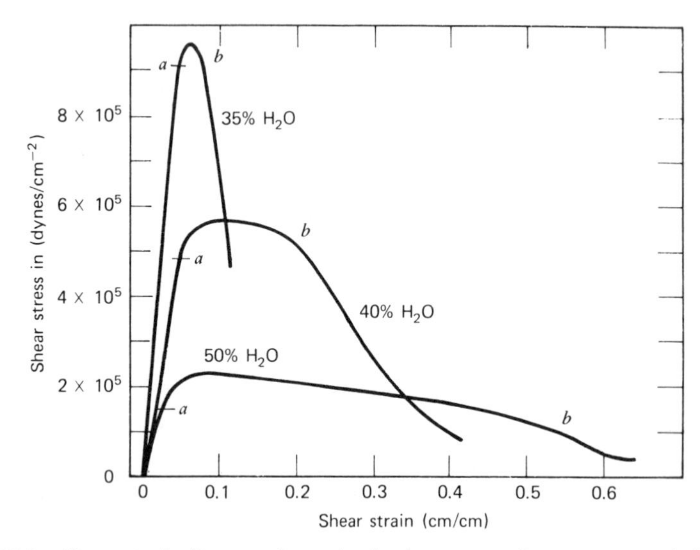

Figure 21.3. Stress–strain diagrams for a plastic clay at several water contents (taken from Norton[2]): a = yield point, b = fracture. Note: percent water is weight percent water M; $C = M/100[(M(1 - X) + X]$ where X^{-1} = specific gravity of clay.

A characteristic difference between a clay–water system and a glass-forming system is that above their glass transition, glasses are viscous and can only support a hydrostatic stress. As seen in Figure 21.3, however, at water contents where $C \gg \theta$, the clay still has a finite modulus of elasticity. When clays are in the plastic range, they can maintain their shape against finite deformation forces, for example, those due to gravity.

Several additional points about Figure 21.3 deserve comment. (1) For a decrease of water concentration ΔC of about 0.30 (which is equivalent to 15 wt %), the modulus of elasticity increases by a factor of about 5. (2) The strain at the yield point remains remarkably constant over the entire range of water contents. (3) The occurrence of cracking as indicated by the points b on the curves occurs at higher strains but lower stress at C increases.

It is well known that so long as $C > \theta$, a clay–water paste will dry at a rate determined entirely by conditions external to the piece[1-3]; that is to say, water flow through the clay paste is always adequate to maintain the flux density of water j consistent with the evaporation kinetics. Typically, the volume of water evaporating per unit surface area per second, that is, the flux j, is given by

$$j = k(p_w - p_0)$$

where k = evaporation constant governed by the flow conditions
p_w = vapor pressure of water at the temperature of drying
p_0 = partial pressure of water in the surrounding atmosphere.

Since k, p_w, and p_0 are often held fixed, this portion of the drying cycle is often called the "constant-rate" period. As the system becomes dryer, water flow within the piece begins to influence the kinetics and causes a "falling rate."

KINETICS OF DRYING

As shown by Philip and Knight,[16,17] the phenomenology of water flow through clay–water systems is best comprehended and related to permeability models when the flow relative to the clay particles is considered. This results in his using the moisture ratio γ, which is equal to V_w/V_s, to describe the water content and substituting a distance of solid variable m, where m is equal to z/C, in place of the usual distance variable z. However, he also shows that there is a direct conversion between the diffusion coefficient appropriate when the (γ, m) variables are utilized and the diffusion coefficient using the (C, z) variables, that is,

$$D(C, z) = (1 + \gamma)^3 \, D(\gamma, m)$$

This leaves us free to write the diffusion equation in the usual way as

$$\frac{\partial C}{\partial t} = \nabla(D \nabla C) \tag{3}$$

Our approach here is to consider solution of this equation for the case of a large slab of thickness $2w$ with an initial uniform water concentration C_0 and an equal drying rate j from both surfaces. Assuming at first that D is constant for all $C > \theta$, we get the solution[22]

$$C(z, t) = C(0, 0) - \frac{jw}{D}\left(\frac{z^2}{2w^2} + \frac{Dt}{w^2} - \frac{1}{6} - \frac{2}{\pi^2} \sum_{n=1}^{\infty} \frac{(-1)^n}{(n)^2} \cos\left(\frac{n\pi z}{w}\right) \exp \frac{-n^2\pi^2 t}{w^2}\right) \tag{4}$$

This solution, displayed as Figure 21.4, reveals that after a short period of time, the form of the concentration profile becomes fixed; dC/dt is equal to $-j/w$, a constant everywhere, and the concentration distribution is parabolic.

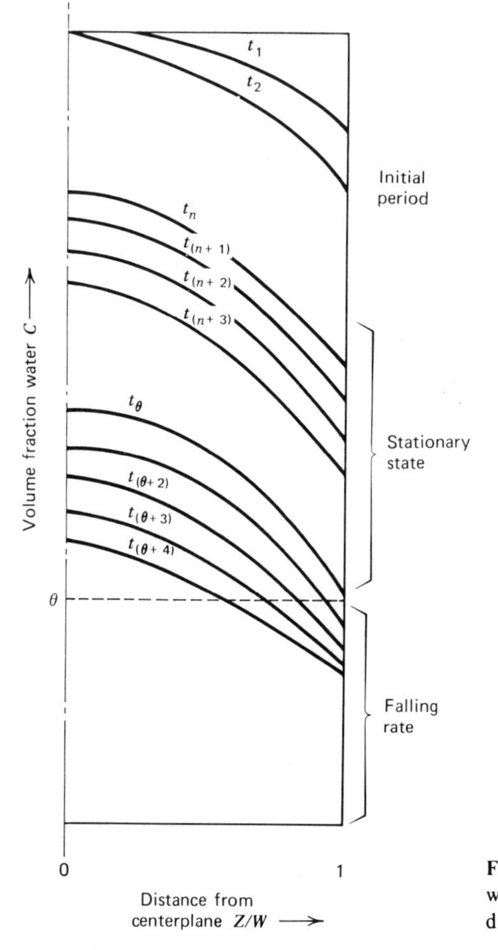

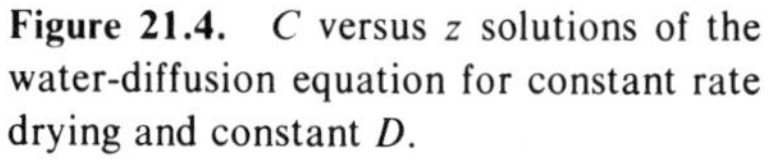

Figure 21.4. C versus z solutions of the water-diffusion equation for constant rate drying and constant D.

This "stationary-state" condition is assymptotically attained. An estimate of the time required before the distribution becomes parabolic is given by the fact that when $t > 0.3w^2/D$, the terms in the summation in Equation 4 always contribute less than 3% of the value of $C(z, t)$ for all z. Even when D is not constant, we may expect that the distribution will tend to the stationary state, that is, dC/dt is equal to $(-j/w)$.

For constant D in the stationary-state regime, it is easy to show that

$$C(w, t) - \bar{C}(t) = -1/3(jw/D) \tag{5}$$

This is also the maximum differential in water content between the average concentration $\bar{C}$ and the concentration at any other point at any time during the drying process with constant drying rate and constant water-diffusion coefficient.

Once the water concentration at the surface reaches θ, the drying rate begins to diminish. Since no volume change occurs in the region of $C < \theta$, there is little interest to us from the point of view of deformation and fracture once the entire specimen satisfies this condition.

While Equations 2 and 3 and Figure 21.4 are for a constant diffusion coefficient, the recent data on $D(\theta)$ for bentonite,[23] as well as the results quoted by Ward (see Figure 21.5), show that D is actually an increasing function of water concentration. This has two important consequences. First, the curves $C(z,)$ during the stationary-state period are no longer simple parabola. Rather, as indicated in Figure 21.6, an increase in the diffusion coefficient in the higher-water-containing interior causes a decrease in the composition differentials over that obtained if the diffusion constant is fixed at its value at the outer surface. Thus if the diffusion coefficient at the surface D_w is substituted for D, Equation 3 still provides an upper

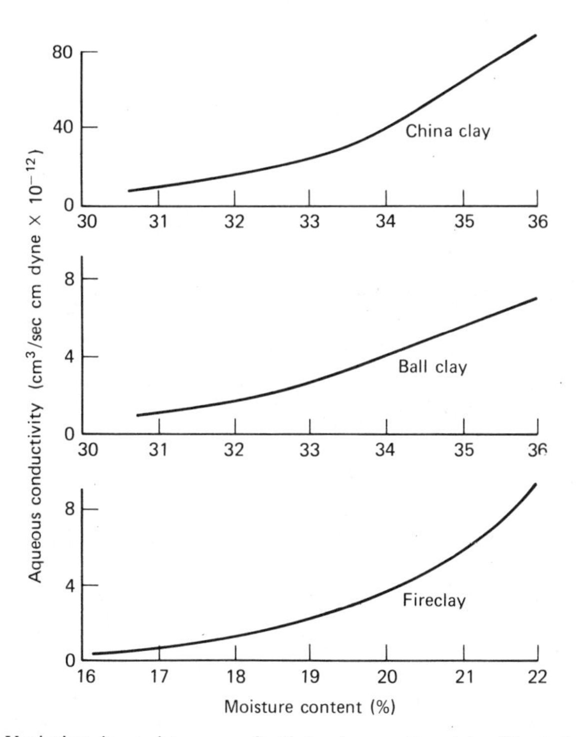

Figure 21.5. Variation in moisture conductivity (proportional to D) of three clays. Note moisture content is weight percent water M; $C = M/[100M(1 - X) + X]$ where X^{-1} is specific gravity of clay. (From Ford,[1] Figure 28, p. 51.)

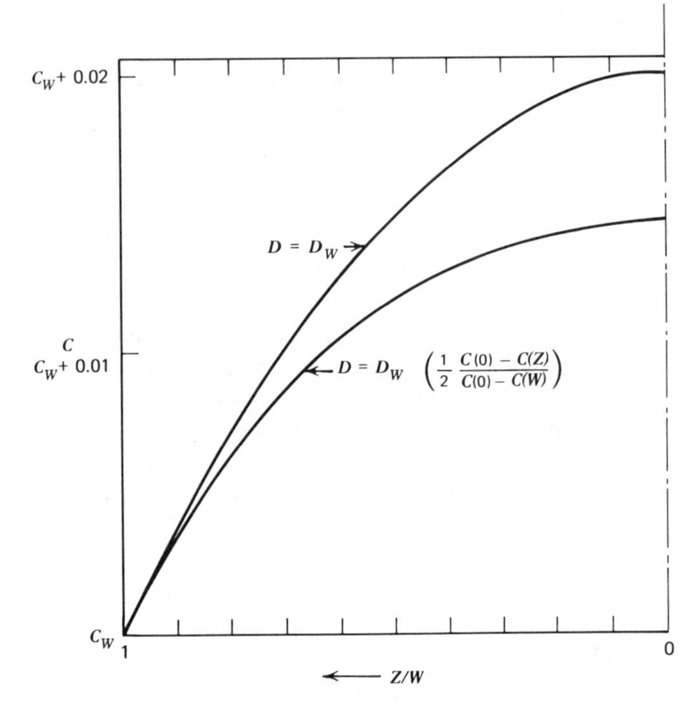

Figure 21.6. Comparison of water distribution for case of constant D (i.e., $D = D_w$) and variable D.

bound for the water concentration differentials. Second, the differentials $[C(w, t) - \bar{C}(t)]$, will increase as time advances during the stationary-state period. Actually, as seen in Figure 21.6, the estimate is quite good for a twofold variation in D across the piece. Because of the decrease of D_w with time, the maximum concentration differential occurs at the time when $C(w) = \theta$.

It is interesting that constant D solutions to Equation 3 for cylindrical or spherical symmetry and constant drying rate also yield stationary-state concentration distributions that are parabolic, just as those in Equation 4 and Figure 21.4, with the only replacement being fractional radius r/R for fractional thickness z/w. Likewise, the differentials are only slightly modified (because of the difference in averaging procedures) to:

$$C(R, t) - \bar{C}(t) = -\frac{1}{4}\frac{jR}{D} \qquad \text{cylinder}$$

$$C(R, t) - \bar{C}(t) = -\frac{1}{5}\frac{jR}{D} \qquad \text{sphere} \tag{6}$$

Drying also causes temperature differentials, as a result of the heat of vaporization, but these are ignored here for simplicity.

THERMAL STRESS THEORY

To understand the consequences of water differentials, such as those given in Equations 4 through 6, we look for an analogy to thermal stress theory. This methodology permits calculation of stresses, strains, deformation, and warping due to the presence of nonuniform temperatures. Explicit solutions exist for simple shapes such as slabs, cylinders, and spheres.[6-8] For example, with constant thermal expansivity a, elastic modulus E, and Poisson's ratio ν, the normal stress σ in the plane of a slab of thickness $2w$ with the origin of the z axis placed at the slab midplane is:

$$\sigma(z) = \frac{aE}{1-\nu}\left(-T + \frac{1}{2w}\int_{-w}^{+w} T\,dz + \frac{3z}{2w^3}\int_{-w}^{+w} Tz\,dz\right) \tag{7}$$

If the temperature distribution T is linear, there is no stress anywhere, that is, $\sigma = 0$. If T is symmetric, that is, $T(z) = T(-z)$, then the last term drops out in Equation 7 and $\sigma(z) = Ea(\bar{T} - T(z)/(1 - \nu)$. (This condition is also true for tangential stress at the surface of spheres and cylinders of radius R, that is, $\sigma(R) \sim [\bar{T} - T(R)]$. In this convention positive values of σ are tension. Hence, during cooling the maximum tension occurs at the surface $z = w$ or $r = R$.

The normal strain in the plane ϵ_{xx} is given by

$$\epsilon_{xx}(z) = \epsilon_{yy}(z) = \frac{a}{w}\int_{-w}^{+w} T\,dz + \frac{3z}{2w^3}\int_{-w}^{+w} Tz\,dz \tag{8}$$

and the curvature ρ by

$$\rho^{-1} \cong -\frac{3}{2}\,\frac{a}{w^3}\int_{-w}^{w} Tz\,dz \tag{9}$$

Notice that when temperature distribution is symmetric, $1/\rho = 0$, that is, a plane slab remains flat.

Materials that are capable of developing residual stresses can exist with a nonuniform temperature distribution T_0 without a stress. If they are quenched to a new temperature regime where stresses can be supported, the stress, strain, and curvature present at the new temperature distribution T are given by substituting $(T - T_0)$ for T in Equations 7 through 9.

A constant function of z is both linear and symmetric. Hence, for T a constant function of z, the stress and curvature are calculable by substituting $-T_0$ for T in Equations 7 and 9. The stresses obtained in tempering glass by quenching are approximately described in this way by letting the zero stress temperature distribution T_0 be that which is present when the center of the slab goes through the glass transition temperature T_g.

APPLICATION OF THERMAL STRESS THEORY TO DRYING

In the region where a clay–water paste exhibits elastic behavior (where $\sigma < \sigma_{yp}$), the equations of thermal stress apply with the substitution $C \rightarrow T$, $C_0 \rightarrow T_0$, and $a_C \rightarrow a$. However, it is not appropriate to use Equations 7 and 8 directly because they apply exactly only for constant properties and, as we see in Figure 21.3, the elastic constant varies markedly with water concentration. Thus we must use an expression[24] that avoids the approximation of constant properties. For an isotropic slab dried at an equal rate from both surfaces, the result is:

$$\frac{\sigma(z, t)}{A(z, t)} = -\int_{C(z,0)}^{C(z,t)} a_C(X)\, dx + \frac{\int_0^w A(\psi, t) \int_{C(z,0)}^{C(\psi,t)} a_C(X)\, dX\, d\psi}{\int_0^w A(\psi, t)\, d\psi} \tag{10}$$

where $A = [E/(1 - \nu)]$ and water expansion a_C are functions of concentration, X is the dummy concentration variable, ψ is the dummy distance variable, and $C(z, 0)$ is a water distribution at which there is no stress. Provided this stress-free distribution is one of uniform concentration, we may replace $C(z, 0)$ with any constant water concentration, for example, $C(w, t)$. Then the maximum tension stress always occurs at the surface and

$$\sigma_{\text{Max}}(t) = \sigma(w, t) = \frac{A(w, t) \int_0^w A(\psi, t) \int_{C(w,t)}^{C(\psi,t)} a_C(X)\, dX\, d\psi}{\int_0^w A(\psi, t)\, d\psi} \tag{11}$$

It is not difficult to evaluate Equation 11 directly for a given case. By making some reasonable approximations, however, we can gain further insight. Since $a_C = \frac{1}{3}\,[(1 - C)]$ for an isotropic material

$$\int_{C(w,t)}^{C(\psi,t)} a_C(X)\, dX = \frac{1}{3} \ln \left\{ \frac{[1 - C(w, t)]}{[1 - C(\psi, t)]} \right\} \tag{12}$$

Further, since

$$\ln \frac{[1 - C(w, t)]}{[1 - C(\psi, t)]} \cong \left(\frac{C(\psi, t) - C(w, t)}{1 - C(w, t)} \right)$$

$$\frac{\sigma(w, t)}{A(w, t)} \cong \frac{1/3[1 - C(w, t)] \int_0^w A(\psi, t) [C(\psi, t) - C(w, t)] \, d\psi}{\int_0^w A(\psi, t) \, d\psi} \tag{13}$$

The quotient of the two integrals in Equation 13 is just the average of the quantity $[C(\Psi, t) - C(w, t)]$ weighted according to the elastic modulus. Since $A(\Psi)$ is a decreasing function of Ψ the water concentrations in the interior will not be weighted as heavily as the surface water concentration. Thus

$$\frac{\sigma(w, t)}{A(w, t)} < \frac{1}{3[1 - C(w, t)]} [\bar{C}(t) - C(w, t)] \tag{14}$$

and substitution of numerical values suggests:

$$\frac{\sigma(w, t)}{A(w, t)_{\text{slab}}} \cong \frac{1}{3} [\bar{C}(t) - C(w, t)] = \frac{1}{9} \frac{jw}{D_w} \tag{15}$$

Thus we see now that the stress is directly proportional to the flux of water from the surface and the width of the sample and inversely proportional to the diffusion coefficient.

For spheres and cylinders of radius R, we can write analogous approximations for the tangential stress:

$$\frac{\sigma_{\theta\theta}(R, t)}{A(R, t)} \cong \frac{1}{12} \frac{j_R}{D_R} \qquad \text{cylinder}$$

$$\frac{\sigma_{\theta\theta}(R, t)}{A(R, t)} \cong \frac{1}{15} \frac{j_R}{D_R} \qquad \text{sphere} \tag{16}$$

The question remains as to what value of σ to select to avoid the development of cracks during drying. From Figure 21.3 it appears that (for the clay measured by Norton) if σ/E remains less than about 0.05, no chance of fracture exists. Also, if this condition is maintained there is likewise no opportunity for irreversible deformation. Then, assuming that no volume change occurs during elastic deformation and hence $\nu = \frac{1}{2}$, we may conclude that for safe drying

$$j_{\text{slab}} \leq \frac{9D}{w}\left(\frac{0.05}{0.5}\right) \cong \frac{D_w}{w}$$

$$j_{\text{cyl}} \leq \frac{12D}{w}\left(\frac{0.05}{0.5}\right) \cong 1.2\frac{D_R}{R}$$

$$j_{\text{sphere}} \leq \frac{15D}{w}\left(\frac{0.05}{0.5}\right) \cong 1.5\frac{D_R}{R} \tag{17}$$

Maintaining drying rates within these restrictions should avoid fracture, provided j, the drying rate, is kept equal at all points on the surface.

Warping is prevented during the drying of an isotropic system during the entire drying cycle if the water concentration distribution is symmetric, as is easily seen by substituting $(C - C_0)$ for T in Equation 9. It also follows from Equation 9 that even if the drying rate is nonuniform, there will be no shape change in a fully dried slab, cylinder, or sphere with an originally symmetric water distribution, provided no stress relaxation occurs during drying. For a material with the properties shown in Figure 21.3, this requires that $\sigma < \sigma_{yp}$. Retention of shape also is achieved if the conditions of Equation 17 are fulfilled.

Since D_w decreases as time of drying increases, Equation 17 suggests that minimum drying time is achieved if j decreases with time during the stationary-state period.

POSSIBILITY OF "TEMPERING" A CLAY DURING DRYING

The analogy of clay drying to glass cooling poses the question, Can residual compressive stresses be created in a dried piece by a method similar to that used to create stresses during rapid cooling of glass?" First this requires the creation of a water distribution that causes some stress relaxation. Such a result can be achieved if $\sigma > \sigma_{yp}$. It is clear from Figure 21.3 that this is most easily obtained at the higher water contents, which suggests drying initially at a rate such that j is greater than that permitted by Equation 17. While this is no doubt a useful idea, it is not clear without analysis beyond the scope of this paper whether the strong dependence of D on C prevents the possibility of obtaining stress relaxation during initial stages of drying without obtaining fracture near the end of the drying cycle.

RELATION TO OBSERVATIONS

The predictions from theory are consistent with the observation that the most critical period for fracture during drying is at the end of the constant-rate period. They are also consistent with the comments of Ward, "In general terms, cracking or warping of an article during drying is due to differential shrinkage of the body, which may have several causes: (1) Differential rate of loss of water from different positions, e.g., surface and interior. (2) Uneven distribution of moisture within the article prior to drying, resulting in non-uniform total shrinkage. (3) Anisotropic shrinkage because of the orientation of particles during shaping. (4) Mechanical restraint of shrinkage at a position where the article is in contact with the surface on which it is resting, particularly with heavy shapes and particles formed and dried on plaster moulds."

It is difficult, however, to obtain a quantitative agreement with the observations of Macey and Wilde,[7] who measured the maximum rates of drying permissible for a brick clay. For example, for a slab of 1.27 cm half width, they found that $j_{\text{Max}} = 1.5 \times 10^{-5}$ cm^3/cm^2 second. According to Equation 17, $j_{\text{Max}} \cong D_w/w = 0.8\ D_w$. Hence, from drying data $D_w \cong 2 \times 10^{-5}$cm^2/second. For comparison, one can make estimates on diffusivity of water in clay from data presented by Ward. Comparing the water flux j with the concentration gradient at the surface, $(dC/dz)_w$ for ball clay (Figure 21.7), one gets $D_w \cong 1.6 \times 10^{-3}$cm^2/second. Likewise, using the aqueous conductivity data, Figure 21.4, and the relation between hydrostatic pressure P and equilibrium water concentration[1] $dP/dC = 1.1 \times 10^9$dynes/cm^2, one obtains $D_w \cong 1 \times 10^{-3}$cm^2/second. A discrepancy of about two orders of magnitude is present.

Several explanations are possible: (1) Factors completely different from the stresses caused by water differentials govern the maximum permissible drying rate. (2) Nonuniformities of body and/or drying cause local stresses far in excess of those considered by our assumption of a uniform isotropic clay body and a uniform drying rate. (3) The drying rate, diffusion coefficient, and deformation characteristics were all measured on different clays and perhaps at different temperatures, and hence good agreement cannot be expected. Some support for the last point is found from the water-diffusion data on Smiles and Harvey[23] on the bentonite–water system. Extrapolating to the range of water concentrations of interest in clay drying, one obtains $D \cong 10^{-6}$ cm^2/second, which is lower than the result obtained from Macey's observed safe drying rates.

Another discrepancy between Macey's results and Equation 17 is that Macey found in the drying of cubes that $j_{\text{Max}} \sim w^{-2.6}$. Although none of the

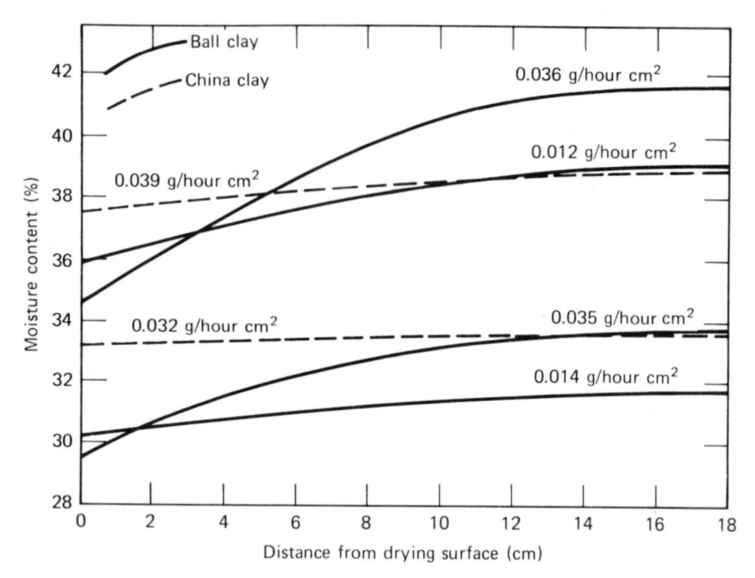

Figure 21.7. Moisture distribution during "constant rate" period for drying of very thick slabs.

simple shapes, spheres, cylinders, and slabs, considered in this paper are cubes, $j_{\mathrm{Max}} \sim w^{-1}$ or R^{-1}.

CONCLUSIONS

It has been shown that by treating water flow as being analogous to heat flow, it is possible to calculate stresses and warping of simply shaped bodies during drying. The approximate results are simple enough to be readily useful. They suggest ways of optimizing the drying process, for example, steadily decrease the drying rate during the period when the body is shrinking; however, at present they lack experimental confirmation. Thus it is hoped that the theory will be useful in (*a*) provoking a more detailed analysis of the process using, for example, the methods of Philip[16] and Parlange,[18] and a more detailed description of the deformation characteristics and (*b*) providing a useful model for design of a series of experiments in drying kinetics of a single body involving determination of water-diffusion coefficients, a detailed determination of the deformation characteristics, and a study of drying fracture and warping.

ACKNOWLEDGMENT

A. H. Heuer critically reviewed the manuscript and made useful and welcome suggestions to improve its unity and clarity.

REFERENCES

1. R. W. Ford, "Institute of Ceramics Textbook Series 3. Drying," McLaren, London, England, 1964.
2. F. H. Norton, *Elements of Ceramics*, Addison-Wesley, Cambridge, Mass., 1952.
3. H. Salmang, *Ceramics Physical and Chemical Fundamentals*, translated by Marcus Francis, Butterworths, London, 1961.
4. F. H. Clews, *Heavy Clay Technology*, Academic, London, 1969.
5. A. J. Dale, *Modern Ceramic Practice*, McClaren, London, 1963.
6. H. H. Macey, "The Relative Safe Rates of Drying of Some Different Sizes and Shapes," *Trans. Br. Ceram. Soc.*, **38**, 464 (1939). H. H. Macey and F. G. Wilde, "Experiments on the Drying of Clay," *Trans. Br. Ceram. Soc.*, **43**, 93 (1944).
7. R. Q. Packard, "Moisture Stress in Unfired Ceramic Bodies," *J. Amer. Ceram. Soc.*, **50**, 223 (1967).
8. H. H. Macey, "The Drying of Clay Ware, A Discussion," *Trans. Br. Ceram. Soc.*, **46**, 207 (1947).
9. B. A. Boley and J. H. Weiner, "Theory of Thermal Stresses," Wiley, New York, 1960.
10. A. R. Cooper and D. A. Krohn, "Strengthening of Glass Fibers: II. Ion Exchange," *J. Amer. Ceram. Soc.*, **52**, 665 (1969).
11. O. Richmond, W. C. Leslie, and H. A. Wreidt, "Theory of Residual Stresses due to Chemical Concentration Gradients," *Trans. American Society for Metals*, **57** (1), 294–300 (1964).
12. A. E. Scheidegger and K. H. Liao, "Thermodynamic Analogy of Mass Transport Processes in Porous Media," *Fundamentals of Transport Phenomena in Porous Media*, Elsevier, New York, 1972.
13. G. Dagan, "Some Aspects of Heat and Mass Transfer in Porous Media," *Fundamentals of Transport Phenomena in Porous Media*, Elsevier, New York, 1972.
14. J. R. Philip, "Hydrostatistics and Hydrodynamics in Swelling Media," *Fundamentals of Transport Phenomena in Porous Media*, Elsevier, New York, 1972.
15. D. E. Smiles, "Infiltration into a Swelling Material," *Soil Sci.*, **117** (3), 140–147 (1974).
16. J. R. Philip, "Recent Progress in the Solution of Nonlinear Diffusion Equations," *Soil Sci.*, **117** (5), 257–262 (1974). (See numerous references to earlier work by this author.)
17. J. R. Philip and J. H. Knight, "On Solving the Unsaturated Flow Equation: 3. New Quasi-Analytical Technique," *Soil Sci.*, **117** (1), 1–13 (1974).
18. J. Y. Parlange, "Theory of Water Movement in Soils: 10. Cavities with Constant Flux," *Soil Sci.*, **116** (1), 1–7 (1973). (This paper gives reference to the nine earlier contributions by this author.)

19. G. Sposito, "Volume Changes in Swelling Clays," *Soil Sci.,* **115** (4), 315–320 (1973).
20. S. Iwata, "Thermodynamics of Soil Water: IV. Chemical Potential of Soil Water," *Soil Sci.,* **117** (3), 135–139 (1974).
21. S. Al-Khafaf, R. J. Hanks, "Evaluation of the Filter Paper Method for Estimating Soil Water Potential," *Soil Sci.,* **117,** 194–199 (1974).
22. H. S. Carslaw and J. C. Jaeger, " Conduction of Heat in Solids," 2nd ed., Clarendon Press, Oxford, 1954, p. 112.
23. D. E. Smiles and A. G. Harvey, "Measurement of Moisture Diffusivity of Wet Swelling Systems," *Soil Sci.,* **116** (6), 391–399 (1973).
24. I. S. Alvarez and A. R. Cooper, Jr., "Thermal Stress Distribution for a Thin Plate with Temperature Variation through the Thickness," to be published by *J. Appl. Mech.*

22
Mineralogy of Curing and Drying of A Refractory Concrete

R. E. Farris
J. S. Masaryk

Tabular alumina aggregate bonded with calcium aluminate cement is a technically important refractory concrete. The mineralogical changes that occur during the curing and drying of this concrete greatly affect many of its properties. One important use of this concrete is in hardening UHF antennae for Minuteman Missile systems. The concrete protects against the heat and shock of a nuclear blast. In this application the dielectric and loss tangent characteristics (at UHF frequencies) must be consistent because the impedance of the antenna is matched. Mismatch causes gain loss, and transmitting power is not sufficient to contact the antenna.

Development studies during the past several years on hardened UHF antennae has led to a better understanding of the mineralogy of curing and drying of the refractory concrete. The phase that develop and their effects on processing and properties have been analyzed. The results of this work are discussed in this chapter.

MATERIALS, PROCESS DESCRIPTION, AND PROBLEMS

The concrete for the hardened antenna consists of a cylindrical mass weighing about 3000 lb. It consists of sized tabular alumina (6 mesh maximum size) bonded with calcium aluminate cement. The cement contains about 80% Al_2O_3 and 18% CaO. The overall concrete contains 94% Al_2O_3.

The process can be divided into four major steps:

1. *Mixing and casting.* The correct ratios of dry concrete and water are mixed and then properly placed in the mold.
2. *Curing.* The cement–water reactions occur and heat is liberated to form the initial bond in the system.
3. *Queueing.* The period where the castings are allowed to reach thermal and reaction equilibrium before the drying cycle commences.
4. *Drying.* Removal of the water to the required level.

Free water in the concrte causes dielectric losses in the antenna and must be removed. With a fairly rigid drying schedule, it was found over a period of time that residual moisture varied between 2.5 and 4.0%. The variable that correlated best to this variation was the cement lot used to manufacture the concrete, and possibly the CaO content of the cement. Other manufacturing variables that can have some effect are "casting water" or "concrete consistency," curing condition, and queueing conditions.

Another concern resulted from postmortem analysis of antennae that split during drying. This analysis revealed dehydration products that are not considered typical for normal low-temperature (275°C maximum) drying of concrete bonded with calcium aluminate cement. These dehydration products have been previously described in the literature, but not under the same conditions of pressure, temperature, and composition encountered in this particular system.

As a result of these concerns, studies were carried out to better define the effect of processing variables on bond-phase mineralogy during drying.

PREVIOUS LITERATURE

To follow the thermal response of the system of water, calcium aluminate cement, and aggregate, one must first establish the hydration reactions and temperatures and then follow the dehydration reactions that occur later. In the discussion of the hydration–dehydration reaction of calcium aluminate cement, the following notations are used: (1) C is for CaO, (2) A is for

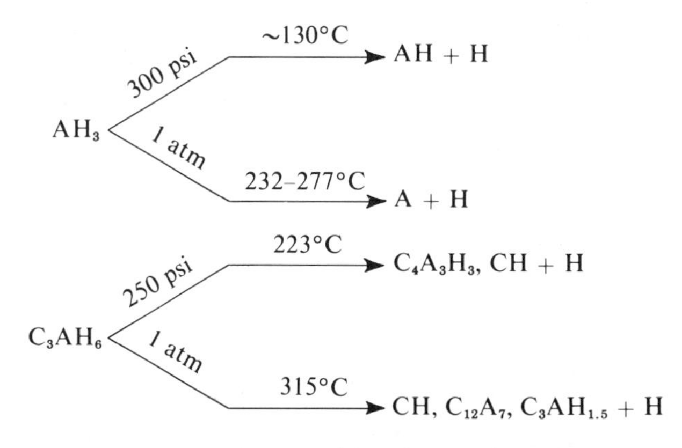

Figure 22.1. Decomposition of AH_3 and C_3AH_6.

Al_2O_3, and (3) H is for H_2O. For example, CA is for $CaO \cdot Al_2O_3$, which is the monocalcium aluminate.

The literature[1] discusses the reaction of CA with water at 1 atm pressure. The reactions are summarized in Table 22.1. The hydration of CA results in CAH_{10} and C_2AH_8. These dehydrate at higher temperatures to C_3AH_6 and finally to $C_{12}A_7$ along with the formation of A. At very high temperatures (600 to 1000°C), the excess A reacts to form various ceramic bonding phases, such as CA_2 and CA_6, along with $C_{12}A_7$.

Majumdar and Roy[2] studied reactions under hydrothermal conditions (high pressure and temperature). They investigated materials initially having C/A ratios of 3:1 and 4:3. In the hydration reaction, C_3A formed C_3AH_6 and the C_4A_3 formed $C_4A_3H_3$. During decomposition the C_3AH_6 formed $C_4A_3H_3$, CH, and H at 223°C and 250 psi. Any AH_3 that formed converted to AH and H at 130°C and 3000 psi. The hydrothermal products are compared in Figure 22.1 with those formed at 1 atm. In general the decomposition of the major phases is lowered about 100°C in systems under pressure.

Table 22.1. Hydration–dehydration of CA cement at atmospheric pressure

Compounds and Reactions	Stable Temperature Range (°C)	Major Decomposition Temperature (°C)
$CA + H \rightarrow CAH_{10} + C_2AH_8$	—	—
$CAH_{10} \rightarrow C_2AH_8 + AH_3$	0–20	20
$C_2AH_8 \rightarrow C_3AH_6 + AH_3$	25–58	58
$AH_3 \rightarrow A + H$	27–350	232 and 277
$C_3AH_6 \rightarrow C_{12}A_7 + H$	58–350	310–320

POSTMORTEM PHASE ANALYSIS OF CONCRETE

Studies on production failures revealed that the predominant phases were AH (bochmite) and $C_4A_3H_3$. The curing was carried out in an ambient of 1 atm. Therefore, the presence of AH and $C_4A_3H_3$ was contrary to what was stated in the literature; these products are produced only under hydrothermal conditions. It was therefore hypothesized that the interiors of the large castings were subjected to hydrothermal conditions as a result of the pressure developed internally by trapped water vapor. To investigate this possibility, autoclave studies were initiated.

AUTOCLAVE STUDIES

In this work it was assumed that during drying an antenna would behave similarly to a "leaky" autoclave. On this basis the maximum pressure in the large concrete mass at any temperature would be that of saturated steam at the temperature in question. The "leaky" portion of the assumption relates to the casting permeability. Casting permeability yields to several processing variables, the work relating to process variables is discussed in the next section.

The autoclave studies were straightforward. Dry concrete, the same as that used in antenna manufacture, was blended with 9% "casting water" and 3 in. diameter × 6 in. high cylinders were cast at about 20 to 24°C. The cylinders were cured for 16 to 24 hours at 20 to 24°C and 90 to 95% relative humidity (RH). Cured cylinders were then placed in a standard autoclave for pressure–temperature treatments as shown in Table 22.2. The standard autoclave was limited to 166°C and 80 psi. For higher conditions the samples were placed in a model bomb over water and heated to the listed temperatures that yield the indicated pressures. Holding times at maximum pressures for 24 hours were found to be adequate to yield reaction products. In all cases the steam was released at temperature and then the bomb or autoclave was cooled. Analysis of mineral phases was made using X-ray diffraction techniques. The relative X-ray intensity of reaction product versus temperature and pressure is shown in Table 22.2 and the results are discussed in the following paragraphs.

Gibbsite (AH_3) decomposes between 115 and 130°C at 10 to 25 psig to form boehmite. Previous work suggested this reaction would occur in this temperature range but at very high pressures (3000 psi). This work and the work of others have shown that at atmospheric pressure, regardless of temperature, the AH_3 to AH reaction does not occur in high-purity

Table 22.2. Relative X-ray intensities of mineral phases in autoclaved concrete–calcium aluminate cement bond

Temperature (°C)	Gauge Pressure (psi)	CA	AH_3	AH	C_3AH_6	$C_4A_3H_3$
115	10	3	16	3	21	—
130	25	—	—	10	21	—
148	50	—	—	13	23	—
166	80	—	—	15	14	—
177	120	—	—	12	15	—
204	233	—	—	25	17	—
232	410	—	—	26	14	—
246	525	—	—	29	3	13
260	665	—	—	41	2	17

concretes. Once formed, however, the boehmite is very stable and becomes a significant bonding phase to temperatures of 510 to 520°C.

Between 148 and 166°C (50 to 80 psig) a decreased intensity for C_3AH_6 was observed. Additional phases were not observed and there should be no tendency for the C_3AH_6 to decrease in crystallinity. The reason for the decrease is not readily apparent. However, an increase in the AH intensity was observed between 117 and 204°C (120 to 233 psig). This may indicate some tendency for the C_3AH_6 to decompose directly to AH and CH under these hydrothermal conditions.

The major decomposition reaction for the C_3AH_6 was seen to occur between 232 and 246°C (410 to 525 psig). The new compound $C_4A_3H_3$ was identified; however, CH was not. The CH was probably amorphous or masked by other reaction products. Longer exposure to the hydrothermal conditions may produce identifiable CH. Boehmite intensity increased again, either indicating greater crystallinity or a direct decomposition of C_3AH_6 to AH.

Work needs to be done to define the minimum temperature–pressure conditions under which the AH_3 and C_3AH_6 decompose. The work reported here was confined relative to those pressures of saturated steam at various temperatures because it was felt that these were the maximum pressures attainable during the drying of the concrete. Discrepancies that exist between this work and that cited earlier are shown in Figure 22.2.

As a result of this work, it was apparent that temperature ranges in the drying schedule that need be accommodated are 120 to 130°C for the AH_3

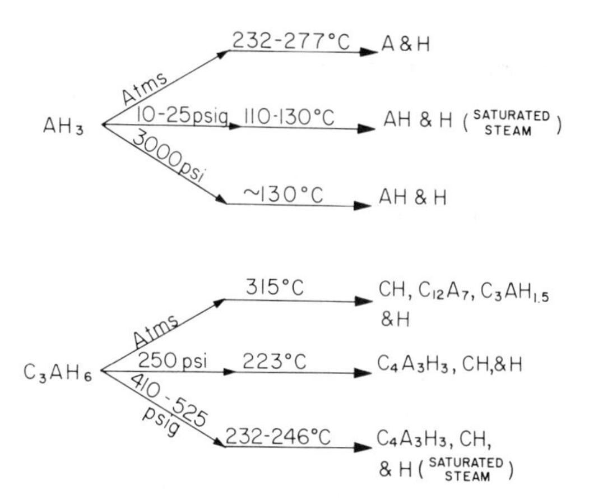

Figure 22.2. Comparison of reported decomposition of AH_3 and C_3AH_6.

and 235 to 245°C for the C_3AH_6 decompositions. These accommodations are not necessarily for the water-release problem but rather for stresses caused by thermal gradients and strains caused by bond-phase changes to smaller volume (shrinkage). In large massive cylinders, center temperatures are depressed relative to surfaces because of the heat-sink effect resulting from the energy required for the decomposition reaction.

CASTING SIZE

During the early part of the program, several of the antennae suffered massive failure during drying. The typical fracture divides the shape in half, generally across the diameter. Samples of concrete obtained from these cross sections were submitted for analysis. Position samples were at about the halfway point vertically and were sampled along the radius surface to

Table 22.3. Relative X-Ray intensity and thermal analysis of mineral phases in dried concrete (typical)

	Relative X-Ray Intensity (Percent of total weight loss)							
Position	C_3AH_6		AH		$C_4A_3H_3$		$CaCO_3$	
Surface	10–15	(40–60)	10–20	(20–30)	—		—	(15–35)
Surface, 5 in.	4–7	(30–40)	20–40	(30–40)	tr-5	(5–15)	5–10	(10–30)
Surface, 10 in.	6–7	(25–35)	20–30	(30–40)	tr-5	(10–15)	5–10	(10–30)
Center	5–6	(30–50)	20–30	(30–40)	tr-5	(5–15)	5–10	(10–15)

center. Data from X-ray and thermal analyses of these samples are shown in Table 22.3. These data confirm the hydrothermal decomposition of the concrete. The thermal analysis data show an additional phase not identifiable by the X-ray method used—calcium carbonate.

The calcium carbonate is of concern because of specifications relative to residual weight after drying. Experiments are currently underway to determine preferred conditions for carbonate formation.

The antennae average about 32 in. in diameter and about 30 in. high above the outer conductor. To establish concrete thicknesses that respond to hydrothermal decomposition versus those that respond to thermal decomposition, three different sizes of castings were made, dried, and analyzed for decomposition products. The sizes were, for diameter and height, 6 × 12 in., 12 × 12 in., and 30 × 30 in. Results of this work are shown in Table 22.4. These data appear to indicate that the outer 1 to 2 in.

Table 22.4. Mineral phases versus casting diameter—relative X-Ray intensity

Size (in.)		Distance from	Relative X-Ray Intensity					
Diameter	Height	Surface (in.)	C_3AH_6	AH	$C_4A_3H_3$	CH	$C_3AH_{1.5}$	CA
6	12[a]	0	6	3	—	—	PT	13
		1½	24	15	—	—	PT	—
		3	27	16	—	—	PT	—
6	12[b]	0	PT	16	3	—	10	—
		1½	PT	16	3	—	14	—
		3	PT	10	5	3	15	9
12	12[c]	0–1	9	10	PT	—	PT	—
		1–2	7	16	10	—	PT	—
		4	6	21	11	—	PT	—
		6	5	18	12	—	PT	—
30	30[d]	0–1	—	9	—	4	—	10
		5	—	24	16	8	—	—
		10	—	56	24	9	—	—
		15	—	75	24	5	—	—
30	30[e]	0–1	13	19	5	—	—	—
		7.5	5	44	18	—	—	—
		15	3	49	20	—	—	—

[a] Drying temperature = 400°F.
[b] Drying temperature = 600°F.
[c] Drying temperature = 540°F.
[d] Drying temperature = 600°F (25°F/hour heating rate).
[e] Drying temperature = 500°F (~5°F/hour heating rate).

of the casting surface respond solely to thermal decomposition, while the inner portions of the casting respond very well to hydrothermal decomposition. We have also found that certain processing variables contribute to the type of decomposition that takes place, and the limited information available is discussed later. Also of importance in this work was the fact that the residual mineralogy of the 12 $\times$ 12 in. casting was nearly identical to that found in production antennae (compare Tables 22.3 and 22.4). This allowed us to decrease our sample weight from $\sim$2000 to $\sim$150 lb.

PROCESS VARIABLES

The physical characteristic of the large concrete mass that must control the pressure in the casting during drying is the permeability. The counterpart to permeability relative to controllable variables would be heating rate during drying. Relatively small specimens with reasonable permeability, if heated rapidly enough, could create enough free water relative to its release rate at the surface to approach autoclave (saturated steam) conditions.

Heating rate has been specified for the concerns of this study; therefore, the variable that needs to be controlled to establish consistent drying behavior is permeability.

Conditions at the time of casting probably exert the greatest influence on permeability. Further, in any lot of concrete with constant aggregate sizing and cement behavior, the controllable variables that allow one to exert some influence on the castware permeability are "casting water" (amount of water required to achieve desired "casting consistency") and raw-material temperature, including casting-water temperature. Fixing material temperature allows one to measure independently the effect of casting water on resultant bond-phase mineralogy and, hence, indirectly on permeability.

As a result of these implications, 12 in. diameter $\times$ 12 in. high castings were made using either 8 or 10% casting water. The normal casting water for this lot of concrete was 9%, dry weight basis. On the basis of thermal diffusivity, a drying schedule was estimated that would simulate the thermal gradient experienced by production-dried antennae. After the units were dried, strength was determined by diametral compression, and samples for mineral-phase analysis were chipped from the diametral cross section. The results of these analyses are shown in Table 22.5. A similar experiment was conducted using 8.5 and 9.5% casting water. Results for 8.5% casting water were similar to those obtained using 8% casting water, and those obtained using 9.5% casting water were similar to those found with 10% casting water. An exact quantitative comparison has not yet been obtained.

Table 22.5. The effect of casting water on bond phase mineralogy during drying to 540°F

Casting Water (%)	Distance from Surface (in.)	Relative X-Ray Intensity				
		C_3AH_6	AH	$C_4A_3H_3$	$C_3AH_{1.5}$	CA
8	0	13	12	3	PT	3
	2	7	27	14	PT	—
	4	7	27	14	PT	—
	(Center) 6	8	32	13	PT	—
10	0	12	10	—	PT	PT
	2	19	14	—	PT	—
	4	14	13	—	PT	—
	(Center) 6	18	15	PT	PT	—

These results indicate that for a particular lot of concrete, variation of casting water by ½ to 1% can alter the bond-phase decomposition path from a strictly thermal to a hydrothermal response. These results also imply that temperature considerations for major thermal decomposition reactions should be adjusted relative to casting behavior or possibly cast density.

Curing and queueing conditions, once established, are held constant and yield relatively consistent mineralogy to commence the drying operation. These early conditions influence to some extent, however, the amount of residual CA in the concrete.

THERMAL DECOMPOSITION TEMPERATURES

Decomposition temperatures for the C_3AH_6 to $C_4A_3H_3$ have been confirmed on production units. Confirmation appears as a temperature lag or depression (endotherm) measured on the surface of the antenna, while heat input to the oven remains constant or is increased. This is noticed while casting surface temperatures are in the 230 to 250°C range.

The conditions have not been established where the casting permeability becomes sufficiently high to allow water transport and release without pressure buildup. This work has been largely confined to temperatures below 316°C (600°F). However, identification of mineral phases has been confirmed by thermal analysis. On the basis of these results, the major decomposition temperatures for several of the hydrate compounds have been measured. This technique has also allowed the detection of certain

Table 22.6. Thermal decomposition temperatures for CA cement bond phases in dried concrete

Compound	Temperature Range (°C)	Peak Reaction Temperature (°C)
AH_3	210–240	~230
C_3AH_6	240–370	~315
$C_3AH_{1.5}$ [a]	465–482	~470
AH	480–565	~525
$C_4A_3H_3$	565–620	~600
$CaCO_3$	650–790	~745

[a] Questionable—when this peak occurs, X-ray results generally indicate the presence of $C_{12}A_7$.

compounds that are difficult to identify by X-ray in this system, specifically calcium carbonate. Table 22.6 gives these phases and their measured decomposition temperature ranges at atmospheric pressure in flowing dry air.

Many times X-ray data indicate the presence of $C_{12}A_7$ when the system temperature has not been sufficiently high to form this compound. When thermal analyses of these samples are made, a second endotherm-decomposition peak occurs at about 470°C. X-Ray diffraction peaks for these compounds are similar; therefore, it is suggested that this TGA peak is that for $C_3AH_{1.5}$. Often some intensification of the AH peak is consistent with this reaction.

CONCLUSIONS

Small variations (0.5 to 1%) in casting water can cause significant variation in hydrated CA cement bond-phase mineralogy during the drying process. These variations result from changes in permeability, causing differences in the internal pressure of the concrete.

Increases in internal pressure cause initial decomposition reactions to occur at lower temperatures, on the order of 100°C. The products of these decomposition reactions, AH and $C_4A_3H_3$, still comprise a hydrate bonding phase that is more tenacious at higher temperatures than that observed with typical low-pressure decomposition.

Additional work is required to better define the pressure–temperature–hydrate phase decomposition interactions. Guidelines or limits for these studies appear to be defined; however, certain inconsistencies are still prevalent.

In addition to casting water and heating rate, cement-to-aggregate ratio and sizing can influence the bond-phase decomposition reactions.

ACKNOWLEDGMENTS

Dr. Kaiser wishes to thank Boeing Aerospace Company, who is the major contractor funding this study. In addition, the authors wish to thank W. H. Boyer for the X-ray analysis results; R. W. Nimmer and W. R. Alder for thermal analysis results; and M. A. Stett, M. E. Green, and J. W. Meeds for carrying out the autoclave and 12 in. diameter cylinder studies.

REFERENCES

1. G. V. Givan, L. D. Hart, R. P. Helich, and G. Maczura, "Curing and Firing High Purity Calcium Aluminate Bonded Tabular Alumina Castable," presented at the Refractories Division meeting of the American Ceramic Society at Bedford Springs, Pa., Oct. 4, 1974.
2. A. J. Majumdar and R. Roy, "The System $CaO-Al_2O_3-H_2O$," *J. Amer. Ceram. Soc.*, **39**, (12) 434–442 (1956).

PART FOUR

GREEN-BODY FORMATION AND MICROSTRUCTURE

The green body is the precursor to the fired body, and consequently the quality of the product depends heavily on the detailed characteristics of the unfired body. In these chapters attention is focused on physical behavior of powders during the formation of green bodies (mixing, packing, and compaction), on the forces of adhesion that provide green strength, and on defects within green bodies that affect the microstructure of the product.

23

Firing—The Proof Test for Ceramic Processing

W. D. Kingery

Processing before firing involves processes used to manipulate fine-particle-size solids and form shapes suitable for firing. After firing the formed shapes must have useful properties meeting more or less stringent requirements. In some cases the properties achieved are less critical than the shape or cost; in other cases the required shapes are simple but requirements imposed on resulting properties are very difficult to achieve. In all cases some minimum level of properties is a processing objective.

Almost all variations in the material as it is prepared for the kiln are amplified during the firing process. As a result, examination of fired ware not only serves as a test for successful processing, but also leads to inferences about the way in which processing must be modified or improved to obtain satisfactory ware.

A few of the ways in which the firing process can be interpreted as indicating the success of, and requirements for, prior processing are examined in this chapter.

A SPECTRUM OF RESULTS

During firing, a formed ceramic body tends toward chemical equilibrium, which is approached by reaction between components, chemical and diffu-

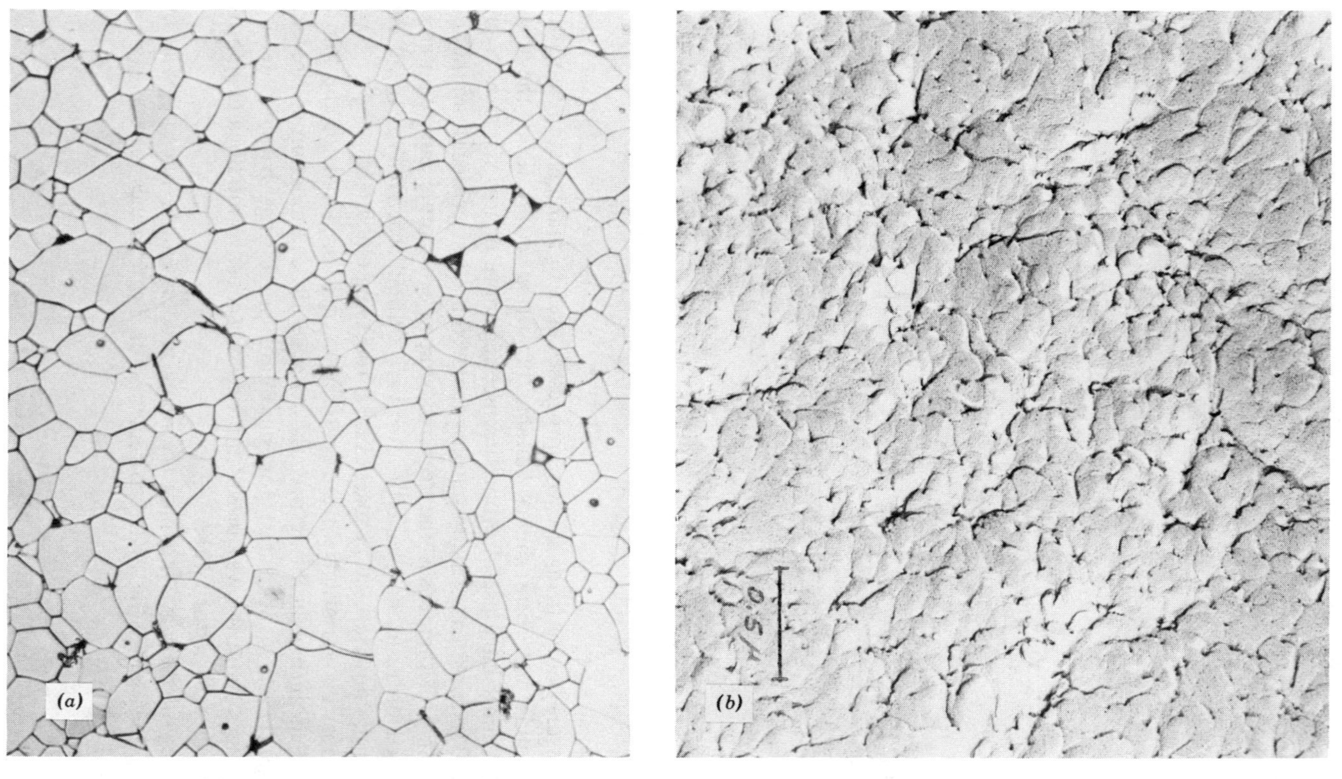

Figure 23.1. (*a*) Polished section of Y_2O_3 + 10% ThO_2 sintered to a uniform grain size nearly pore-free transparent ceramic. 100× (courtesy C. Grescovich and K. N. Woods). (*b*) Fine-grain yttria-stabilized zirconium oxide ceramic hot-pressed at 1300°C. 30,000× (courtesy T. Vasilos).

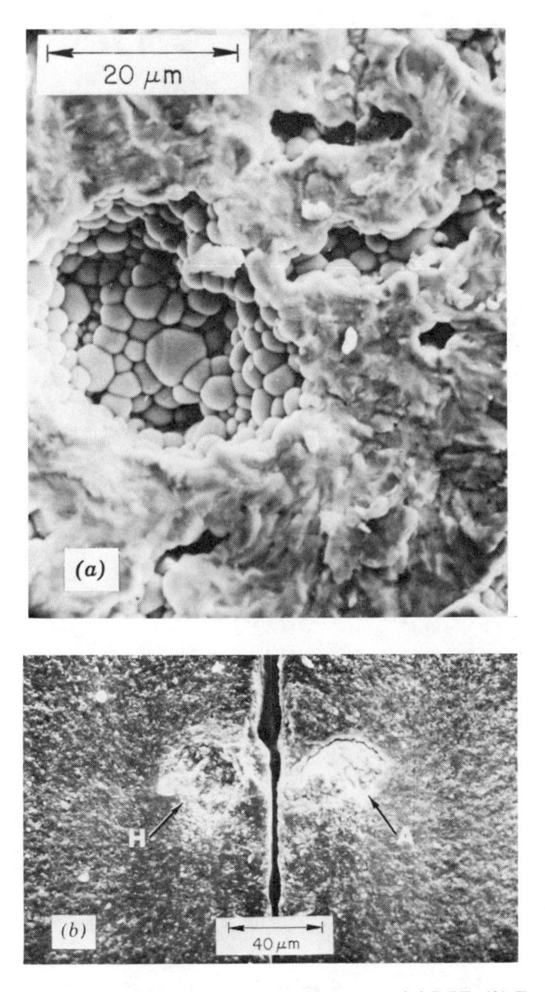

Figure 23.2. (*a*) Large internal pore in a sample of commercial PZT. (*b*) Foreign agglomerate in a sample of experimental Al_2O_3. (Fractographs, courtesy Roy W. Rice.)

sion, and phase transformations; it also tends toward physical equilibrium, a minimization of the surface energy and strain energy, approached by grain growth, pore elimination, and phase separation on cooling. When a material is processed such that the final composition ready for firing is a uniform homogeneous fine-particle-size mixture, these processes occur at a relatively low temperature uniformly throughout the sample. The result for two single-phase compositions is shown in Figure 23.1 for a completely dense pore-free uniform-grain-size optically transparent sintered yttrium

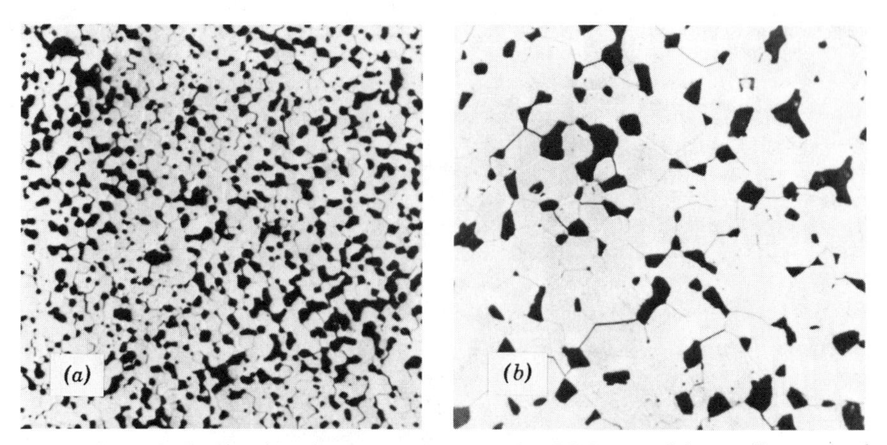

Figure 23.3. Grain growth and pore growth in sample of UO_2 containing small amount of carbon after (*a*) 2 minutes and (*b*) 5 hours at 1600°C. 312× (Francois et al.[2]).

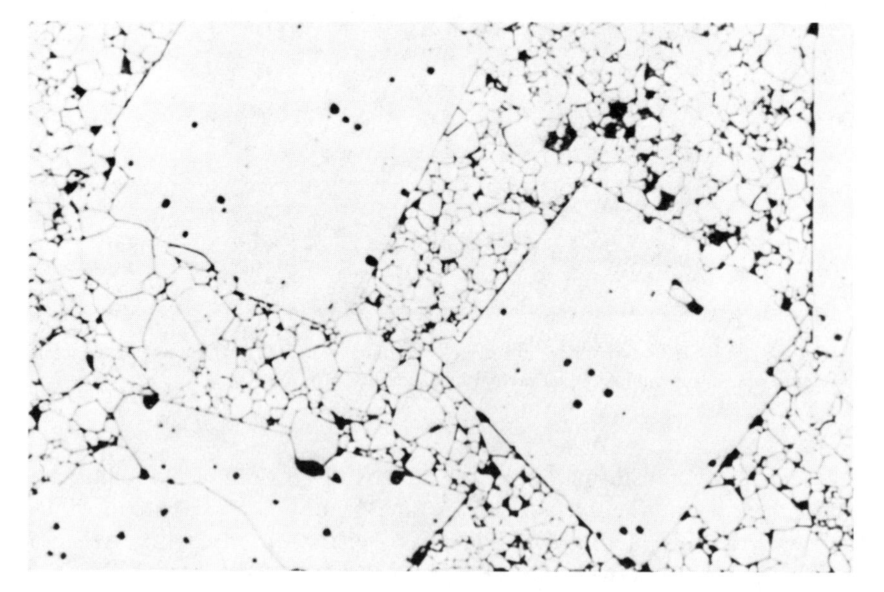

Figure 23.4. Idiomorphic grains in polycrystalline spinel containing a small amount of liquid phase. 350× (courtesy R. L. Coble). From W. D. Kingery, *Introduction to Ceramics,* Wiley, New York, 1960.

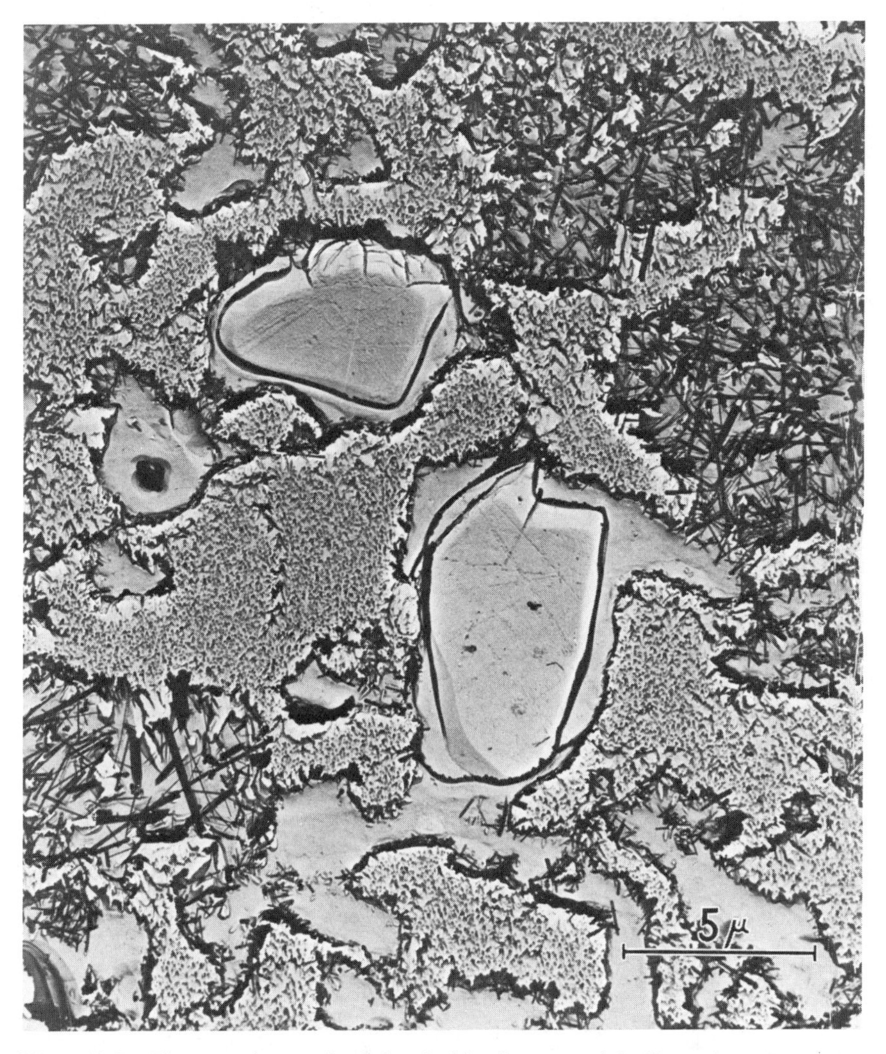

Figure 23.5. Electron micrograph of electrical insulator porcelain illustrating solution rims and cracks associated with large quartz grains. 1245× (courtesy S. T. Lundin). From W. D. Kingery, *Introduction to Ceramics,* Wiley, New York, 1960.

oxide and for a hot-pressed zirconium oxide. When samples are prepared from uniform-particle-size material without agglomeration, nearly complete densification can be achieved at remarkably low temperatures as compared with ordinary firing requirements.

On the other hand, many samples are characterized by gross defects. Large internal voids and unconsolidated foreign agglomerates, such as shown in Figure 23.2, are very common indeed; so common, that fracture of commercial ware almost always is found to originate at this kind of macroscopic defect.

COMPOSITIONAL CONTROL

One of the most spectacular examples of firing as a proof test for processing comes for ware that has been insufficiently dried. Many of us have had the experience, or seen the results of an overeager potter, firing poorly dried ware with excess water present. The steam generated literally explodes the product.

A related but somewhat less spectacular result was observed by Laurent and Bernard[1] in the preparation of alkali halide samples for diffusion measurements. They sintered imperfectly dry samples by rapidly heating to a temperature near the melting point; the residual porosity was substantially greater than in perfectly dry samples, and the measured diffusion charac-

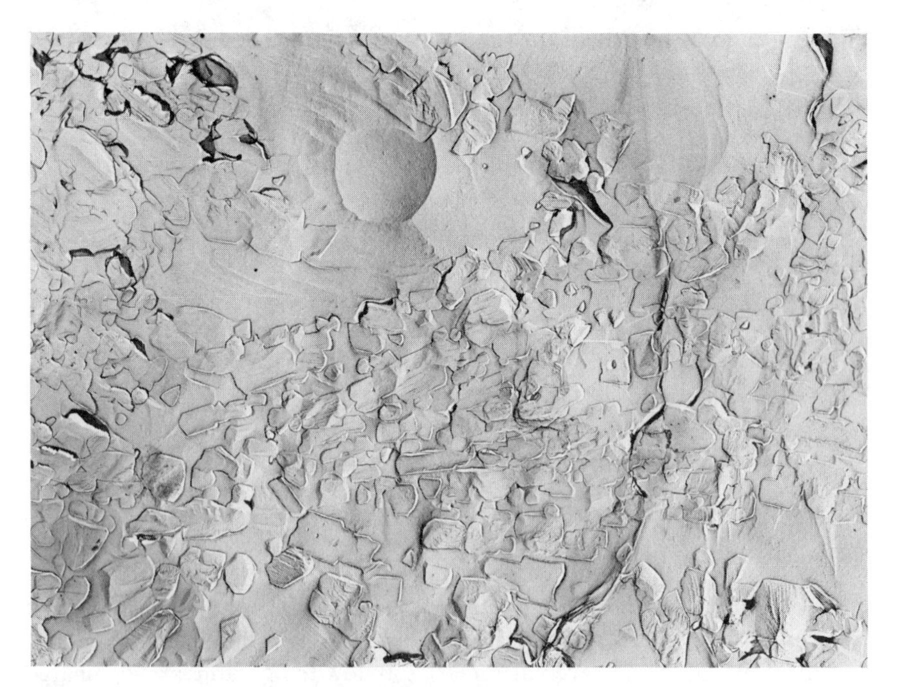

Figure 23.6. Irregular distribution of protoenstatite crystals in a steatite body. Major glass phase etched on a fractograph surface. 2700× (courtesy K. H. Schuller).

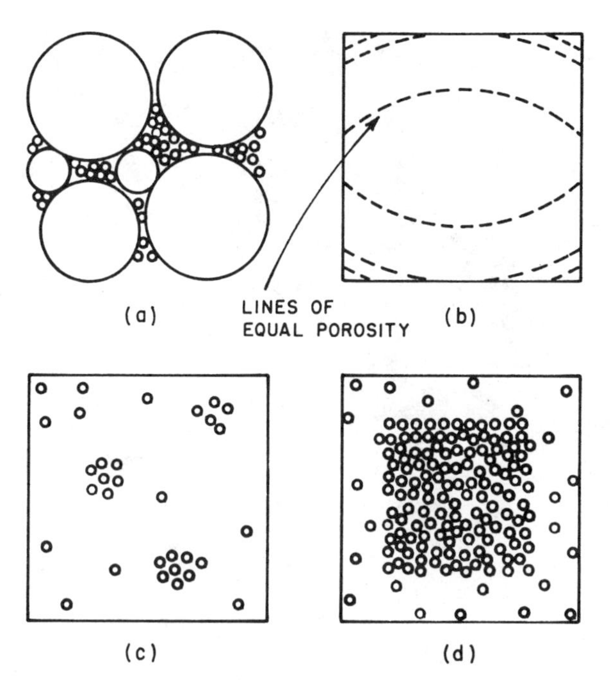

Figure 23.7. Pore concentration variations resulting from (*a*) a variation in grain sizes, (*b*) die friction, (*c*) local packing and agglomeration differences, and (*d*) more rapid pore elimination near surfaces.

teristics were markedly affected by the residual water content and its influence on grain boundary behavior. In a more definitive study, Francois et al.[2] investigated rapidly heated samples of uranium dioxide containing a small fraction of carbon. The surface was sealed during firing, with the result that gas generation in the pores led to a limitation of density that persisted during long heating; the pore configuration was such that pore growth occurred along with grain growth, essentially all the porosity remaining as intergranular pores (Figure 23.3).

Similarly, the control of composition, including minor constituents, has a major influence on the development of microstructure. If a small amount of wetting liquid is present, idiomorphic grain growth can take place, making it impossible to achieve a fine uniform grain size; the liquid phase formed at the boundaries of a largely crystalline material causes selected growth of large individual grains (Figure 23.4).

The results of compositional variations and the requirements for processing controls are classically illustrated by the influence of large grains of

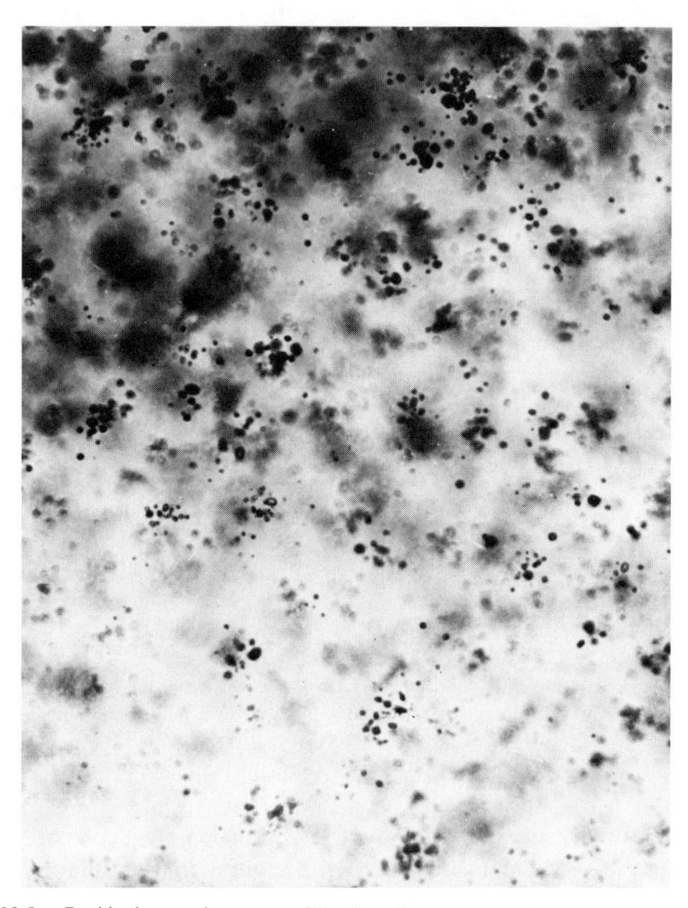

Figure 23.8. Residual pore clusters resulting from improper powder processing in a sample of 90 mole % Y_2O_3–10 mole % ThO_2. Transmitted light, 137× (courtesy C. Greskovich and K. N. Woods).

quartz in procelain bodies. The rate of dissolution of the quartz grains is the slowest process taking place during chemical equilibration. As a result, for usual particle sizes the quartz will completely dissolve to form a glass only after long high-temperature firing. At lower temperatures a solution rim that has very different thermal-expansion characteristics forms around quartz grains and gives rise to stresses on cooling, often cracks as shown in Figure 23.5. Dramatic increases in the strength are obtained by using fine-particle-size silica as a body constituent or by replacing the quartz with another material such as alumina.

UNIFORMITY

In analytical treatments of the firing process it is implicitly or explicitly assumed that the starting material has been processed in such a way that it is physically uniform with regard to the distribution of components and the distribution of porosity. In products where particle sizing has been used to

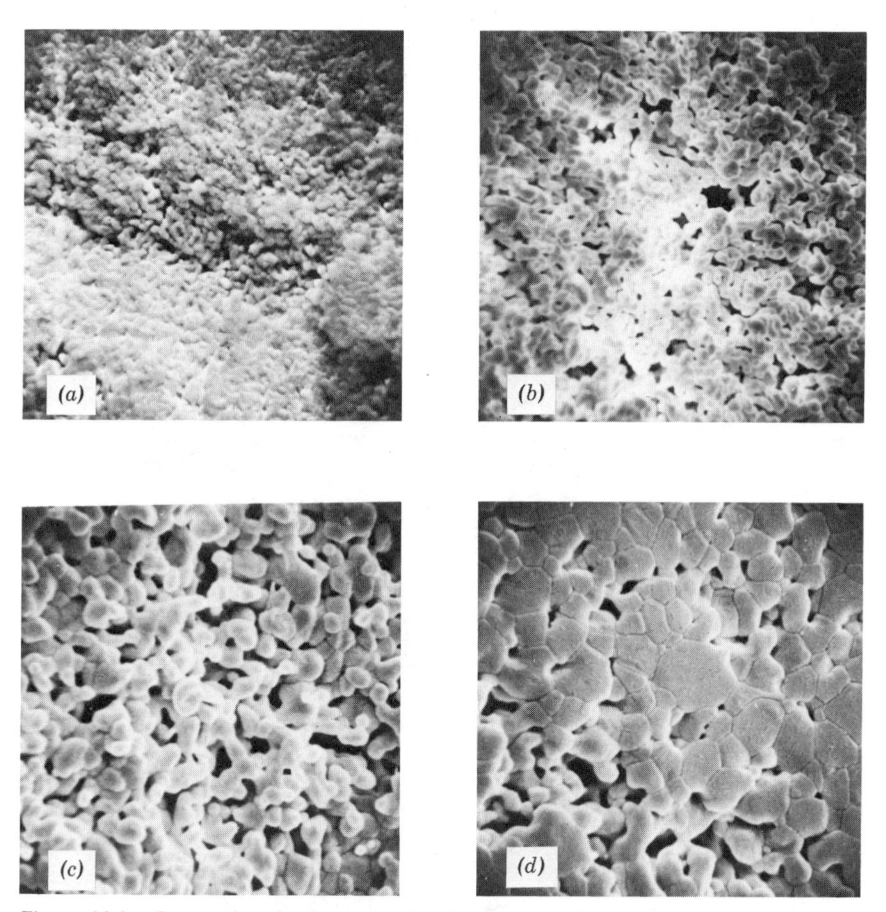

Figure 23.9. Progressive development of microstructure in Lucalox alumina. Scanning electron micrographs of (*a*) initial particles in the compact (3125×) (*b*) after 1 minute at 1700°C (3125×), (*c*) after $2\frac{1}{2}$ minutes at 1700°C (3125×), and (*d*) after 6 minutes at 1700°C (3125×). Note that pores and grains increase in size, that there are variations in packing and pore size, and that pores remain located between dense grains. (Courtesy C. Greskovich and K. W. Lay.)

maximize the green density, this is clearly, and intentionally, not true. Accidental and intentional variations of particle size and material distribution have strong consequences in the resulting structure.

For chemical equilibration processes the rate-limiting process is usually diffusion, and the time required is related to the nondimensional constant Dt/a^2, where D is the diffusion coefficient (cm^2 second^{-1}), t is the time (seconds), and a^2 is a characteristic dimension (cm^2). When this parameter is nearly unity, diffusively limited processes are approximately complete. For firing processes the maximum reasonable time is limited by economic considerations and by the boundaries of normal human patience. The diffusion coefficient depends on the particular system and the temperature used for the firing process. The characteristic dimension over which diffusion occurs depends on processing parameters, particularly the uniformity and the grain size of the constituents. Large grain-size materials do not reach chemical equilibrium, as illustrated for the quartz particles shown in Figure 23.5. A result of nonuniformity is shown in Figure 23.6.

In addition to chemical equilibration, the physical processes of pore elimination and grain growth are controlled by diffusion, either in the bulk

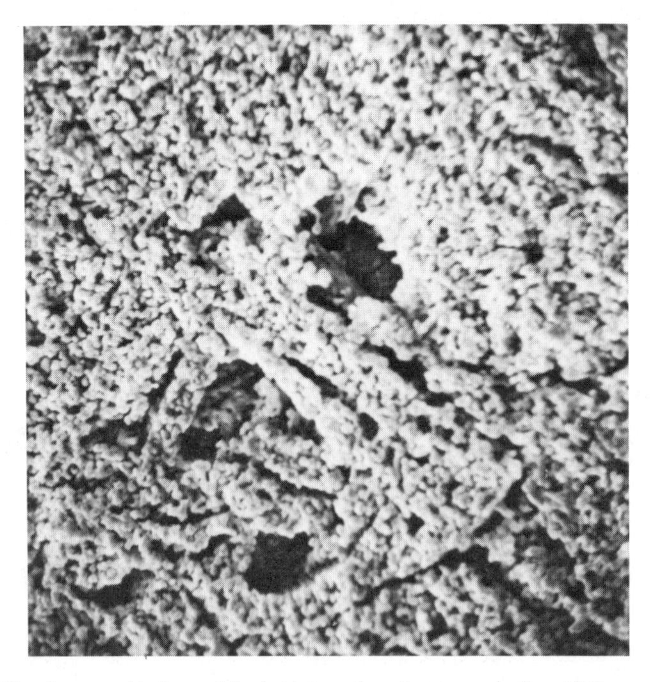

Figure 23.10. Large voids formed by bridging of agglomerates in fine Al_2O_3 powder viewed with scanning electron microscope at 2000× (courtesy C. Greskovich).

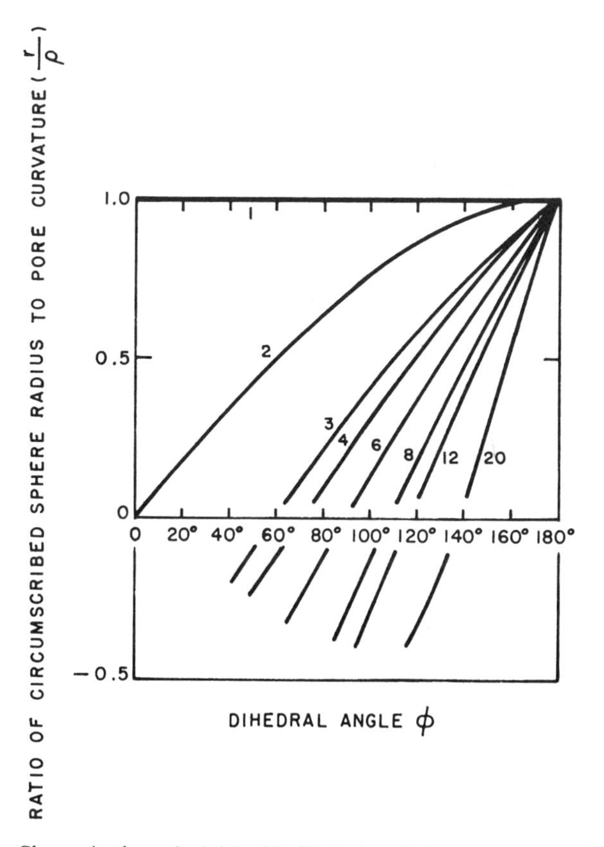

Figure 23.11. Change in the ratio (r/ρ) with dihedral angle for pores surrounded by different numbers of grains as indicated on individual curves.

or in the grain boundaries; a similar nondimensional parameter, Dt/a^2, describes the time required for equilibration to occur. For an isothermal completely uniform mixture of uniform pore sizes, a corresponds approximately to the pore-separation distance. For processing variables that give rise to changes in pore distribution, as shown in Figure 23.7, the characteristic distance required for pore elimination corresponds to a much larger value related to the spacing between pore clusters or the distance for high-porosity regions to the surface. As a result, pore elimination is not feasible, and microstructures with residual pore clusters result, as shown in Figure 23.8.

The exaggeration of any initial nonuniformity during the firing process is seen in Figure 23.9, which shows the progressive development of pores and grains for a sample of Lucalox alumina prepared from Linde A material.

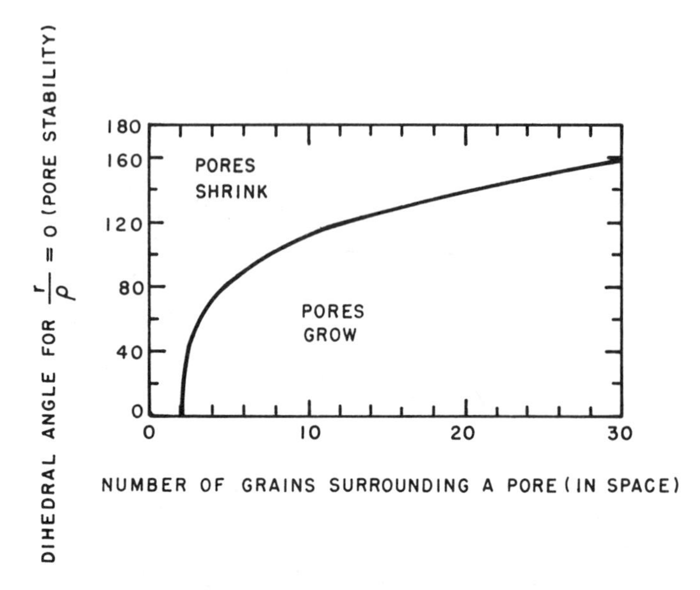

Figure 12.12. Conditions for pore stability.

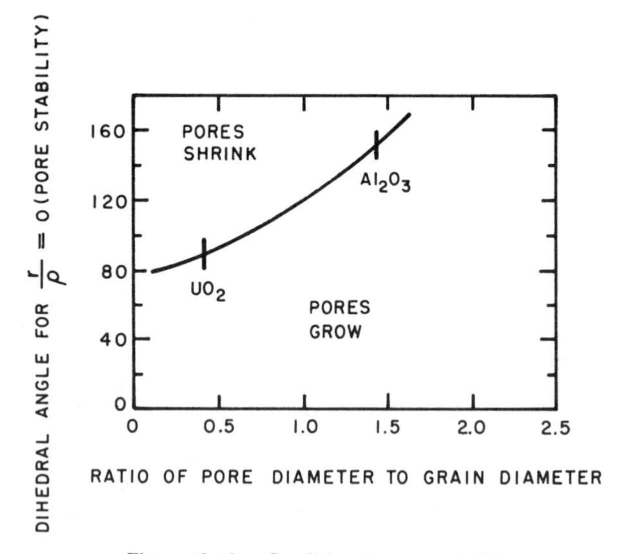

Figure 12.13. Conditions for pore stability.

Initially the structure appears quite uniform in the compact. However, after 1 minute at 1700°C there is an increase in grain size and also an increase in pore size. In areas where agglomerates come together, the pores remain located between dense grains in which both grain size and pore size are increasing. This condition is substantially greater in some samples than others, depending on the effectiveness of processing. A sample containing large voids formed by bridging of agglomerates is shown in Figure 23.10.

When a pore is surrounded by grains, the surface curvature between the pore and the grains depends both on the value of the dihedral angle at the grain boundary intersection with the pore and on the number of surrounding grains. If we take r as the radius of a circumscribed sphere around a polyhedral pore, the ratio of this value to the radius of curvature of the pore surface is shown in Figure 23.11. When the ratio of pore size to radius of curvature (r/p) decreases to zero, the interface will be flat and there will be no tendency for pore shrinkage or growth. Pores will be stable, as shown in Figure 23.3 for cubic pores in UO_2 surrounded by six grains with a dihedral

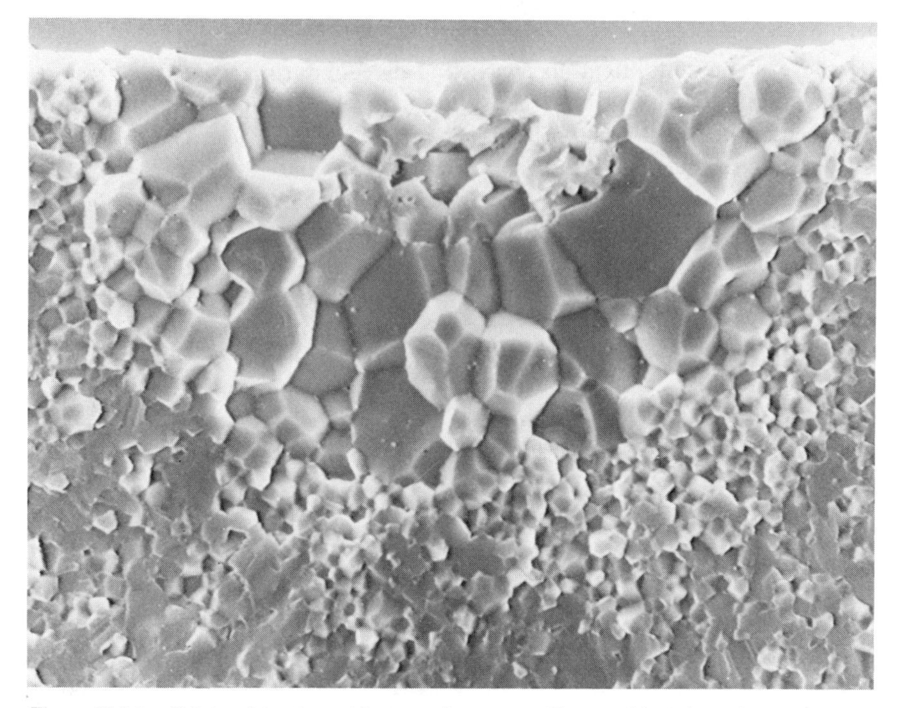

Figure 23.14. Origin of fracture at large grains surrounding an oblong impurity zone in hot-pressed alumina. 460× (courtesy W. H. Rhodes).

angle of about 90°. When (r/p) becomes negative, there is a thermodynamic tendency for pore growth rather than shrinkage; that is, the direction of the change in pore shape depends on the ratio of pore size to grain size. This is shown for a range of dihedral angles and number of grains surrounding a pore in space in Figure 23.12, where a line of thermodynamic equilibrium separates regions of pore shrinkage (the usual case) from regions of pore growth. This is shown in a different way in Figure 23.13.

This relation between pore behavior and the pore size/grain size ratio accounts for most of the increase in pore size during the early stages of sintering fine powders and explains the extreme sensitivity of residual porosity to agglomerate formation and seemingly minor variations in processing. Large changes in structure occur in the early stages of firing (Figure 23.9) and strongly affect subsequent events. The same underlying phenomena accounts for the internal structure of the large pore shown in Figure 23.2*a*.

CONSEQUENCES

Because of the amplification of existing defects during firing, the arrangement and uniformity of body constituents resulting from processing before ware enters the kiln is the primary factor that determines many properties of ceramic products. As shown in Figures 23.2 and 23.14, fracture events are almost always initiated at regions of physical or chemical inhomogeneity. This is also true of other properties dependent on failure events, such as thermal shock resistance, dielectric strength, and high-temperature deformation. Defects can also be introduced by inadequate firing processes—warping, bloating, overfiring, black core, dunting, and so forth. However, even when these faults are avoided and the firing process itself is carried out to perfection, resulting properties are limited almost wholly by the ware that enters the kiln. Cases in which firing serves to correct inadequacies of processing are rare.

ACKNOWLEDGMENT

Work was supported by the U.S. Atomic Energy Commission under contract AT(11-1)2390. A fuller account of many of the principles involved is given in W. D. Kingery, H. K. Bowen, and D. R. Uhlmann, *Introduction to Ceramics,* 2nd Ed., Wiley, New York, 1975.

REFERENCES

1. J. F. Laurent and J. Bernard, *J. Phys. Chem. Solids,* **7,** 218 (1958); J. Cabance, *J. Chim. Phys. Physico Chim. Biol.,* **59,** 1123, 1135 (1962).
2. B. Francois, R. Delmas, G. Cizeron, and W. D. Kingery, *Mem. Sci. Rev. Metall.,* **64,** 1079 (1967).

24

Effect of Process Optimization Properties of Alumina Sintered Under Rate Control

T. M. Hare
H. Palmour, III

INTRODUCTION

Systematic studies of the cumulative effects of prefiring variables on firing processes and subsequent properties have not been generally available in the open literature even for a single material. Prefiring variables include milling practice, binder and/or lubricant additions, forming procedures, burnout methods, and resulting fractional green densities.

Our work over a number of years[1-10] with several different high-purity aluminas of chemical- or conventional-Bayer-process origin has shown that improvements in firing methods are related to improvements in the prefiring processing of those materials.[8] The importance of various process variables upon the bend strength of finished alumina bars[10] is shown in Figure 24.1.

The correlations between processing and strength were found to be related to the flaws in the samples[9]; the effectiveness of a given preparative step is inversely related to the characteristic size of flaws that it generates in the final microstructure. An example of such process-related flaws is shown in Figure 24.2.

The current study is an extension of these earlier investigations of processing variables and their relationships to rate-controlled densification processes.[1-10] It is more quantitative and represents an attempt (within our experimental limits) to achieve for alumina a systematic optimization of the whole sequence of process steps that precede firing.

The apparatus used in rate-controlled sintering studies is a specially instrumented high-temperature dilatometer, described elsewhere,[1,3,4,7] which permits monitoring and control of the shrinkage-time firing profile. Since linear shrinkage in any given direction is proportional to volume shrinkage, and hence to fractional density D, it is possible to preprogram the entire densification profile in terms of preselected densification rates $dD/D\ dt$, which are maintained over specific segments of the total density range to be traversed. Once such a program (usually comprised of three progressively

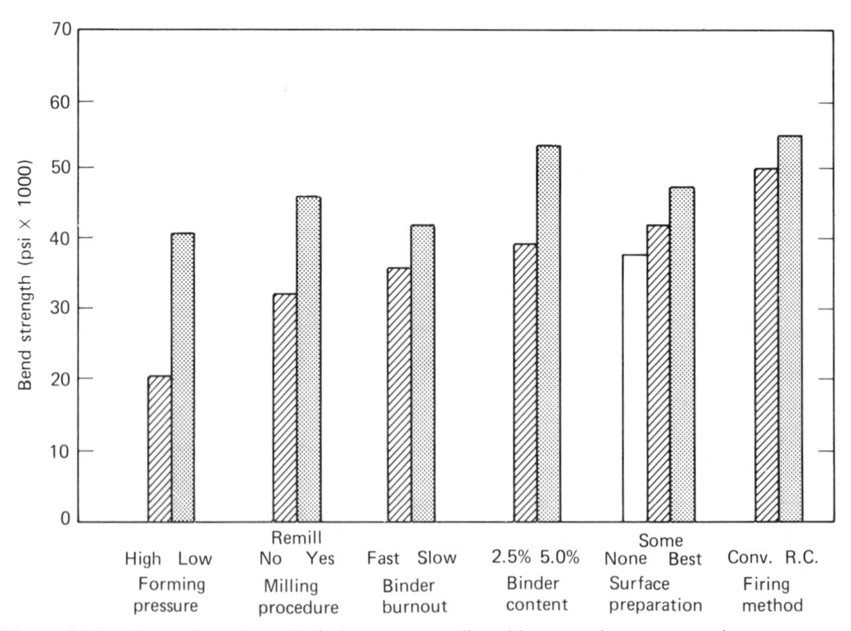

Figure 24.1. Examples of strength increases attributable to various process improvements (After Palmour et al.[8,11]).

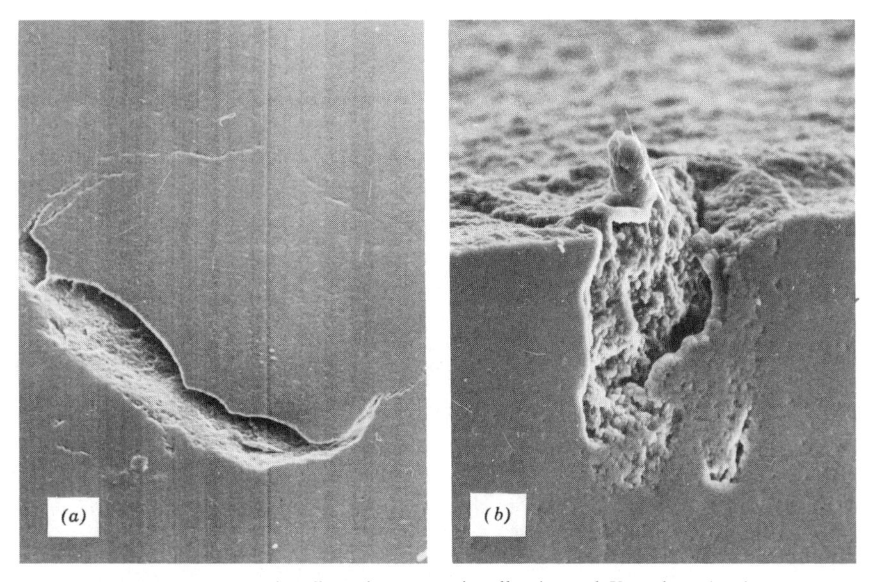

Figure 24.2. Gross processing flaws in conventionally sintered Kemalox alumina: (*a*) SEM 83× and (*b*) SEM 250×. After Palmour et al.[8,10].

slower rate regimes) is designed, the dilatometer provides an instrumental means for assuring that the small test specimen follows the desired program from the onset of shrinkage D_0 at some elevated temperature T_0 (normally attained at some constant heating rate dT/dt) to completion of the firing at D_f, T_f. During the rate-controlled portion of the firing, the ceramic specimen itself provides its own feedback by means of the high-temperature dilatometer, calling for more or less power as required to maintain the preprogrammed shrinkage–time (and hence density–time) curve. During this phase, temperature becomes a monitored but wholly dependent variable; the resulting temperature–time trace is a consequence of the energy requirements of the specimen during sintering as it conforms to the programmed density–time profile.

When a temperature–time profile is determined in the dilatometer it may thereafter be used to program reproductions of the original rate-controlled firing.[5] This makes it possible to achieve the desired density–time profile either in the dilatometer under temperature control or in conventionally instrumented kilns or furnaces. This ability to scale up, which has been successfully demonstrated for alumina processed in laboratory gas-fired periodic kilns,[8] obviously is important in developing practical industrial applications for rate-controlled sintering technology.

EXPERIMENTAL PROCEDURES

The equipment and methods used in characterizing and processing the alumina samples have been described in detail elsewhere.[8]

An organic binder is necessary for use with these alumina materials to assure sample integrity after dry pressing and to attain adequate levels of fractional green density. The binder utilized in our study was Carbowax 4000 (Union Carbide). Although the waxy binder provides some degree of lubricity to aid in forming an additional lubricant additive, isostearic acid has proved effective. On the basis of prior studies, the level of each of these additives was fixed at either 0 or 5 wt % unless otherwise noted.

All milling and remilling reported in this study was carried out at ~60% of critical speed in a 3 gallon polyurethane-lined mill containing ~1.9 cm diameter 99% Al_2O_3 balls with a 20:1 ball-to-charge ratio. Before-and-after spectrographic analyses of alumina materials did not show any significant impurity increases attributable to dry milling (18 hours), remilling if any (18 hours), or organic binder or lubricant additives.

In those experiments calling for a fluid vehicle for binder suspension and distribution, the binder was premixed in an appropriate quantity of a selected solvent (e.g., 600 ml/200 g Al_2O_3), which was then slurried (under agitation) with previously dry-milled alumina powder. The mixed slurry was dried at room temperature in large Pyrex trays prior to (*a*) crushing, screening, and forming or (*b*) remilling, screening, and forming.

For dry processing, the as-received flaky binder material was added directly to previously dry-milled alumina powder in a clean dry mill along with lubricant, if any. After the first several hours of remilling, the mill was emptied and scraped down with a plastic spatula to remove and redistribute any adherent, binder-rich material. After reloading, remilling was continued to complete the 18 hour schedule. This redistribution step was found to be extremely important, particularly if the batch formulation called for a (liquid) lubricant.

Rectangular bar specimens measuring ~0.25 × 0.30 × 1.4 in. were formed by dry pressing a preweighed charge of prepared powder (~3.75 g) in a hardened steel die in an air-operated hydraulic press. Unless otherwise noted, dry pressing was carried out at 2000 psi for materials containing binders and at 1000 psi for binder-free materials. When repressing was needed, the formed bars were placed in small rubber balloons, evacuated, tied off, and repressed isostatically at 40,000 psi, unless otherwise noted. The repressed bars displayed very little shape distortion.

On the basis of prior thermogravimetric (TGA) and dynamic differential calorimetry (DDC) studies of binder oxidation and removal, a binder burnout schedule was designed. Formed bars were heated from 100 to

200°C over a 4 hour period and from 200 to 500°C over an additional 4 hour span and were cooled to room temperature in the oven.

Starting materials included several different lots of chemically prepared Kemalox aluminas (W. R. Grace & Co.), both undoped (KA-100) and MgO-doped (~0.09%, KA-210), together with two types of Alcoa superground, low-soda Bayer-process aluminas, including a moderate purity grade (A-16 SG) and a higher purity MgO-free grade (XA-139 SG). Detailed characterizations of these several materials along with descriptions of sintering behavior, microstructural features, and mechanical strengths have been described elsewhere.[8-10,12]

Exploratory phases of the experiments reported here were carried out with the same lot of A-16 SG described above. Once the basic optimization of a refined preprocessing sequence was established with A-16 SG, the integrated process was applied to Linde A alumina (Union Carbide Corp.). This material was used both in the as-received undoped form (MgO-free) and with doping (0.1%) added as MgO during initial dry milling.

PREFIRING PROCESS OPTIMIZATIONS

Dry Milling

The importance of intensive dry milling in the sintering of alumina has been demonstrated previously.[3,13-17] As in the earlier study, an 18 hour dry-milling operation was utilized with Linde A to break up and disperse agglomerates and to introduce annealable excess energy (internal strain). The superground Alcoa aluminas had already been subjected to intensive milling; they were advanced to the next process step in the as-received condition.

Binder Additions with Fluids

In a previous study[8] binder additions were made in the form of methyl alcohol slurries, using procedures already described. In this investigation other solvents (water, cyclohexane) were included with methyl alcohol in a more detailed study of binder distribution using Alcoa A-16 SG material. A dry binder milled directly in the batch was included as a control.

Water added in various ways was effective in increasing green density but was less effective than methyl alcohol in improving fired density. As vehicles, methyl alcohol and water were found to have a small concentration dependence of their effect on green and fired densities, with a 3:1 vehicle-to-solid ratio being somewhat better than 1:1. The anhydrous vehicle, cyclohexane, yielded poorer green densities but better fired densities than either water or methyl alcohol. However, the most promising fired densities were obtained with the directly added dry-remilled binder, even though green densities were somewhat low.

Results of this study emphasize the desirability of anhydrous processing of alumina. These findings are in good agreement with a recent study of deleterious H_2O–Al_2O_3 reactions.[16]

Dry Binder and Lubricant Additions

Experiments using dry additions of binder were carried out with Alcoa A-16 SG material. The influence of lubricant additions and isostatic repressing to increase green densities is summarized in Table 24.1.

The lubricant level had a much greater effect on green density than did binder level, both at constant weight volatiles (batches 1, 2, 5) and at constant binder content (batches 3, 4, 5). The highest green and fired densities were obtained with isostatic repressing and 2.5% lubricant; the 5% binder level yielded slightly higher values than 2.5%. At the highest lubricant level the conventionally pressed bars had nearly the same green density

Table 24.1. The effect of several processing variables on densities of A-16 SG alumina

					Fired Density	
Batch	Binder Content (%)	Lubricant Content (%)	Press Method	Green Density[a]	Conventional 1530°C + 2½ hr	Rate Controlled to 1550°C
1	2.5	2.5	Conv[c]	0.6010	0.9895	0.9904
			Iso[d]	0.6077	0.9910	0.9912
2	3.75	1.25	Conv.	0.5888	0.9885	
			Iso	0.5993	0.9887	0.9897
3	5.0	2.5	Conv	0.6061	0.9902	0.9904
			Iso	0.6111	0.9902	0.9916
4	5.0	1.25	Conv	0.5825	0.9902	0.9904
			Iso	0.5927	0.9904	0.9912
5	5.0	0	Conv.	0.5700	0.9898	
			Iso	0.5931		

[a] Volatile-free basis.
[b] Density–time profile optimized for earlier form of processing.
[c] Conventional dry pressing at 20,000 psi.
[d] Isostatic repressing at 40,000 psi.

Table 24.2. Green density as a function of pressing variables for milled, undoped, Linde A alumina, remilled with binder (B) and lubricant (L).

Initial Pressure (psi)	Iso Repressing (psi)	Fractional Green Density	
		B 2.5%; L 2.5%	B 5.0%; L 5.0%
1000	25,000	0.5938	0.5955
	50,000	0.6022	0.6039
2000	25,000	0.5992	0.6030
	50,000	0.6086	0.6130
4000	25,000	0.5985	0.6079
	50,000	0.6110	0.6190

as those repressed isostatically. When well-processed A-16 SG was sintered under rate control with a reasonably well optimized density–time profile, a fractional density of 0.992 was achieved. For conventional firings the density was limited to ~0.990.

With finer, higher-surface-area Linde A, similar results were obtained (Table 24.2), but a 5% binder, 5% lubricant addition was found to be slightly better than only $2\frac{1}{2}$% lubricant. Green density increased with both initial forming pressure and isostatic repressing pressure. An initial forming pressure of 2000 psi and a final isostatic pressure of 40,000 psi was chosen as a standard condition.

For milled, undoped Linde A, use of optimized processing conditions (remilled with 5% binder, 5.0% lubricant, dry pressed at 2000 psi, followed by 40,000 psi isostatic repress and controlled slow binder burnout) produced bars that attained fractional green densities of 0.6055 (volatile-free basis). When fired conventionally (1530°C, $2\frac{1}{2}$ hours), a density of 0.9880 was attained.

PREPROCESSING EFFECTS ON SINTERING OF LINDE A ALUMINA

Five batches of Linde A alumina were prepared by several methods to compare the effects of process variables on sintering characteristics. Table 24.3 shows some characteristics of the prepared materials.

Batches were characterized primarily by the fractional green density obtained during formation under an initial pressure of 2000 psi, followed by isostatic repressing at 40,000 psi. In the case of the binderless batch, an initial pressure of 1000 psi was used because of cracking problems at the 2000 psi level.

Batches were dry milled for 18 hours under previously described "optimum" conditions except for batch 3, which had less initial dry milling. The fractional green density obtained was a function of the organic additives employed, as previously discussed.

The "best" processing method (batch 1) utilized 5% binder and 5% lubricant. "Poorly processed" material (batch 2) employed no additives, although it was well milled (18 hours + 18 hours remilled). The fractional green density obtained after slow burnout of organics was 0.05 higher than that obtained using the binder alone and ~0.10 higher than that obtained without the use of organics, when all were milled for a comparable time.

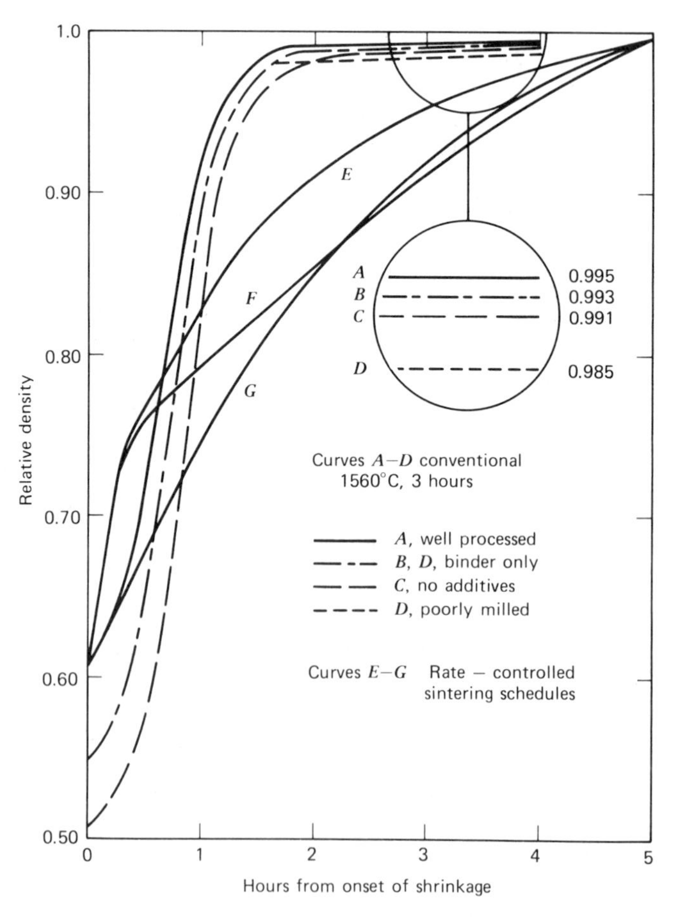

Figure 24.3.

Table 24.3. Processing parameters investigated in sinterability study for Linde A alumina

Batch	Condition	Milling Time (hours)	Remilling Time (hours)	% Binder	% Lubricant	Fractional Geeen Density
1	Doped (0.1% MgO)	18	18	5	5	0.6075
2	Doped	18	18	—	—	0.5055
3	Doped	~2	18	5	—	0.5502
4	Doped	18	19	5	—	0.5600
5	Undoped	18	19	5	5	0.6055

Differences in sinterability between batches 3 and 4 were not apparent from green density alone.

Figure 24.3 shows the relative importance of the prefiring process variables and the basic difference between the conventional sintering schedules and the rate-controlled sintering profiles employed in this study. The four conventional curves *A* through *D* represent typical densification paths for specimens conventionally fired at 8.7°C/minute to 1560°C and held for 3 hours. After 3 hours at 1560°C, the relative-density differences existing between various batches are small but significant (see inset). Curves *A* through *C* represent batches that received equal milling treatment; their final densities rank in the same order as their fractional green densities. Curve *D* shows the density–time path of the poorly milled batch with binder only (batch 3), which did not go beyond the 0.985 relative density level. The chief differences between conventional and rate-controlled densification profiles are the slower densification rates maintained for the latter between 0.75 and about 0.90 and the somewhat faster sintering rates near the end of the firing (curves *E* through *G*).

PROCESSING AND SINTERING EFFECTS ON MICROSTRUCTURE OF LINDE A ALUMINA

These studies have attempted to achieve high density and fine grain size, in keeping with an overall objective of increasing mechanical strength. This does not necessarily mean firing to extremely high density and achieving a high degree of translucency. Grain growth encountered in raising density from about 0.992 to 0.995, for example, certainly would reduce the bend strength by virtue of the increased grain size, probably offsetting any

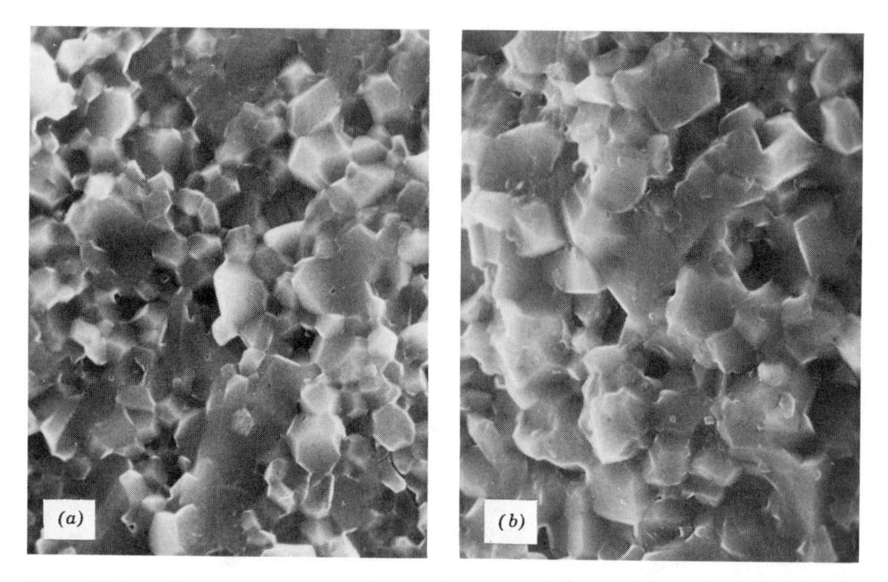

Figure 24.4. Microstructural comparison of MgO-doped Linde A sintered-conventionally at 1560°C to about equal densities, SEM 1840×. (*a*) best processing: 1 hour soak, $D = 0.992$; (*b*) poor processing: 3 hours soak, $D = 0.991$.

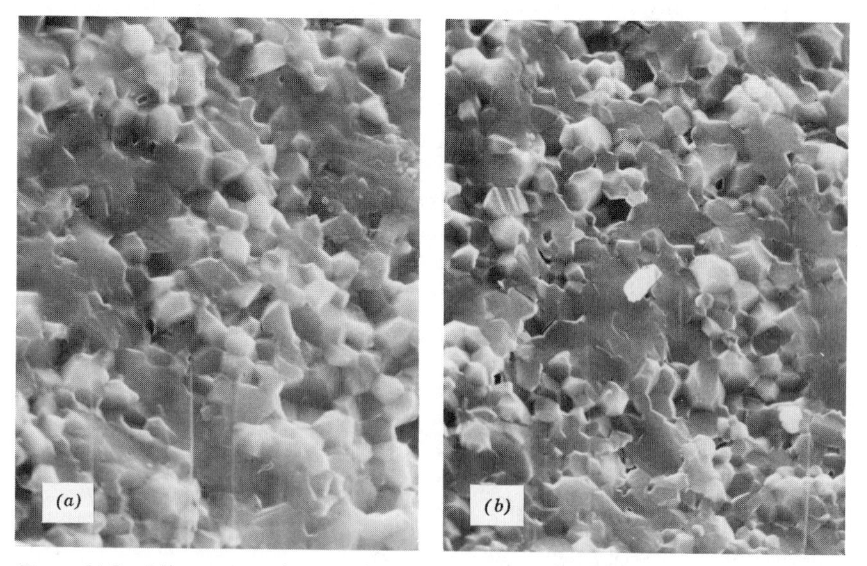

Figure 24.5. Microstructural comparison of well-processed, MgO-doped Linde A using different sintering methods, SEM 1800×. Note: about equal grain size. (*a*) conventional 1450°C, 3 hours, $D = 0.986$ and (*b*) rate controlled to 1560°C, $D = 0.993$.

advantage gained by reducing the porosity by that small amount. The exact level of a strength-maximizing density has been difficult to determine in view of confounding effects attributable to the many process-related flaws that become relevant as stress concentrators at very fine ($< 2\ \mu$) grain sizes.

For alumina in the intermediate sintering range (< 0.95) it has been well established that, given a starting fractional green density, grain size is a function of the final density attained. One of the objectives of this and related research has been to show that this simple grain size–density relationship is not strictly true for final-stage (> 0.95) densification of MgO-doped alumina. Until recently the facts have been largely masked by variability attributable to processing.

This study has shown the importance of controlling preprocessing variables and firing schedules in the control of grain size. Figure 24.4 shows that well-processed material sintered for 1 hour at 1560°C had about the same density (within 0.1%) as poorly processed material sintered for 3 hours at the same temperature; however, the well-processed material had a considerably finer grain size.

Figure 24.5 shows an important comparison between a well-processed sample sintered according to rate-controlled profile *G* (Figure 24.3), and a

Figure 24.6. Low-density region in MgO-doped, well-processed Linde A fired conventionally to 1450°C, 3 hours. SEM 3040×.

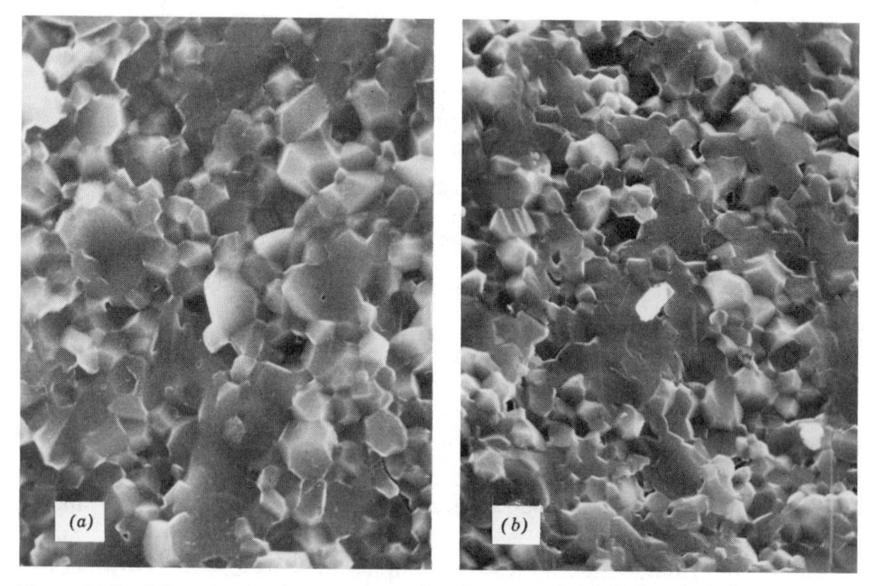

Figure 24.7. Microstructural comparison of well-processed, MgO-doped Linde A using different sintering methods, SEM 1840×. Note: comparable densities. (*a*) conventional 1560°C, 1 hour, $D = 0.992$ and (*b*) rate controlled to 1560°C, $D = 0.993$.

sample fired conventionally at 1450°C for 3 hours. Although the rate-controlled specimen did spend some time 1500°C, its grain size is not appreciably different from that of the conventionally sintered material. However, the rate-controlled specimen showed a considerably higher fractional density. Figure 24.6 shows a region of low density typical of the 1450°C conventional firing, illustrating that even the best processing obtained in this study is not yet fully optimized. Apparently a major advantage of rate-controlled sintering is that it permits such low-density regions to densify uniformly before pore-channel closure isolates them prior to final-stage sintering. Further evidence of this is seen in Figure 24.7. Specimens rate-control sintered according to profile *G* (Figure 24.3) had a finer grain size when compared at nearly equal density with conventionally sintered material.

ACKNOWLEDGMENTS

This work sponsored by the U.S. Navy, Naval Air Systems Command, under Contracts N00010-73-C-0134 and N00019-74-C-0265. We gratefully acknowledge technical assistance by M. L. Huckabee, R. K. Boozer, E. M. Gregory, J. P. Kirkland, R. M. Roberts, Wanda K. Johnson, Betty L. Smith, E. H. Kendall, Jr., and R. Glenn Simpson, and secretarial assistance by Ann S. Ethridge and Anne Y. Gregory.

REFERENCES

1. H. Palmour III and D. R. Johnson, "Sintering and Related Phenomena," G. C. Kuczynski, N. A. Hooton, and C. F. Gibbon, eds., Gordon and Breach, New York, 1967, pp. 770–791.
2. H. Palmour III, R. A. Bradley, and D. R. Johnson, *Kinetics of Reactions in Ionic Systems,* T. J. Gray and V. D. Frechette, eds., Plenum Press, New York, 1969, pp. 392–407.
3. R. A. Lawhon, "Rate Controlled Sintering of Polyphase Ceramic Bodies by Direct Instrumentation," Master of Science Thesis, North Carolina State University, Raleigh, North Carolina, 1970.
4. D. R. Johnson, "Densification Kinetics of Spinel Sintered under Rate Control," Ph.D. Dissertation, North Carolina State University, Raleigh, North Carolina, 1970.
5. M. L. Huckabee and H. Palmour III, "Rate Controlled Sintering of Fine-Grained Al_2O_3," *Amer. Ceram. Soc. Bull.*, **51** (7), 574–576 (1972).
6. H. Palmour III and M. L. Huckabee, "Studies in Densification Dynamics," *Sintering and Related Phenomena,* G. C. Kuczynski, ed., Materials Science Research Ser., Vol. 6, Plenum, New York, 1973, pp. 275–282.
7. H. Palmour III and M. L. Huckabee, "Process for Sintering Finely Divided Particulates and Resulting Ceramic Products," U.S. Patent Application filed April 26, 1973, Serial No. 354,515.
8. H. Palmour, T. M. Hare, and M. L. Huckabee, "Optimal Densification of Ceramics by Rate Controlled Sintering," Final Technical Report, Contract No. N00019-73-00139, Naval Air Systems Command, March 1974.
9. H. Palmour III, T. M. Hare, and M. L. Huckabee, "Conventional vs. Rate Controlled Sintering of Alumina: I. Microstructural Comparisons," presented at the 77th Annual Meeting of the American Ceramic Society, Washington, D.C., May 3–8, 1975.
10. T. M. Hare, H. Palmour III, and M. L. Huckabee, "Convention vs. Rate Controlled Sintering of Alumina: II. Strength Comparisons," presented at the 77th Annual Meeting of the American Ceramic Society, Washington, D.C., May 3–8, 1975.
11. D. E. Witter, R. A. Bradley, D. R. Johnson, and H. Palmour III, "Use of Computers in Ceramic Research," *Ceram. Age,* **84** (11), 28–31 (1968); *Errata, ibid.*, **85** (1), (1969).
12. T. M. Hare, M. L. Huckabee, and J. P. Kirkland, "Intermediate Stage Densification Kinetics for Alumina Sintered under Rate Control," presented at Fall Meeting, Basic Science Division of the American Ceramic Society, Pittsburgh, Pennsylvania, Sept. 24, 1973.
13. D. Lewis and M. W. Lindley, "Enhanced Activity and the Characterization of Ball Milled Alumina," *J. Amer. Ceram. Soc.*, **49** (1), 49 (1966).
14. A. Pearson, J. E. Marhanka, G. Maczura, and L. D. Hart, "Dense, Abrasion Resistant 99.8% Alumina," *Amer. Ceram. Soc. Bull,* **47** (7), 654–658 (1968).

15. D. E. Niesz, R. B. Bennett, and M. J. Snyder, "Strength Characterization of Powder Aggregates," Technical Report Number 1, Contract Number N00014-68-0347, Office of Naval Research, Oct. 1970.
16. R. B. Bennett and D. E. Niesz, "Comminution of a Reactive Alumina Powder," Technical Report Number 2, Contract Number N00014-68-C-0342, Office of Naval Research, Oct. 1971.
17. R. B. Bennett and D. E. Niesz, "Strength Characterization of Powder Aggregates," Technical Report Number 3, Contract Number N00014-68-C-0342, Office of Naval Research, Oct. 1972.

25

Dynamic Particle Stacking

F. N. Rhines

It is widely recognized that the mode of stacking of particles in a powder compact greatly influences the properties of the green body, the course of subsequent sintering, and the properties of the sintered product. Efforts to explain this dependence, through an understanding of the nature of powder stacking, have often taken the form of constructing assemblies of spheres, either of identical or mixed sizes. This approach assumes that it is possible to represent a stack of powder in terms of a typical arrangement of particles, which is repeated many times to produce the stack.

The origin of the idea of modeling particle assemblies with spheres has been credited to Rene J. Haüy (1789),[1] who was concerned with the atomic arrangement in crystals. A century later W. Barlow[2] used the face-centered-cubic and close-packed-hexagonal models to describe particle stacking. In the 90 years since then, more than a score of investigators have contrived stacking models by the use of real or imagined spheres, of like or variable size, systematically or randomly stacked. One of the most comprehensive of such analyses was that of L. C. Graton and H. J. Fraser.[3]

Considering the great amount of effort that must have gone into these studies, as well as the length of time over which the results have been known to the public, it seems surprising that there are still no recognized laws of particle stacking, or even generally accepted industrial procedures based on these models. Thus this writer concludes that the unit-cell approach is incapable of leading to a useful result in this type of natural process. This is not to imply that the work has been barren. A number of important trends and a few relationships that promise to apply to the general case of natural powder stacking have emerged, but these cannot be woven into a general understanding of stacking until a valid base for the subject has been established.

Except for the rare case in which the particles are so perfectly identical that they can assume a "crystalline arrangement,"[4] the assumption of repeated, typical arrangements of particles is not valid for hard particles. For most powders the particles differ individually in the detail or their shapes and sizes.[5] Since the particles can come to rest only through mutual contact, these differences result in an infinity of local arrangements, which cannot be represented by any iterative model. The stack, nevertheless, has a distinctive character that reflects the nature of the powder or power mix that has gone into its composition and the manner in which the stack has been formed, for example, poured, shaken down, or slip cast. Thus, to describe a stacking realistically it is necessary to characterize the powder and also the mode of stacking in terms of parameters that can be evaluated independently of the specific particle shape and size, as well as the difference from site to site in the structure of the stack. In other words, it is necessary to characterize the stack as a whole, rather than particle by particle.

STACKING PARAMETERS

Two appropriate parameters that are capable of evaluation without reference to particle size or shape are: (1) the number of particles P in the stack and (2) the number of interparticle contacts C in the stack. Both these numbers are subject, in principle, to direct and exact evaluation. The numerical difference between them is a measure of the topological connectivity G of the stacking[6]:

$$G = C - P + 1 \tag{1}$$

In this view the powder stack may be thought of as a three-dimensional network in which the individual particles represent nodes that are joined by interparticle contacts that serve as branches of the net. The connectivity of such a net is defined as the number of branches (interparticle contacts) that would have to be severed to reduce the net to a single chain of particles. As a chain there is one more particle than there are contacts; hence the connectivity is one more than the difference between the number of contacts and the number of particles. It is obvious that porous particles are already multiply connected and that their connectivity becomes part of the connectivity of the system as a whole. Where porous particles are involved, a more elaborate method for evaluating the topological parameters of the system is required. In some cases it might be possible to determine an average connectivity of the powder particles by serial sectioning and to add this to the main-network connectivity. It is doubtful that the result would justify the great effort required.

DeHoff and coworkers[7] have demonstrated the important principle that the course of geometric change in second-stage sintering is determined by the connectivity of the system, which is often the same as the connectivity of the original powder stack. The ratio of the surface area to the volume of the porosity in second-stage sintering remains constant and is proportional to the connectivity. This makes it possible to design and construct a sinter body of specified structure and properties. The same authors have shown also that the permeability of the porous body is a simple function of connectivity. Other properties that display a qualitative relationship to the connectivity, but which have not yet been studied in detail, include powder flow rate, sintering rate and, perhaps, electrical and thermal conductivity. In a somewhat less direct fashion it appears that the connectivity may prove to be involved in some of the mechanical properties of both green and sintered bodies.

The number of particles P in a powder mass may be determined directly by the use of a Coulter Counter, or equivalent device. Such a count must be made without distinguishing the sizes and shapes of the particles. It is necessary, however, to count the particles separately, not as disintegratable clumps. The particles should be counted in the form in which they will appear as individuals in the ultimate stack. If they will break up into pieces in the stack, then they should be disintegrated before counting. It is also required, in principle, that *all* the particles in the stack be counted, because any kind of segregation in the stack would tend to distort the count. In practice, it is impractical to make total counts, for which reason rather sophisticated sampling techniques may be required.

The number of interparticle contacts C in a powder mass has been counted directly by a number of investigators, of whom the first seem to have been Smith et al.[8] They used lead shot wetted with acetic acid, which left a white deposit of lead acetate at each point of contact. The white marks were then counted as the stack was disassembled. Such methods are cumbersome at best and become impractical as the particle size becomes very small. Fortunately, it is possible to estimate C through a knowledge of apparent density.

Apparent density is another property that applies to the powder mass as a whole. It has been found that the apparent density is simply related to the number of interparticle contacts, at least over the range from half to fully dense (Figure 25.1).[9] This important relationship, which makes possible the

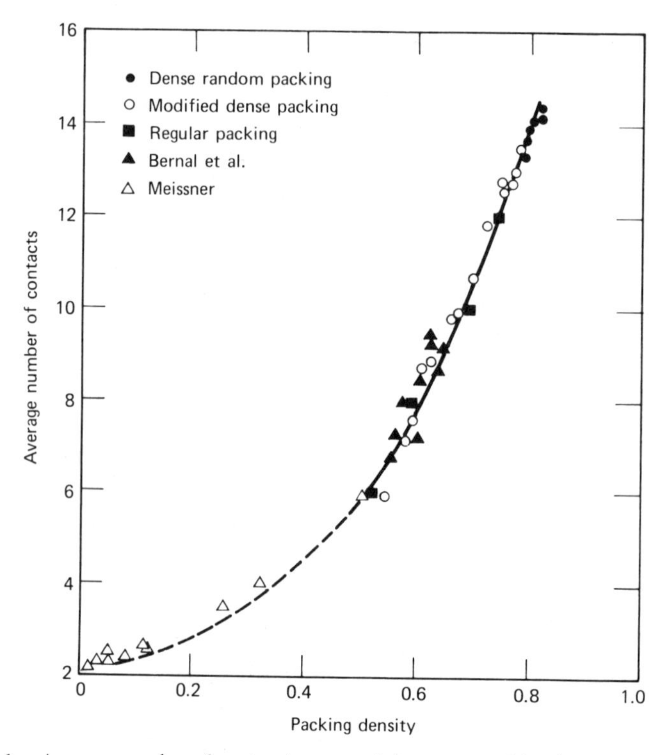

Figure 25.1. Average number of contracts per particle versus packing density. From Norman and Maust[49].

estimation of the number of contacts from one simple measurement, was first reported by Smith et al.[8] and has since been confirmed by others.[10,11] The relationship displayed in Figure 25.1 is valid irrespective of particle size, but the slope of the curve varies with particle shape. That the relationship must differ with particle shape may be illustrated with a thought comparison of spherical versus dendritic particles, wherein it is evident that the apparent density of a stack of dendritic particles must be lower than that of a stack of spheres having the same total number of interparticle contacts. Thus powders of different shapes require special calibration for determination of the number of contacts through the apparent density.

EXPERIMENTAL INVESTIGATIONS

Having means by which the stack and the powder can be characterized in an exact and meaningful manner, it becomes of interest to inquire into the question of how the characteristics of the stack can be controlled. It serves, for the present, to control the number of particles and the number of interparticle contacts. The number of particles per gram, or per unit volume, can be manipulated through the particle size and particle-size distribution. Since the quantities involved are directly additive it does not appear that the control of the number of particles requires special consideration at this time.

Control of the number of interparticle contacts can be had in two rather obvious ways, namely, through the orientation of the particles so as to contact a maximum number of neighbors and through the mixing of particle size and shape. Particle reorientation and relocation can be accomplished by mechanical agitation of the stack. The controlled mixing of powders involves suitable classification and blending of powders from selected sources. In so doing the apparent density of the stack provides an index of the number of contacts.

About 20 years ago, a study of the packing of powders became the subject of a series of theses at the Carnegie Institute of Technology sponsored by a luncheon club of Pittsburgh powder metallurgists. The results, which have not been published up to now, are summarized in this chapter and form the basis upon which a new analysis of powder stacking is proposed. The authors of these theses are John A. Brown (1950), R. J. MacDonald (1951), and Richard H. Vogt (1958). Although some aspects of their work have been duplicated by others, there remains much that is new.

A mild steel powder was made for the research by Arthur H. Grobe at the Vanadium Alloy Steel Company in Latrobe, Pennsylvania. The particles were typical of atomized powders, being ovate in form and spanning a large range of particle size. A water quench, which was part of the atomizing process, gave the particles a shiny, rust-resistant surface, which remained clean after 10 years of storage. The powder was first sized by taking cuts with consecutive sieves of a new Tyler Screen set (Table 25.1). Each cut was then tested for its apparent density. Duplicate measurements were made at four laboratories, namely, Kennametals, Westinghouse, Gibson Electric, and Carnegie Tech. In all cases the determination was made by the use of a Scott Volumeter. The results, which were mostly in good agreement, are summarized in Figure 25.2. It is to be noted that the apparent density diminished approximately in proportion to the particle size over most of the range. This is an effect that has been noted by nearly everyone who has carried out a similar experiment. The less-dense packing of the finer particles has been ascribed to bridging, or clumping, resulting from interparticle adherence that is more persistent the lighter the particle. The loss in density at the coarse end of this series is not commonly found. It is believed to be associated with a departure from the common particle

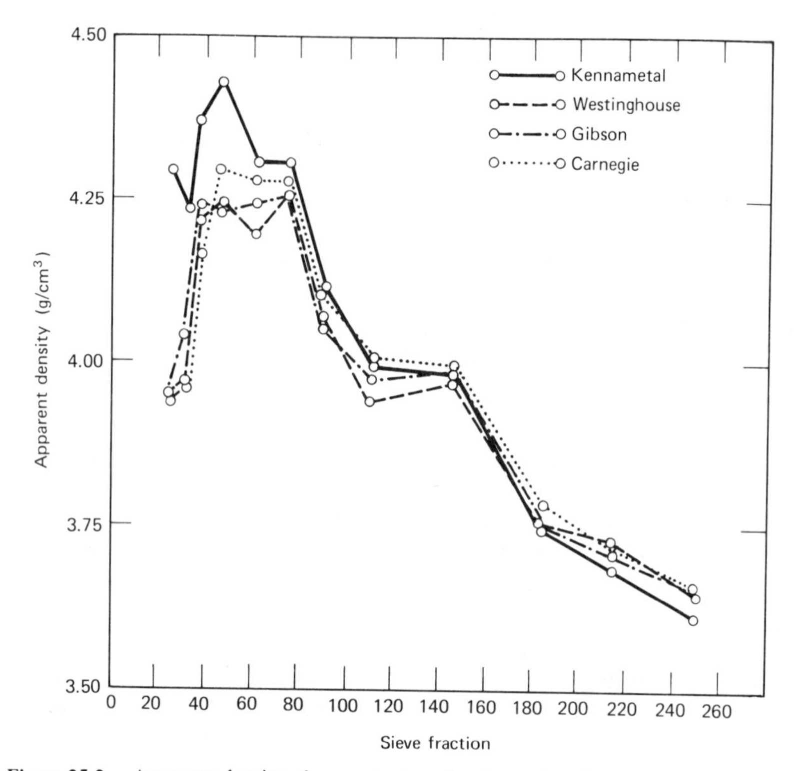

Figure 25.2. Apparent density of separate sieve fractions of steel powder. From Brown.

shape among the largest particles, for which reason the coarsest cuts were excluded from subsequent experiments.

Brown measured the apparent densities of binary mixtures of steel powders of different sizes. Weighed fractions of two cuts were blended by quartering and rolling on glossy paper, as in assaying practice, and then were passed through a Scott Volumeter to determine the weight of 1 in.^{3} of the mixture. Two typical results are displayed in Figure 25.3. Here the 408 μ powder is mixed, respectively, with 163 and 68 μ powder, giving size ratios of approximately 3:1 and 6:1. Both curves arch to a maximum density with slightly less than one-third of the finer cut present in the mixture. In all the nine series studied by Brown, the density attained a similar flat maximum near 30% of the finer cut in the mixture.

The same result as that described above has been obtained by essentially everyone who has conducted comparable experiments. Westman and Hugill proposed an explanation of this phenomenon in terms of the packing volume Figure 25.4). In this diagram volume is plotted vertically and com-

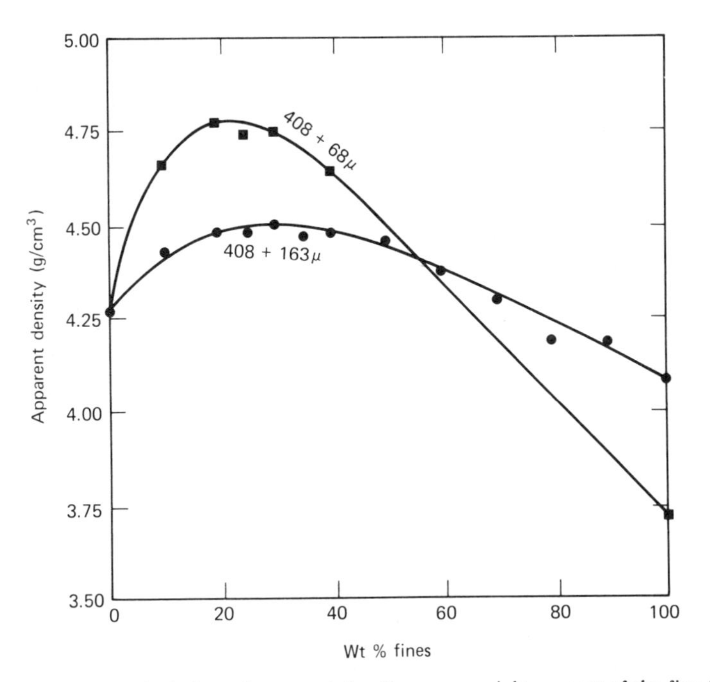

Figure 25.3. Some typical plots of apparent density versus weight percent of the fine fraction in binary mixtures: 408 μ powder mixed with 163 μ powder; 408 μ powder mixed with 68 μ powder. From Brown.

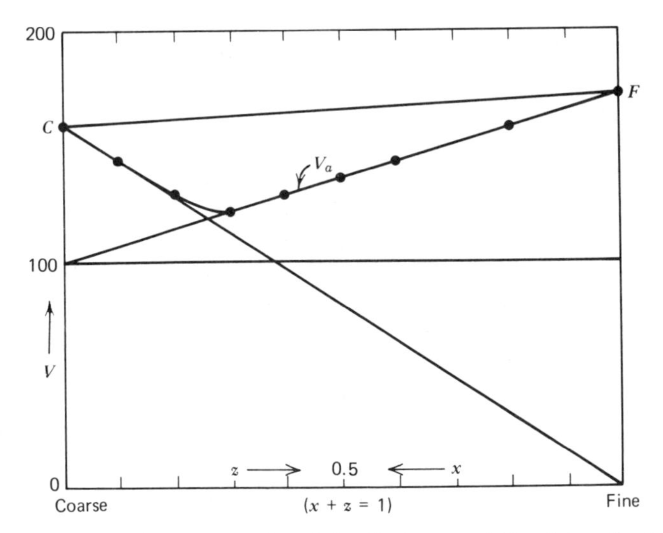

Figure 25.4. Diagramatic representation of the Westman and Hugill hypothesis to account for the apparent density of binary mixtures of powders.[7]

Table 25.1. Mean particle size, apparent density and flow rate of unmixed sieve fractions of atomized steel powder (Brown)

Lot No.	Sieve Fraction	Mean Particle Size (μ)	Apparent Density (g/cm^3)	Flow Rate (seconds)
0	−20 + 30	715	3.94	None
1	−30 + 35	505	3.96	None
2A	−35 + 40	460	4.16	34.6
2B	−40 + 50	408	4.29	29.8
2C	−50 + 70	253	4.27	24.6
2D	−70 + 80	199	4.27	22.8
3	−80 + 100	163	4.10	23.2
4	−100 + 120	137	3.98	23.1
5	−120 + 170	106	3.97	22.5[a]
6	−170 + 200	81	3.78	24.0[a]
7	−200 + 230	68	3.73	24.3[a]
8	−230 + 270	58	3.65	24.7[a]

[a] A smooth flow was not observed. Stacking occurred.

position of the mix is plotted horizontally. The horizontal line at $V = 100$ represents the ideal volume of the fully dense material. The line CF represents the packing volume of mixtures, obtained by adding the packing volume of the unmixed components. The diagonal lines terminating at C and F are the contributions of the two cuts to the total volume, based on the concept that the coarse fraction is of primary importance in determining the skeleton of the stack and that the fine fraction first fills the void spaces among the coarse particles and, when the void spaces are filled, begins to add to the total volume of the stack. Accordingly, to the left of the intersection of the diagonal lines the portion connecting the point C represents the packing volume as determined by the coarse fraction alone, while the portion at the right, connecting to point F, consists of the fully dense contribution of the coarse fraction added to the normal volume of the fine fraction. As is seen later in this chapter, and as was in fact recognized by Westman and Hugill, this analysis constitutes an oversimplification. It has the charm, however, of providing a rather good simulation of the variation of apparent density with composition.

The maximum in the apparent density of the composition series becomes higher as the difference in particle size increases, up to at least a ratio of coarse to fine of 6:1 (Figure 25.5). A similar finding has been reported by

McGeary,[12] except that he carried the study to larger ratios and found very little change in the maximum density above a 6:1 ratio.

Flow rate is a property that may be expected to vary with the number of interparticle contacts in the system, because the displacement of a particle requires the rupture of contacts. With the number of interparticle contacts in a powder stack proportional to the apparent density, it is expected that the flow rate will be slower the greater the apparent density. Brown measured the flow rates of all the powder mixtures in the same nine series, using a Hall Flow Meter, ASTM Standard B 213-46T, and found the expected relationship (Figure 25.6). In each case the composition of slowest flow corresponded nearly with that of maximum density. Where deviations occurred, the minimum flow rate was at slightly larger content of fines in the mixture.

Ternary mixtures of the steel powders were examined by MacDonald. He found a flat maximum in the apparent density, higher than that of any of the associated binary maxima, located at about two-thirds of the coarsest fraction. This result has since been duplicated by other investigators.[13]

Segregation according to particle size became apparent as the foregoing studies progressed. In an attempt to deal with this factor, Vogt examined the effect of vibratory packing upon the apparent density of binary mixtures. To observe directly any tendency for the particles to segregate according to size, he employed a white silica beach sand and dyed the finer cuts each a different color. The particle size range was somewhat shorter than that used by Brown (Table 25.2), but the particle shapes were similarly

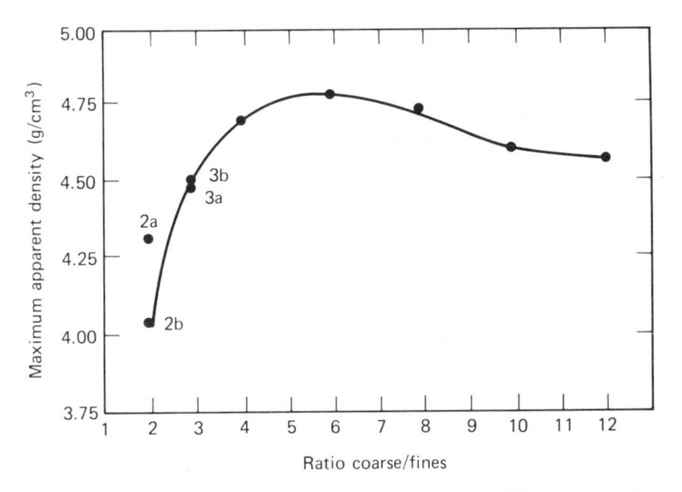

Figure 25.5. Maximum apparent density (composition not specified) versus the ratio of the sizes of the coarse and fine particles. From Brown.

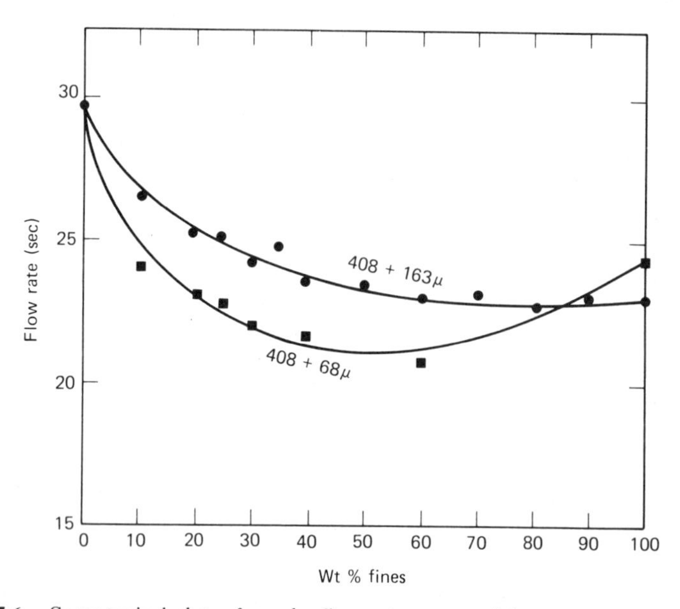

Figure 25.6. Some typical plots of powder flow rate versus weight percent of the fine powder in the mixture (same series represented in Figure 25.3). From Brown.

ovate and smooth. Instead of using a Scott Volumeter, Vogt poured weighed portions of the powder into a 100 ml glass graduate cylinder and used the filling volume to calculate the apparent density. At the same time the glass container made it possible to observe segregation visually. The glass cylinder was strapped to a wooden base with rubber bands, and the base was vibrated by means of a small motor with an eccentric fly wheel

Table 25.2. Mean particle size, apparent density of as-poured and vibrated sieve fractions of white sand (Vogt)

Lot No.	Sieve Fraction	Mean Particle Size (μ)	Apparent Density As-Poured (g/cc)	Apparent Density Vibrated g/cc)
A	$-20 + 30$	715	1.348	1.462
B	$-30 + 40$	506	1.306	1.466
C	$-40 + 50$	358		
D	$-50 + 70$	254		
E	$-70 + 100$	180	1.158	1.392
F	$-100 + 140$	127	1.073	1.322
G	$-140 + 200$	89	0.994	1.268

rotating at a speed of 1800 rpm. The cylinder was filled and the as-poured volume was measured before the vibrating motor was started. Thereafter, the vibrator was operated and readings of apparent density were made periodically, until a steady state was indicated by constant readings.

The powder was blended in two ways. One method, similar to that used by Brown, consisted of quartering and rolling the weighed portions of the mixture upon glossy paper and then pouring the blend directly into the glass cylinder. Particle-size segregation was apparent during blending, in the as-poured stack, and became more prominent as the powder was left unvibrated in the cylinder. In an effort to overcome this separation, the two powders of the binary combination were introduced into the cylinder through two funnels so arranged that the two streams of powder combined in fall. The mixing appeared, at first, to be more homogeneous, but after 24 hours of quiescent standing, gross separation of the fines to the bottom of the cylinder was evident. No homogeneous stack was ever achieved in these experiments.

It was first noticed that the apparent density increases with the height of the powder stack in the cylinder (Figure 25.7). The effect persists, moreover, through long vibration. For this reason a constant weight of powder was used in all subsequent experiments. The same effect has been

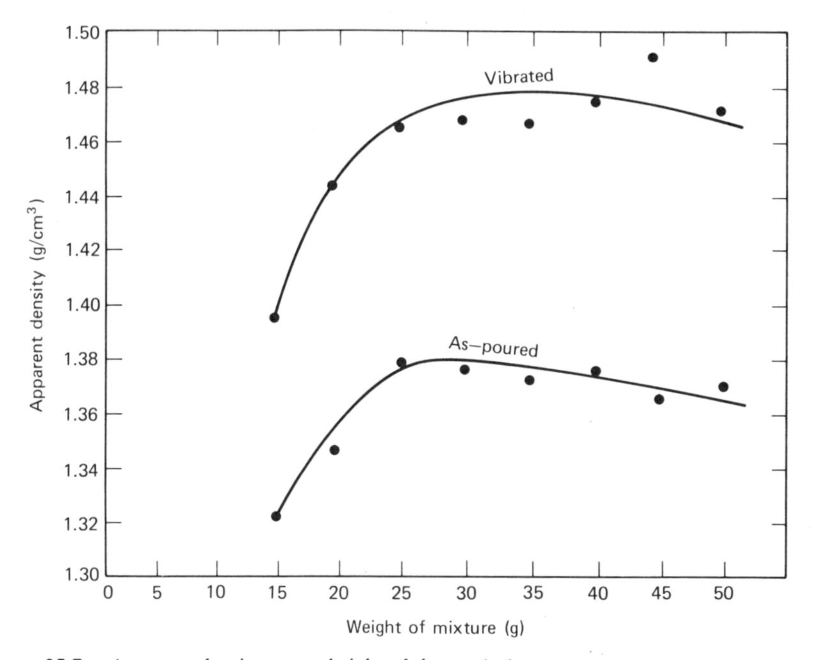

Figure 25.7. Apparent density versus height of the stack (i.e., weight of powder). From Vogt.

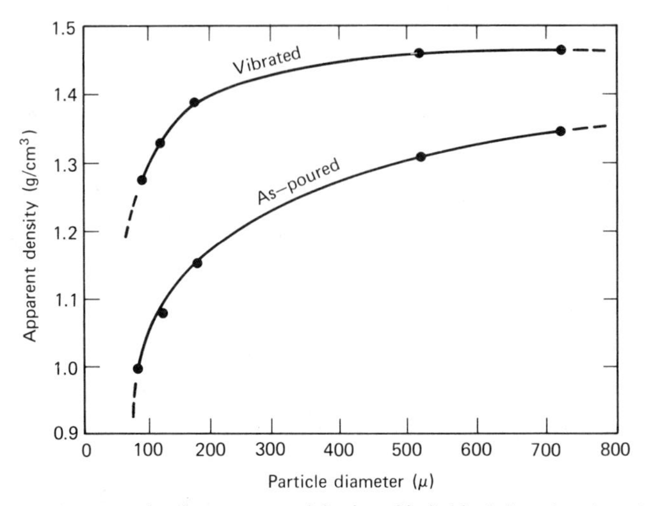

Figure 25.8. Apparent density versus particle size of individual sieve fractions. From Vogt.

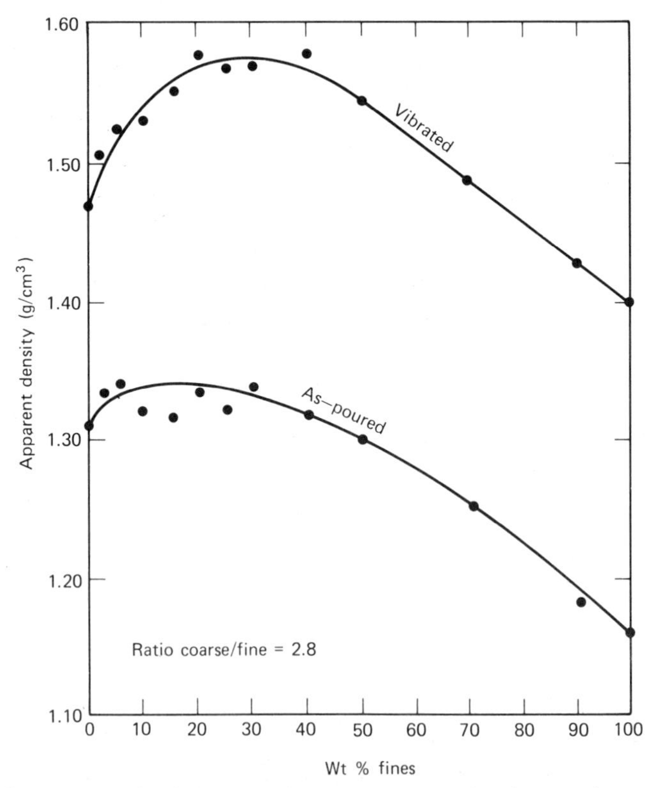

Figure 25.9. Apparent density versus weight percent of fines in mixtures of two powders (506 and 180 μ) with size ratio of 2.8 : 1. From Vogt.

found by other experiments; for example, Bell et al.[14] used an additional loading on the powder to secure increased densification. It may be surmised that the larger load tends to assist the breaking of bonds between particles, reducing bridging and clumping.

Vogt also found the now-familiar effect of particle size on the apparent density of single-size cuts (Figure 25.8). Again it can be seen that this effect persists in spite of vibratory packing, although the finer cuts densify most during vibration.

Vibratory settling of binary mixtures was found to increase the apparent density of all mixtures progressively, until no further change would take place (Figures 25.9 through 25.12). The composition of maximum density is preserved, although it tends to occur at slightly higher fines content after vibration. No reversal in densification, such as that reported by Gugel and Norton,[15] was found. This may appear surprising in view of the additional fact that the size separation became distinct during vibration. Intuitively, it had been supposed that the most dense state, toward which the stack would

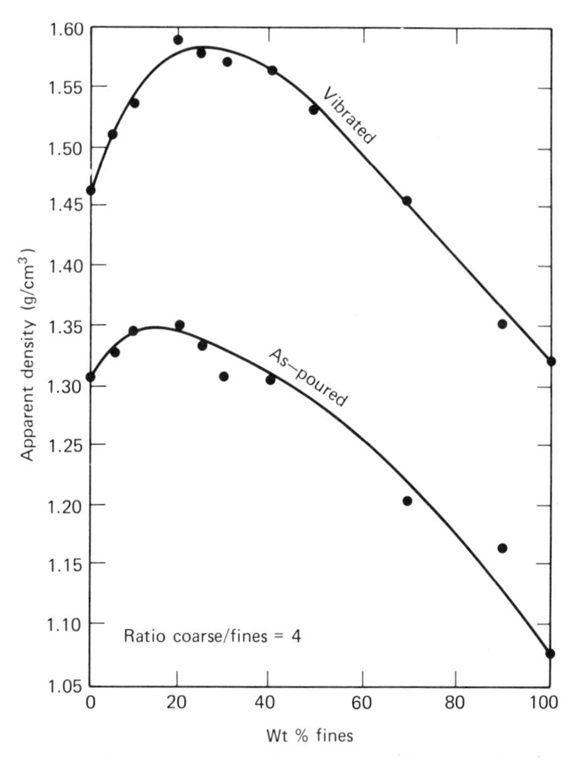

Figure 25.10. Apparent density versus weight percent of fines in mixtures of two powders (506 and 127 μ) with size ratio of 4:1. From Vogt.

tend, would be one in which the smaller particles would be held in interstices between larger particles. It had been supposed, accordingly, that any segregation of the two sizes would result in an expansion corresponding to the return of each cut to its individual stacking density. Clearly, these suppositions were not fulfilled.

The color distribution in the stack suggested a gradation in the mixture, somewhat as indicated schematically in Figure 25.13. Such a gradation does not exclude interstitial packing, but it makes the degree of interstitial placement a function of vertical position in the stack. It seems, therefore, that the steady-state stacking must be a dynamic equilibrium wherein the particles are kept in a state of partial mixing by the mechanical energy of vibration, acting against a natural tendency to segregate, under the influence of gravity.

The effect of particle-size difference on segregation was examined by

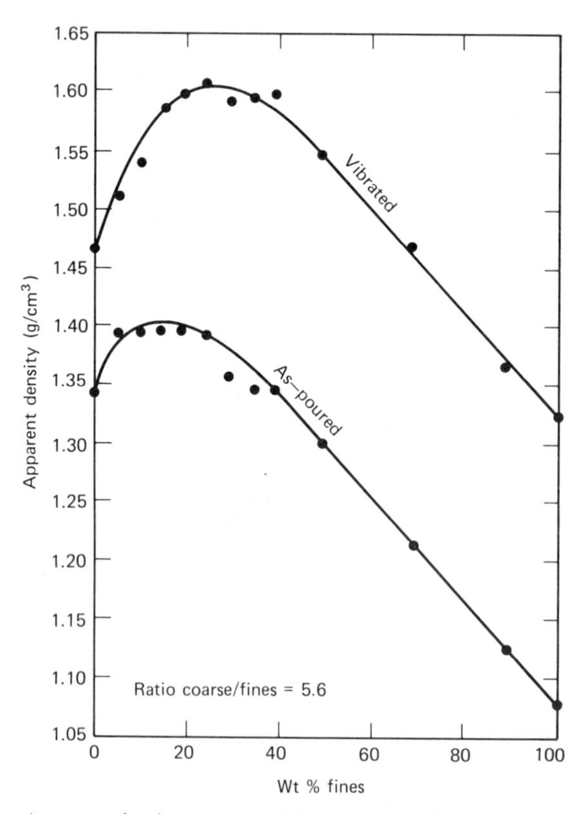

Figure 25.11. Apparent density versus weight percent of fines in mixtures of two powders (715 and 127 μ) with size ratio of 5.1 : 1. From Vogt.

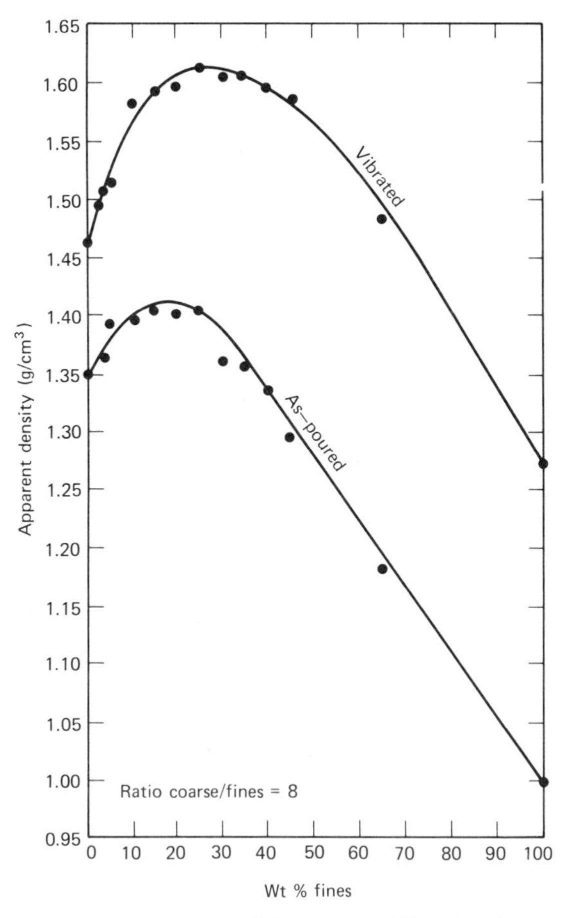

Figure 25.12. Apparent density versus weight percent of fines in mixtures of two powders (715 and 89 μ) with size ratio of 8 : 1. From Vogt.

sampling the extreme top layer of each of the mixes after stabilization by vibration. The particles of the sample were spread on a glass slide and were counted under a microscope. The ratio of the number of coarse to fine particles in the original mixture is plotted as a function of the ultimate ratio in the top layer of the stack after stabilization by vibration (Figure 25.14). Each line of this graph corresponds to a different size ratio of the powder mixture. The 45° line, labeled "stable mixture," represents the ideal case of no segregation. There was no case in which any of the mixtures approached this ideal line. The slopes of the experimental lines, which represent the degree of segregation in each case, are plotted as a function of the particle-

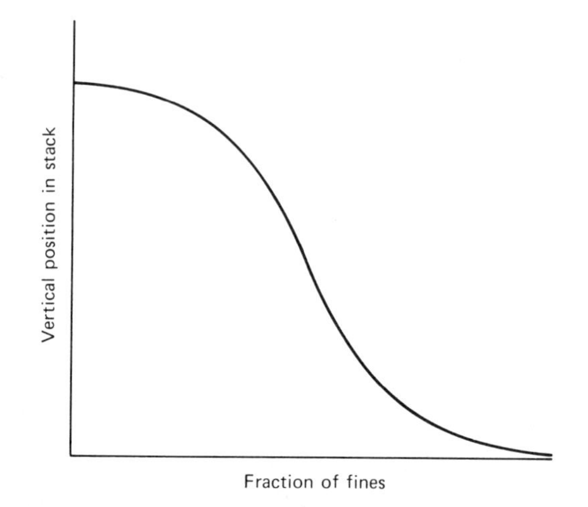

Figure 25.13. Schematic representation of the fraction of fines as a function of vertical position in the stack.

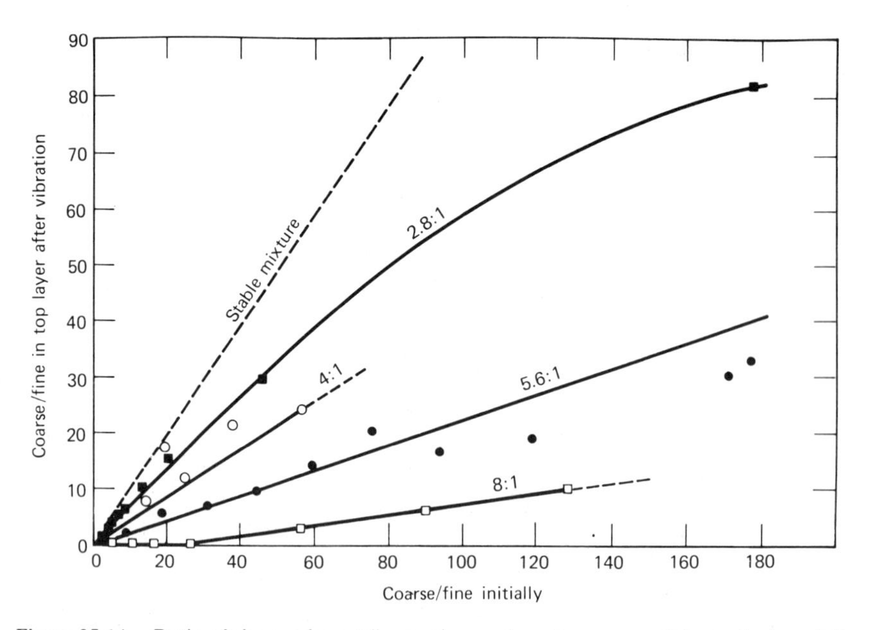

Figure 25.14. Ratio of the number of fine to the number of coarse particles in the top of the stack after vibrating versus the original ratio in the mixture. If there were no segregation all plots would superimpose on the line labeled stable mixture. Departure from homogeneity increases with the ratio of the particle sizes. From Vogt.

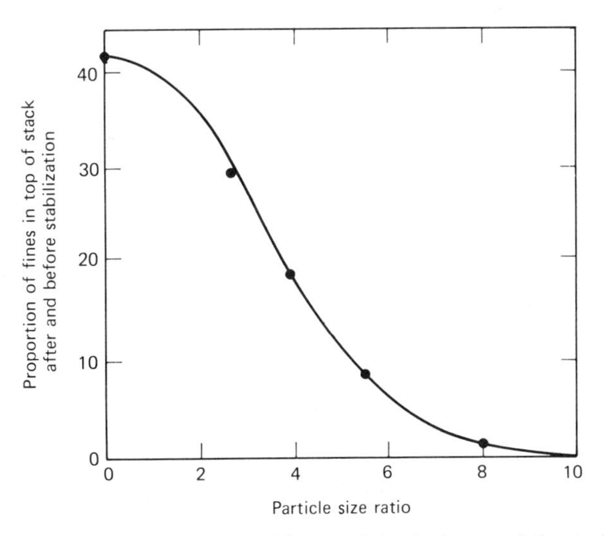

Figure 25.15. Ratio of fine to coarse particles remaining in the top of the stack after vibrating plotted as a function of the particle-size ratio of each mixture.

size ratio in Figure 25.15. Here it is seen that increasing size difference acts progressively to increase segregation, until the separation becomes almost complete, in the top of the stack, with a size ratio in excess of 8:1. It appears that there is no tendency for any particular size of particle to be most stable in the stack. The most stable finer cut was that nearest the size of the coarse fraction. This was indeed an unexpected result!

DISCUSSION

The picture of stacking that thus emerges is a dynamic one in which the system is seeking maximum density of packing for a given particle-size mixture and a given mechanical energy input. To be sure, the dynamic behavior may be thwarted by anchoring the particles in one way or another, but the tendency for the stacking to adjust itself dynamically must always be present, waiting for an opportunity to act.

Particle size varying, as it does, from top to bottom of the stack denies the usual simple ideas of interstitial filling as a direct reason for the higher density of mixed aggregates. Also, the progressive manner in which the density increases with size difference in binary mixtures does not argue for the conventional concept of hole filling by the finer particles. The effect of size mixing on the apparent density is something that involves the whole stack of powder, not merely local sites. Moreover, it affects the as-poured stack and the vibrated stack in like fashion and amount. This seems to limit rather severely even the general idea of best fit as an explanation for the existence of a composition of maximum density. The only solid ground that seems to remain is the correspondence between number of contacts and apparent density. The experiments indicate simply that the number of contacts tends to maximize when about two-thirds of the particles are of the coarser variety, irrespective of homogeneity of mixing, or perfection of fit.

Segregation according to particle size appears to be a gravity effect. The finer particles filter through the coarser particles until they are arrested by coming to a stratum of finer particles through which passage is difficult. The powder stack might be compared, in this respect, to a set of graded sieves, composed of the particles themselves. Some mechanical agitation seems necessary to give the particles mobility and to release particles from interstices in which they may be trapped. The same mechanical energy is expected, however, to promote mixing. The steady state to which the mixture tends during vibration amounts to an equilibrium between size separation and mixing, the mixing being the more pronounced the more energetic the mechanical input.

Densification by vibration has been shown by Bell et al.[14] to proceed very rapidly at the beginning, most of the ultimate change in density being realized in a matter of 5 seconds or less. This argues for densification by the rotation of the particles, virtually *in situ,* to achieve more contacts, rather than by the time-consuming random diffusion of particles to sites of best fit. Indeed, it appears that the most significant result of particle migration must be size segregation. This too must result in an increase in the total number of contacts in the system and an increase in the stability of the structure.

Stability may be thought of as the resistance of the powder mass to change of stacking. This property is illustrated, for example, by resistance to flow, or by resistance to further densification under the influence of vibration, such as is displayed by the finer-size powders. In all cases the stability seems to increase with number of interparticle contacts, perhaps more fundamentally with the connectivity of the system.

The strength of the bond at the contact is also a stabilizing factor. The bond always has some strength, even when it consists of mere van der Waals forces, so that some mechanical force is needed to break it and to set readjustment of the stack in motion. Where a binder, such as a thin film of water, is present, the surface tension can provide a strong enough bond so that considerable force is required to produce particle motion. When the number of contacts and the strength of their bonds has become sufficient to immobilize the powder stack, no further densification occurs.

Bonding strength can also be diminished by chemical means. If a powder stack is filled with a wetting medium, so that the only internal surface becomes that between the solid particles and the liquid, the bonds may become relatively weak and spontaneous particle migration may be initiated. A typical example is that of water added to dry sand. An almost instant shrinkage of the sand mass occurs, because the weakening of the interparticle bonds allows particle rotation. Indeed, dynamic particle stacking appears to attain its ultimate perfection in a slowly moving aqueous medium. This is illustrated by the case of the formation of opal, which consists of a close-packed stacking of identical silica spheres, the color of the opal depending on the sphere size.* Stratification and stacking of the silica spheres develops in thin crevices in rocks where sand and slowly moving water interact. The large number of contacts of the close packing stabilizes the structure, but, as many a wearer has found to her dismay, water and agitation can destroy the stacking and with it the opal.

Particle shape has had much too little study as a variable in particle stacking. There can be no doubt of its importance, but systematic means for dealing with so complex a variable are not yet available. Fortunately, many of the particles of industrial interest are individually dense, approximately equiaxed, and nearly convex in form. For such particles it is possible to visualize some principles of stacking. When a particle falls on a stack, it tends to come to rest in contact with three other particles and, as the stack grows, three more particles are likely to rest on the first one, giving a total of six contacting neighbors, or an average of three contacts per particle. It is readily apparent that, by rotation, the subject particle might find a position in which it would contact more than six neighbors. Perfect spheres of identical size can assume a close-packed arrangement, like that of opal, in which they have up to 12 neighbors. Owe Berg et al.[16] among others, have succeeded in causing a stack of steel spheres to assume close packing under the influence of three-dimensional vibration. It is geometrically possible for identical particles of any shape to be arranged in close packing, but orientational alignment, as well as spatial positioning, must be attained and this is not an event of high probability. It has been assumed that close packing sets a limit to the number of contacts that can occur in a stack. That this is an erroneous assumption is shown by the fact that the grains in a polycrystalline material average 14 neighbors (i.e., 7 contacts per particle). Still more contacts per particle seem possible where the particle shape is branched, and especially where the particles are not rigid. For example,

* According to recent studies at the University of Melbourne CSIRO, under the direction of Alan Moore.

there scarcely could be any limit to the number of contacts in a bucket of octopi. At the other end of the scale, a pile of straws may be expected to average considerably less than six contacting neighbors.

In real powder stacks, and especially under the influence of vibration, the increase in apparent density, above that corresponding to the three contacts per particle stack, may be thought of as a measure of the frequency of occurrence of higher-contact configurations. Similarity of particles should increase the likelihood of close-packed configurations, wherefore a size-segregated stack may consist of strata containing not only particles of like size, but a larger fraction of close-packed configurations than would be possible for unlike neighbors. These matters invite further investigation.

Characterization

It must now be apparent that the deductions drawn initially with respect to the kind of description that would be needed for a stack of powder were well justified. No kind of iterative geometry could describe the structure of the stack as it has revealed itself through experiment. The only parameters that can be applied are those that are totals for the entire stack. Number of particles, number of interparticle contacts, apparent density, and connectivity all apply to the stack as a whole. It is true that these parameters do not describe the stack completely, but there are additional parameters available. Total surface area and total curvature of surface are but two of a number of other parameters that can be evaluated for the stack as a whole and add to the detail of the description. Progress with the correlation of these geometric parameters with physical properties of the powder stack has been gratifying. Where correlations exist they have usually been found to be simple relationships, easy to apply in practice.

REFERENCES

1. R. J. Haüy, "Exposition abrégé de la teorie de la structure des cristaux," *Anal. chem.* 1–28 (1789).
2. W. Barlow, *Nature* **29,** 186–188, 205–207 (1883).
3. L. C. Graton and H. J. Fraser, "Systematic Packing of Spheres with Particular Relation to Porosity and Permeability," *J. Geol.*, **43,** 785–909 (1935).
4. R. M. Koerner, "Limiting Density Behavior of Quartz Powders," *Powder Tech.*, **3,** 308–312 (1969/70).
5. J. E. Davis, V. G. Carithers, and D. R. Warson, "Practical Method for Using Particle Size Distribution of Calcined Aluminas to Predict their Compacting and Sintering Properties," *Amer. Ceram. Soc. Bull.*, **11,** 906–912 (1971).

6. F. N. Rhines, "A New Viewpoint on Sintering," *Trans. Metal Powder Association,* **1958,** 91–101, *Plansee Proc.,* **3,** 38–53 (1958).
7. R. T. DeHoff and F. N. Rhines, Final Report to AEC, University of Florida, 1973.
8. W. O. Smith, P. D. Foote, and P. F. Busang, "Packing of Homogeneous Spheres," *Phys. Rev.,* **34,** (11), 1271–1274 (1929).
9. Yu. E. Pivinskii, "Increasing the Density of Particle Packing in Forming Ceramics," *Steklo Keram.,* **21** (9), 25–29 (1969), Translation, Consultants Bureau of Plenum Publishing Corp, New York, 1970.
10. A. E. R. Westman, "The Packing of Particles and Particle Shape," *J. Amer. Ceram. Soc.,* **20,** 155–166 (1937).
11. J. J. Hauth, "Vibrationally Compacted Ceramic Fuels," U.S. Atomic Energy Commission HW 67777, 44, 1961.
12. R. K. McGeary, "Mechanical Packing of Spherical Particles," *J. Amer. Ceram. Soc.,* **44,** (10), 513–522 (1961).
13. F. F. Pownes, "The Effect of Sand Grain Size on Packing Density," *Br. Foundryman,* **54** (11), 346–468 (1961).
14. W. C. Bell, R. D. Dillinder, H. R. Lominac, and E. G. Manning, "Vibratory Compacting of Metal and Ceramic Powders," *J. Amer. Ceram. Soc.,* **38** (11), 396–404 (1955).
15. E. Gugel and F. H. Norton, "High Density Fire Brick," *Ceram. Bull.,* **41** (1), 8–11 (1962).
16. T. G. Owe Berg, R. L. McDonald, and R. J. Trainor, Jr., "The Packing of Spheres," *Powder Tech.,* **3,** 183–188 (1969/1970).

26

Particle Compaction

O. J. Whittemore, Jr.

The largest number of polycrystalline ceramic products are formed by pressing. Products include insulators for electronic equipment and electrical appliances, refractory brick and other shapes, some building brick, grinding wheels, nuclear fuel, and many other special ceramics. Product sizes may vary from 1 mm to grinding wheels 2 m in diameter. The particle or grain sizes within the products may vary from 0.05 μm to 1 cm. The pharmaceutical and powder metallurgy industries also employ pressing as their principal forming process. Pressing processes include: (1) dry pressing, where less than 2% water is present; (2) semidry pressing, with 5 to 20% water; and (3) isostatic pressing, with less than 2% water. Advantages of pressing are the rapid molding rates (as high as 5000 pieces/minute in rotary presses) and the precision (molded tolerances on the order of 0.1 mm). The final product precision, however, depends on many factors, including powder mixing and sintering control.

EQUIPMENT

A variety of presses are available for compacting ceramics. Their designs display consideration of compact size and shape, material compaction characteristics, molding rate, product density, and powder abrasiveness.

The compacting force is applied by mechanical, hydraulic, or pneumatic means and sometimes by a combination of these. The force can also be applied repeatedly in some designs, in this way overcoming the rheological resistance to flow.

Mechanical

Many presses for molding small precise shapes compact by action of cams against push rods or plungers.[1,2] Cams can be provided both above and below, and also in parallel so that secondary motion will achieve uniform compacting with different shape thicknesses. Developed originally for pharmaceuticals, these "pill" presses can be provided with hydraulic or pneumatic assists to equalize pressure. Rotary presses have a rotary table on which a number of dies are arranged in a circle, each being filled, compacted, and emptied in sequence. These presses are available with maximum pressures to 100 tons, although most are in the range of 1 to 20 tons.

Most refractory brick is formed in large mechanical toggle presses capable of pressures up to 800 tons. The toggle mechanism operates much like the elbow in a human arm.[3] Toggles can also be driven hydraulically to provide pressures up to 1300 tons and to form deeper shapes (see Figure 26.1).

Friction presses have long been used to press tiles, and so forth.[2] The plunger is attached to a large vertical screw, which is driven down by a large friction wheel. By the time the plunger compacts the fill, it has high velocity. The impact can be repeated rapidly and from three to five impacts are normally applied. Large friction presses are now being employed in molding large thick refractory shapes.[4] Although the press is rated at 500 tons the density achieved is close to that obtained on 1000 ton mechanical toggle presses (see Figure 26.2).

Cam and toggle presses close to a given volume. Therefore, although the size is assured, the density achieved is determined by the mold fill. Friction presses close to that point where the inertial energy is absorbed.

Hydraulic

Presses compacting by fluid pressure against a piston are generally used in the abrasives industry[5] and occasionally in other industries. Hydraulic press sizes vary from a few hundred kilograms to 5000 tons. Here, in contrast to mechanical presses, the density is determined by the pressure while the size is determined by the mold fill. An objection to hydraulic presses is the generally slower rate.

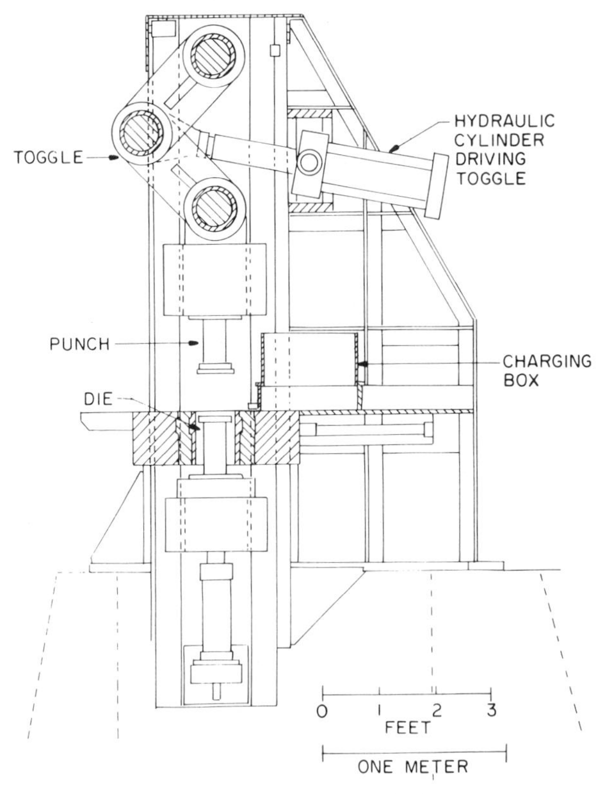

Figure 26.1. Hydraulic toggle press, 1300 ton capacity.

Pneumatic

Special refractory shapes are frequently compacted by hand using air hammers. Uniform high densities can be achieved with skill. Impact presses are constructed with one or more air hammers driving the mold plunger onto the fill. The assemblage is held together by low hydraulic pressure. Impact presses also can produce denser and more uniform products than toggle presses, but their operation is slower.

Isostatic

When a powder mixture is placed in a flexible rubber mold and hydrostatic pressure is applied to the mold exterior, this forming process is called isostatic or hydrostatic pressing.[6,7] The two variations used are "wetbag," where the powder is loaded into a portable mold outside the pressure vessel, and "dry-bag," where that part of the flexible mold that contacts the pressurizing liquid is an integral part or lining of the pressure vessel. In the latter, the mold is usually sufficiently rigid to hold its shape when filled. Products range from spark plug insulators (see Figure 26.3) to sewer pipe 2 m long and massive refractory blocks for glass furnace linings.

Roll-Compacting

If feed material is passed between two horizontal smooth rolls, a continuous sheet of product may be compacted.[8,9] If indentations are cut in the surface of the rolls, briquets are formed that may be granulated for subsequent compacting or may be sintered for refractory grog.

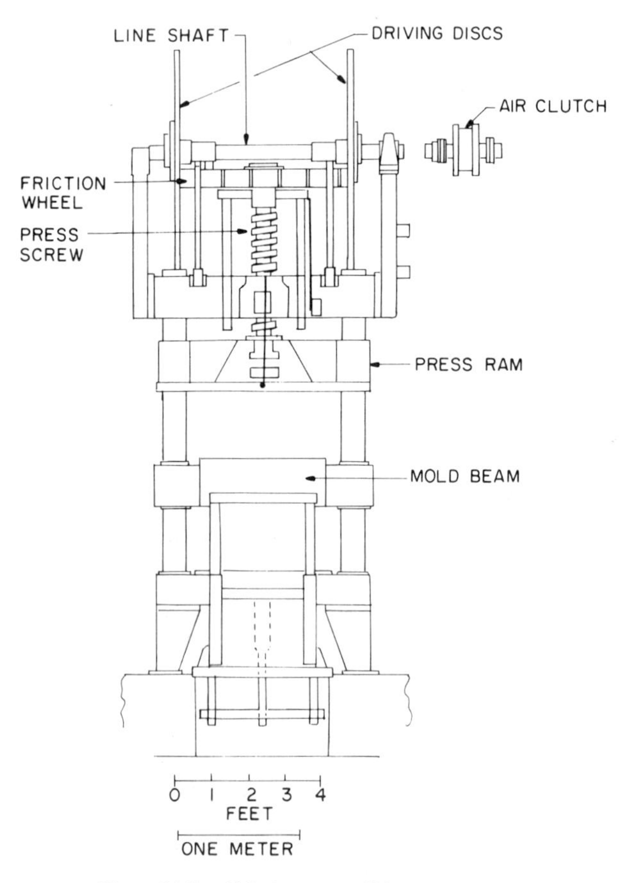

Figure 26.2. Friction press, 500 ton capacity.[5]

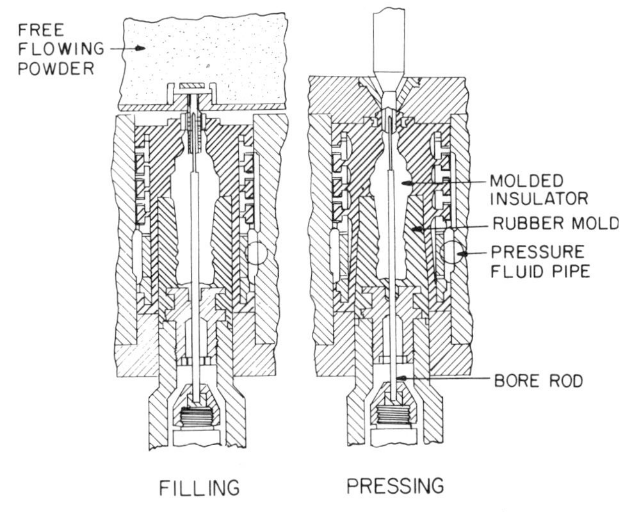

Figure 26.3. Isostatic press for spark plug insulators.[8]

Vibratory Compacting

When particles are poured into a mold that is then vibrated, considerable compaction can occur if the particle sizing is well graded and high accelerations of vibration are applied. Nuclear fuel rods have been produced at 90% of theoretical density.[10] Experimentally, densities as high as 95% of theoretical have been produced.[11] These products are not subsequently fired, as they are not bonded sufficiently to be removed from their containers.

BINDERS AND LUBRICANTS

To assist the compaction process, a number of materials are employed as binders or lubricants.[12] If the ceramic composition includes clay in sufficient amount, other additions are not necessary. For example, foundry molds can be compacted readily of sand with only 2 to 3% of bentonite clay. However, many compositions preclude the inclusion of clay so a number of organic additives are used. These perform several functions.

Binders

When compact strength is required for subsequent handling or other uses, a number of organic materials are used, usually as water solutions. Examples are the lignosulfonates (residues from the sulfite paper process), dextrines, starches, celluloses, and poly(vinyl alcohol). Amounts added may be as low as 0.5% and as much as 5% for coarse-grained mixtures.

Internal Lubricants

Some plastic flow may occur during compaction if a wax such as paraffin or poly(ethylene glycol) is added. These additives also contribute some strength but less than the binders. A large amount can be added and is contained in the void volume of the particle mass.[13] This was illustrated by preparing three series of mixtures of aluminas with poly(ethylene glycol) wax (Carbowax 4000, Union Carbide Corp.).[14] The aluminas used were 38900, 7 μm average size (Norton Co.); XA16, 0.5 μm average size (Aluminum Company of America); and Linde A, 0.3 μm average size (Union Carbide Corp.). The first two were well divided and the last contained aggregates. After pressing at 15,000 psi, densities were determined, samples were fired at 800°C where no shrinkage occurred, weight losses were determined, and pore-size distributions were determined by mercury porosimetry.

Volume percent of alumina in the compacts is plotted in Figure 26.4 versus weight percent wax and shows little densification of the alumina. The mid pore sizes determined by mercury porosimetry are also indicated in Figure 26.4 and show little change for the 7 and 0.5 μm aluminas with wax contents up to 10 wt % (27 vol %). The wax thus filled the interstices between particles up to this amount. At higher wax contents the particles were being held apart, as evidenced by larger pore diameters. No difference was noted for the 0.3 μm alumina up to 16 wt % wax (39 vol %).

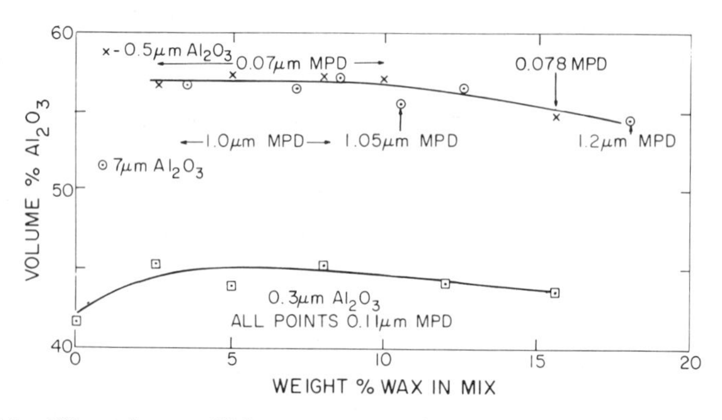

Figure 26.4. Effect of wax addition on compacted alumina density (midpore diameters, MPD, by mercury intrusion).

Of interest is the ratio of 7 between particle size and mid pore size for the 7 and 0.5 μm aluminas. The lower ratio for the 0.3 μm alumina is probably caused by the aggregates present; a bimodal pore distribution was noted, with the peak of smaller pores occurring at 0.04 μm, which also is a ratio of 7.

Mold Lubricants

To reduce the frictional effects at the mold walls, lubricants such as stearic acid and colloidal graphite are employed. These may be wiped or sprayed on the mold wall or pumped through porting.[15]

GRANULATION

Fine powders usually have low bulk densities and high compaction ratios (ratio of bulk density to compact density). Fine powders also do not flow well when the die is filled. To alleviate these problems, some form of granulation is often employed, usually a form of compaction. One method of granulating is spray-drying,[16] where drops of liquid containing powder are sprayed into a chamber and fall through rising hot gases that evaporate the liquid. Surface tension of the liquid holds the drops in spherical form, which, when dry, are free flowing, feed uniformly into the die, and have little dust.

Granulation is also accomplished by pressing the powder into compacts that are chopped or crushed and then screened to obtain the desired granule size for die filling. A large reduction in the compaction ratio can be realized by precompacting and more uniform final products result. However, although high granule density will reduce the compaction ratio, coarse weak bodies may result, with greater density variations within the compact.[17]

COMPACTION EQUATIONS

For a basic understanding of powder compaction, an analytical equation of state would be desirable. However, information required includes the flow properties of powders under stress, stress or pressure distribution within compacts, the distribution of particle-to-particle stresses, and the strength distribution of ceramic particles. Some information has been determined on the last two factors on glass spheres.[18]

Because of the unsolved analytical problems, several empirical equations have been proposed that have value in understanding compaction. Walker[19] determined pressure–volume relationships in compacting several powders and observed the plot of logarithm of pressure versus volume to be linear. Balshin[20] obtained a similar relationship assuming that incremental pressure is proportional to incremental energy expended.

$$\log p = VL_0 + C$$

where p is pressure, V_0 is relative volume, C is a constant, and L is a constant called the pressing modulus, considered analogous to Youngs' modulus used in Hooke's law. The pressing modulus is a function of the initial material properties but was observed to vary during pressing. Balshin showed near linearity in plots of log p versus relative volume for pressing graphite and sodium carbonate, and Huffine and Bonilla[21] later showed linearity for mixed sizes of salt.

An equation describing the compaction behavior of metals was proposed by Heckel,[22]

$$\ln\left(\frac{1}{1-D}\right) = Kp + A$$

where D = relative density at pressure p
K = material constant
$A = \ln [1/(1 - D_0)] + B]$ where D_0 is the initial relative density

He described three compaction stages as (1) densification by die filling, ln $[1/(1 - D_0)]$, (2) densification by particle rearrangement, B; and (3) densification by particle deformation. However, the equation did not describe the compaction behavior of alumina. Leiser[23] later found compaction of fused pure magnesia conformed to Heckel's equation but that of alumina, mullite, and glass did not. As fused pure magnesia has some ductility, this expression appears valid only for ductile materials.

A probablistic approach was proposed by Cooper and Eaton[24] with an expression relating fractional volume compaction V^* with applied pressure P;

$$V^* = \frac{V_0 - V}{V_0 - V_\infty} = a_1 \exp\frac{-k_1}{p} + a_2 \exp\frac{-k_2}{p}$$

where V_0 = initial volume of compact
V = volume of compact at pressure P
V_∞ = compact volume at infinite pressure or theoretical density
a_1, a_2, k_1, k_2 = constants calculated to give agreement with experiment

The two terms assume a two-stage process, described as (1) rearrangement of particles within the compact with accompanying slight fracture or elastic deformation so as to fill voids about the size of the original particles and (2) major fragmentation and filling of small voids. The initial stage of compaction, or die filling, is eliminated by definition of V^*. Sized alumina, silica, magnesia, and calcite particles that vary in hardness on Mohs' scale from 3 to 9 were studied. As hardness increased, the relative compaction at pressure decreased.

The coefficients a_1 and a_2 indicate the fraction of compaction occurring in the two terms. When their sum is not unity, other processes become operative before the theoretical limit of compaction is reached. As hardness increased, the sum of a_1 and a_2 decreased from unity for calcite to 0.85 for alumina. The coefficient a_1 is chosen by data at the lower pressures and its value was found[23] to be dependent on the lower pressure chosen. Oudemans[25] studied compaction of fine alumina and iron oxide particles and was unable to apply either Cooper and Eaton's expression or that of Balshin.

Kawakita and Ludde[25] compared equations of 14 other investigators with his equation:

$$C = \frac{V_0 - V}{V_0} = \frac{abP}{1 + bP}$$

where C = degree of volume reduction
V_0 = initial volume
V = volume at pressure P
a and b = constants characteristic of the powder

This equation can be rearranged as:

$$\frac{P}{C} = \frac{1}{ab} + \frac{P}{a}$$

and if there is a linear relation between P/C and P, the constants can be evaluated. The linear relation held for soft, fluffy, and pharmaceutical powders and also for stainless steel and copper oxide powders except for deviation at the lower pressures. Linearity of the P/C versus P plot was not found to be exact using Leiser's data[23] for magnesia and alumina $-35 + 42$ mesh powders. Kawakita's equation has also been applied[25] to tapping compaction and vibratory compaction, replacing pressure with tapping number and vibrating time, respectively.

Although compaction equations have been criticized as only curve fitting and none have been found to be generally applicable, they all focus attention on important considerations, such as the stages of compaction, the mechanisms, and the variables.

MECHANISMS OF COMPACTION

Where the stages of ceramic powder compaction are die filling, particle rearrangement, and particle fracture, the mechanisms are considered to be particle sliding, elastic deformation, and fragmentation. Describing compaction can be difficult because two stages may be occurring simultaneously and mechanisms predominating during a stage cannot be clearly defined.

Particle fracture was shown by Gormly[5] to occur in pressing a 24 grit grinding wheel. Although there was only a reduction from 50 to 40% retained on a 25 mesh sieve, this amount could result in collapse of many particle bridges. Kingery[26] also showed reduction in size of fine fused alumina when pressed and repressed. Calkins[18] studied the compaction of 250 μm diameter glass spheres that had been vibrated to minimum volume. After compaction, he determined the pore size distribution by mercury porosimetry of the compact without removal from the mold. The void frequency is shown in Figure 26.5 for the as-packed state, after 20,000 psi compaction (23% fracture) and after 40,000 psi compaction (45% fracture). The filling of voids or void-size reduction thus is accompanied by an increase in small voids.

Fracture is accentuated, however, when single-size particles are compacted as in this latter study. When mixtures of two particle sizes are compacted, Chattopadhyay[27] found much less fracture of the coarser size. When −16 + 20 mesh fused-alumina particles were pressed at 24,000 psi, only 23% survived. When 70% of −16 + 20 mesh particles with 30% of one of

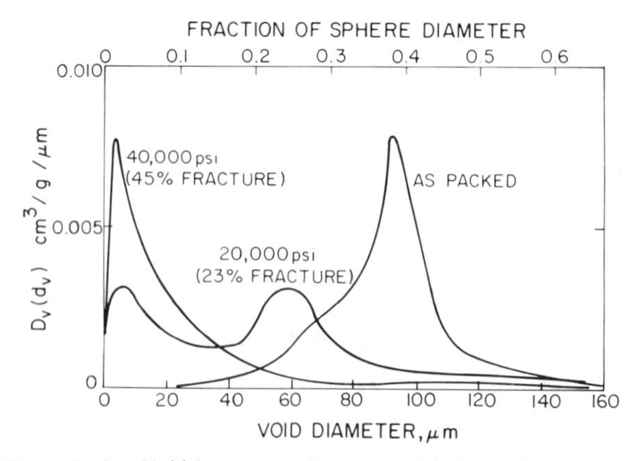

Figure 26.5. Void frequency of compacted 250 μm glass spheres.

three finer sizes was compacted, much greater amounts of the coarse size survived. The greatest amount of coarse-particle survival occurred with the finest of the finer sizes. With the finest particles, the number of contacts increase, the stress per contact point decreases, and less fracture of coarse particles occurs.[28]

Most compaction is not conducted at pressures where a large amount of particle fracture takes place. Instead, we rely on flow of the mixture of fine fractions together with the plasticizers present and some entrainment of the coarser particles to produce rearrangement of particles to obtain higher densities. Since we usually start with compaction ratios of 2 or larger, much of the compaction occurs at low pressures.

Particle sliding is more difficult to observe. Hartmann[28] microscopically observed scratches on coarse particles after compaction. The orientation of the layered silicates (clays and talc) under pressure was demonstrated by Berry et al.[17] The distinction between particle sliding and rearrangement is not easy to make for these minerals.

STRESS DISTRIBUTION

A common method of determining stress distribution within a compact is to measure density distribution and correlate stress with density. In pressing grinding wheels, a characteristic X pattern of slightly higher density is found in the cross section of the wheel.[5] Lower densities are found on the center exterior of faces and ends. Grinding wheels are pressed from both ends. When a compact is pressed from one end, the density decreases from the top to the bottom.

Actual measurements of stress within a compact during pressing were made by Train[29] by means of wire resistance gauges. The isobars of highest pressure in pressing from one direction were found to originate where the face of the moving plunger met the die wall and to pass diagonally through the compact meeting near the center. This work was done on magnesium carbonate, a relatively soft material.

Indirect "end point" measurements have been done by McRitchie[30] in large compacts by incorporating small pressure cells constructed as Brinell hardness testers. Cooper and Goodnow[31] evaluated the deformation of lead grids placed within compacts of alumina and talc by radiography.

Analytically, Schwartz and Weinstein[32] computed the stress distribution from a knowledge of the cohesion and the angle of internal friction of the powder mass as related in the Coulombs' yield criterion. The predicted results were compared with measured results on urania powders.

REFERENCES

1. H. Thurnauer, "Controls Required and Problems Encountered in Production Dry Pressing," *Ceramic Fabrication Processes,* W. D. Kingery, ed., M.I.T. Press, Cambridge, 1958, pp. 62–70.
2. F. H. Norton, *Fine Ceramics,* McGraw-Hill, New York, 1970, pp. 376, 401–403, 427, 437, 439, 442.
3. F. H. Norton, *Refractories,* 3rd Ed. McGraw-Hill, New York, 1949, pp. 130–139.
4. G. L. Barna, "Automatic Friction Press," presented at Pacific Coast Regional Meeting, American Ceramic Society, Oct. 24, 1974.
5. M. W. Gormly, "Technical Aspects of Vitrified Grinding Wheel Manufacture: III Mixing and Molding," *Amer. Ceram. Soc. Bull.,* **37** (4), 189–192 (1958).
6. B. A. Jeffery, "Method of and Apparatus for Shaping Articles," U.S. Patent 1,863,854 (1932) (to Champion Spark Plug Co.).
7. B. A. Jeffery, "Method of and Apparatus for Molding Materials," U.S. Patent 2,152,738 (1939) (to Champion Spark Plug Co.).
8. C. Orr, Jr. *Particulate Technology,* MacMillan, New York, 1966, pp. 400–427.
9. R. C. Ragan, "Method for Continuous Manufacture of Ceramic Sheets," U.S. Patent 3,007,222 (1961) (to Gladding-McBean & Co.).
10. R. B. Holden, *Ceramic Fuel Elements,* Gordon and Breach, Science Publishers, New York, 1966, pp. 27–49.
11. R. K. McGeary, "Mechanical Packing of Spherical Particles," *J. Amer. Ceram. Soc.,* **44** (10), 513–522 (1961).
12. S. Levine, "Organic (Temporary) Binders for Ceramic Use," *Ceram. Age,* 39–42 (January 1960) and 25–36 (February 1960).
13. C. A. Bruch, "Problems in Die Pressing Submicron Size Alumina Powder," *Ceram. Age,* 44–47 (October 1967).
14. O. J. Whittemore, Unpublished research, University of Washington, Seattle.
15. A. W. Mohr, "Pressing Problems Not Unique," *Ceram. Age,* 24 (December 1970).
16. B. A. Jeffery, "Preparing and Molding Material," U.S. Patent 2,251,454 (1941) (to Champion Spark Plug Co.).
17. T. F. Berry, W. C. Allen, and W. A. Hassett, "Role of Powder Density in Dry Pressed Ceramic Parts," *Amer. Ceram. Soc. Bull.,* **38** (8) 393–400 (1959).
18. D. J. Calkins, "Ceramic Powder Compaction by a Brittle Fracture Mechanism Using a Glass Sphere Model," Ph.D. Thesis, University of Washington, Seattle, 1970.
19. E. E. Walker, "The Compressibility of Powder," *Trans. Faraday Soc.,* **19**, 73 (1923).
20. M. Y. Balshin, "The Theory of the Process of Pressing," *Vestnik Metalloprom,* **18** (2) 124–137 (1938).
21. C. L. Huffine and C. F. Bonilla, "Particle-Size Effects in the Compression of Powders," *Amer. Inst. Chem. Eng. J.,* **8,** (4), 490–493 (1962).
22. R. W. Heckel, "Analysis of Powder Compaction Phenomena," *AIME Trans.,* **221** (5) 1001–1008 (1961).
23. D. B. Leiser and O. J. Jr., Whittemore, "Compaction Behavior of Ceramic Particles," *Amer. Ceram. Soc. Bull.,* **49** (8), 714–717 (1970).
24. A. R. Jr. Cooper and L. E. Eaton, "Compaction Behavior of Several Ceramic Powders," *J. Amer. Ceram. Soc.,* **45, 99,** 97–101 (1962).

25. K. Kawakita and K. H. Ludde, "Some Considerations on Powder Compression Equations," *Powder Technol.*, **4,** 61–68 (1970/1971).
26. W. D. Kingery, "Pressure Forming of Ceramics, *Ceramic Fabrication Processes* W. D. Kingery, ed., M.I.T. Press, Cambridge, 1958, pp. 55–61.
27. A. K. Chattopadhyay and O. J. Jr. Whittemore, "Powder Compaction in Systems of Bimodal Distribution," *Amer. Ceram. Soc. Bull.*, **52** (7), 575–577 (1973).
28. H. S. Hartmann, "Packing and Compaction in Dies of Spherical Particles," Sc.D. Thesis, Massachusetts Institute of Technology, Cambridge, 1964.
29. D. Train, "Transmission of Forces Through a Powder Mass During the Process of Pelleting," *Trans. Inst. Chem. Eng.*, **35** (4), 258–266 (1957).
30. F. H. McRitchie, "A Device for Determining Pressure Distribution in Dry Pressing Refractories," *Amer. Ceram. Soc. Bull.*, **43,** (7), 501–504 (1964).
31. A. R. Jr. Cooper and W. H. Goodnow, "Density Distributions in Dry-Pressed Compacts of Ceramic Powders Examined by Radiography of Lead Grids," *Amer. Ceram. Soc. Bull.*, **41** (11), 760–761 (1962).
32. E. G. Schwartz and A. S. Weinstein, "Model for the Compaction of Ceramic Powders," *J. Amer. Ceram. Soc.*, **48** (7), 346–350 (1965).

27

Adhesion Forces in Agglomeration Processes

H. Rumpf
H. Schubert

Agglomeration occurs naturally in powders because of adhesion forces that always act between fine particles. In some important cases agglomerates (granules) are formed intentionally by the addition of a liquid or binder. Because the behavior of powders is so strongly affected by agglomeration processes, this chapter focuses on the nature of agglomeration forces and the factors that affect these forces.

GENERAL CONSIDERATIONS

For particles larger than 1 cm, gravity forces acting on particles are much larger than the natural adhesion forces between particles. If the particle size is reduced, the gravity force decreases very rapidly (by the third power of particle diameter), while natural adhesion decreases approximately by the first or second power of diameter. With 1 μm particles, for example, van der Waals adhesion forces are about six orders of magnitude larger than gravity. Figure 27.1 shows this feature. Gold spheres of 10 μm diameter are adhering to a glass surface by van der Waals forces only. They stick as they come together, with negligible influence of gravity force. This is the world of fine particles; their behavior depends much more on surface phenomena than on volume forces.

To overcome adhesion forces by volume forces, we have to apply high centrifugal forces of about 10^5 to 10^6 times gravity. The result of such an experiment is shown in Figure 27.2. Gold particles of about 10 μm diameter, adhering to an originally plane anthracene surface, were submitted to high centrifugal forces until some of them flew off. Before flying away they plastically deformed the anthracene surface by van der Waals adhesion forces only.

Other adhesion mechanisms are realized by liquid or solid bridges. In the example shown in Figure 27.3 the solid bridges were formed during drying by crystallizing salts.

The principal bonding mechanisms of agglomerates are as follows:

1. Electrostatic.
2. Van der Waals.

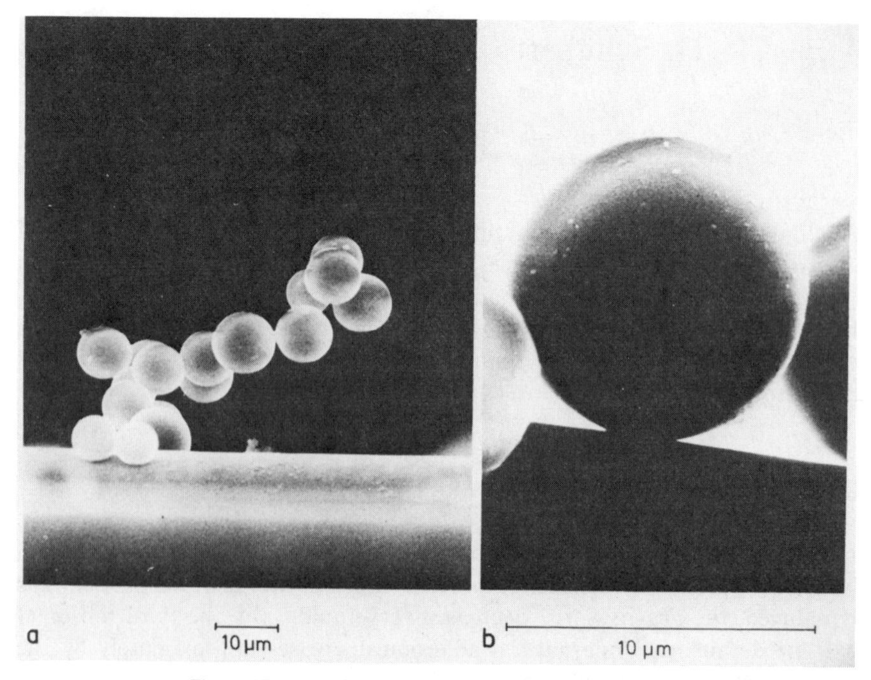

Figure 27.1. Gold spheres on glass fiber (Stereoscan).

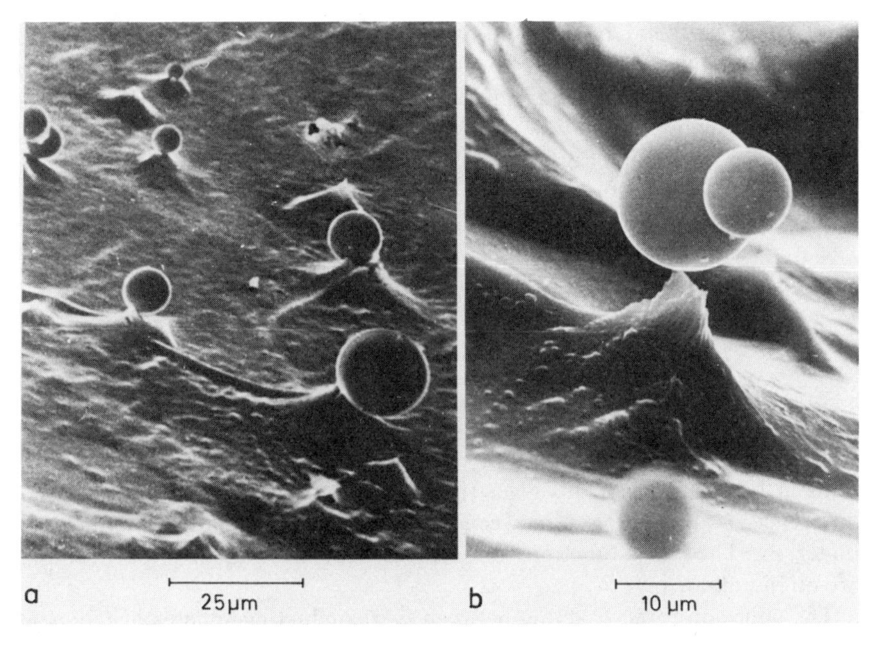

Figure 27.2. Gold spheres on an anthracene surface (Stereoscan).

3. Liquid bridges.
4. Capillary liquid.
5. Viscous binders.
6. Solid bridges.

As an approximate rule, the adhesion forces increase in order from 1 to 6. Chemical bonds are not specifically listed because they normally do not

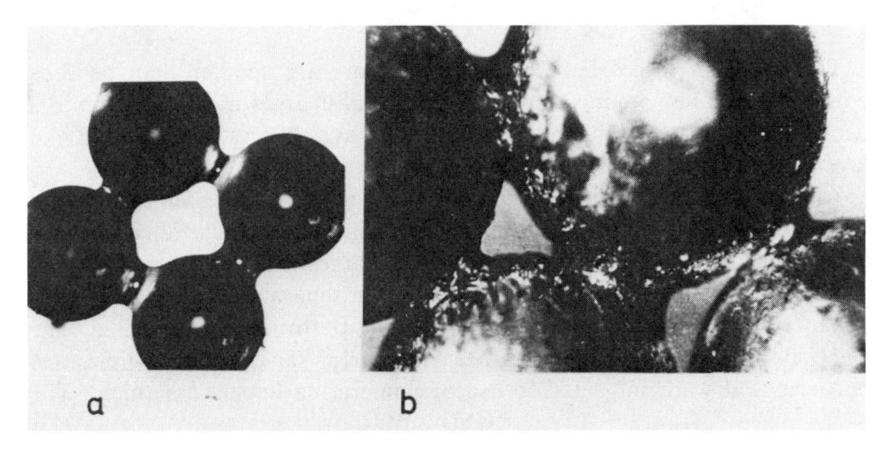

Figure 27.3. Bridges between spheres: (*a*) liquid bridges and (*b*) salt-crystal bridges.

occur unless a solid bridge is formed. An exception is paper, where the fibers adhere by hydrogen bonds. Chemical bonds of course can be effective between solid surfaces and liquid and can cause adhesion by viscous binders. With liquids of low viscosity, surface tension and capillary phenomena are responsible for adhesion. The bonding mechanisms listed above are discussed in detail in the sections that follow.

BONDING MECHANISMS

Electrostatic Forces

When two different materials come into contact, electrons tend to flow from one to the other because of differences in the electronic work functions at both surfaces. This gives rise at equilibrium to a contact potential difference U that ranges between 0 and 0.5 V, depending on the two materials. The work functions of surfaces depend on local impurities and are often unknown.

The adhesion force F acting between two conducting spheres of radius R is given by[1]

$$F = \frac{1}{2} \epsilon \epsilon_0 \pi \frac{U^2 R}{a_0}$$

where ϵ = dielectric constant of the gas
ϵ_0 = absolute dielectric constant of vacuum
a_0 = distance between the two surfaces in contact

a_0 is assumed to be on the order of interatomic dimensions (e.g., 4×10^{-8} cm). Similarly, expressions exist for sphere–plane surfaces and plane–plane surface interactions.[1]

The adhesion forces between nonconductors are smaller than between conductors. In nonconductors the accumulated charges may extend up to a depth of about 1 μm, while conductors may have charges concentrated in a layer of a few angstroms at the surface.

After a quick separation of particles, half of the charge remains on the separated surfaces, giving a surplus charge that may be neutralized in time according to the conductive conditions. The largest surface charge densities are realized if particles are contacted and separated many times, for instance, by impact in pneumatic conveying or by impact grinding. The electrical charge distribution has been measured[2] for quartz and limestone particles that were subjected to a deagglomerating impact treatment. The maximum surplus charge density found was around 100 elementary charges per square micrometer, which corresponds to a potential gradient of 20,000 V/cm at the surface. Fracture also produces surplus charges on the new surfaces.

Spherical particles of opposite surplus charge densities σ_1 and σ_2 attract each other according to Coulomb's law:

$$F = \frac{\pi}{4\epsilon_0} \frac{\sigma_1 \sigma_2 x^2}{[1 + (a/x)]^2}$$

where x = diameter of the spheres
a = distance of separation between the spheres

With ideal insulators, this formula also gives the adhesion forces in the case of contact.

Van der Waals Forces

When a chemist speaks of van der Waals forces, he thinks of small bonding forces between molecules that depend on the distance according to the -7 power law. We speak of the same forces, the London-van der Waals dispersion forces, which are due to fluctuating dipoles and which are found with every substance. These forces can be superimposed to a first approximation. Consequently, if two solid bodies are separated by a distance a between their surfaces, the van der Waals molecular forces can be integrated over the whole bodies. This yields the van der Waals adhesion force, according to the Hamaker theory. If the bodies are spheres with radii R, the van der Waals adhesion force F is inversely proportional to the square of the distance a between their surfaces according to the formula:

$$F = \frac{\hbar\bar{\omega}}{16\pi} \frac{R}{a^2}$$

where $\hbar\bar{\omega}$ is the Lifshitz-van der Waals constant, which ranges between 1 and 10 eV depending on the materials in contact. Thus van der Waals adhesion forces are not short-range chemical bonds between the surface molecules only. They depend on the whole solid continuum near the surfaces and also on the properties of a fluid phase between the solid bodies. They are still measurable up to distances a of about 500–1000 A (far above the effective range of single chemical bonds). Lifshitz has derived a so-called macroscopic theory of van der Waals adhesion forces, applying Maxwell's equations to the electrodynamic fields in the solid continua. A study of the van der Waals interaction is given by Krupp.[1]

Liquid-Bridge Forces

The best information is available for the liquid-bridge bonding mechanism. The shape of a liquid bridge is such that the capillary pressure in the bridge is the same in the whole volume (if gravity is negligible). This means that the mean radius of curvature is the same at all points of the surface. We have evaluated the exact solution of the corresponding differential equation for all radially symmetrical geometries of the contacting bodies with variable distances, bridge volumes, and contact angles. The results are available in the form of a set of diagrams.[3,4]

Figure 27.4 gives the results of calculation for the case of two spheres where the contact angle δ is zero. Here, F is the adhesion force, γ is the surface tension, and x is the diameter of the spheres. The quantity $F/(x \cdot \gamma)$ is a dimensionless adhesion force F_H that can be conveniently plotted as a function of a/x, where a is the distance between the two spheres. Curves for different V_l/V_s ratios are plotted, where V_l and V_s are the volumes of liquid and the two spheres, respectively. The maximum F_H is equal to π, when a/x is zero and V_l/V_s is zero. With increasing V_l/V_s the force–distance curves become less inclined. For V_l/V_s greater than 0.1, the force is nearly independent of distance.

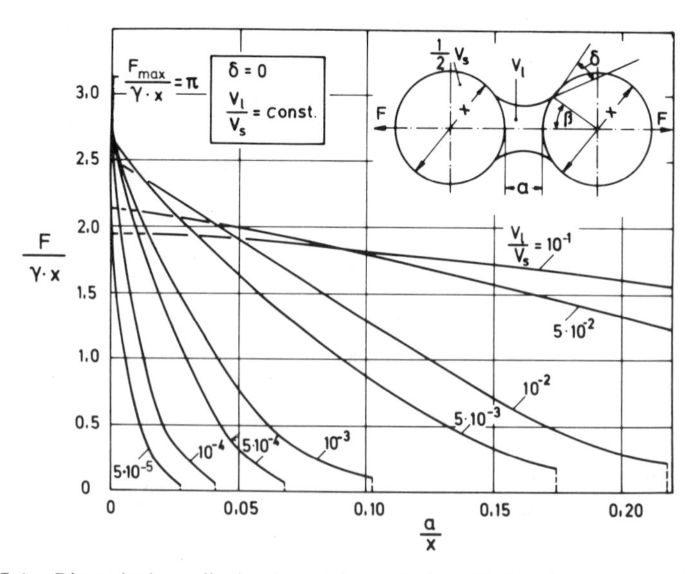

Figure 27.4. Dimensionless adhesion force $F/\gamma \cdot x$ of a liquid bridge between two spheres as a function of the distance ratio a/x.

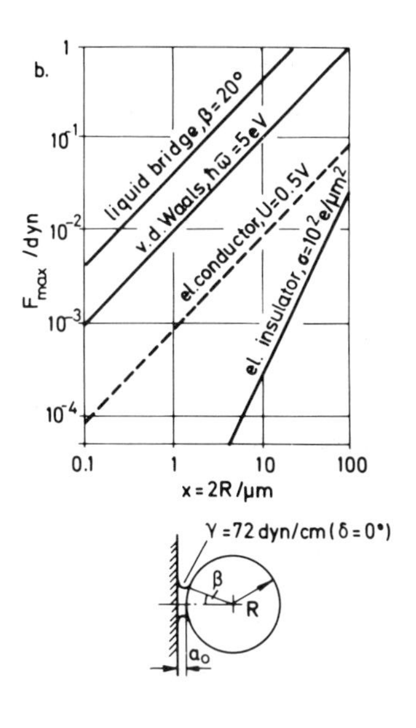

Figure 27.5. Adhesion force F_{max} of liquid bridge, van der Waals, and electrostatic interaction between a smooth sphere and a half space as a function of the sphere diameter x. The surfaces of the two interacting bodies are in contact, $a_0 = 4 \times 10^{-8}$ cm.

Comparison of Adhesion Forces

In Figure 27.5 the calculated adhesion force is plotted against sphere diameter for a sphere adjacent to a plane surface. The Lifshitz-van der Waals constant $\hbar\bar{\omega} = 5$ eV is relatively high; the contact potential difference $U = 0.5$ V and the surface charge density $\sigma = 100$ C/μm^2 are both maximum values; the liquid-bridge angle $\beta = 20°$ gives a medium value of the liquid-bridge force. The liquid-bridge forces are about four times as large as the van der Waals forces, which are by an order of magnitude greater than electrostatic adhesion forces between conductors, due to contact potential, and these are again about 100 to 10 times as great as adhesion forces due to maximum surplus charges of opposite sign.

Solid Bridges

Solid bridges are formed by crystallizing salts or by sintering. The strength of agglomerates due to crystallizing salt depends sensitively on the drying conditions. Pietsch's experiments[5] have illustrated rather comprehensively the different tendencies. We are presently investigating these effects under still more defined drying conditions.

Sintering is a field of research of its own. The different sintering mechanisms have been studied intensively by many researchers. We are

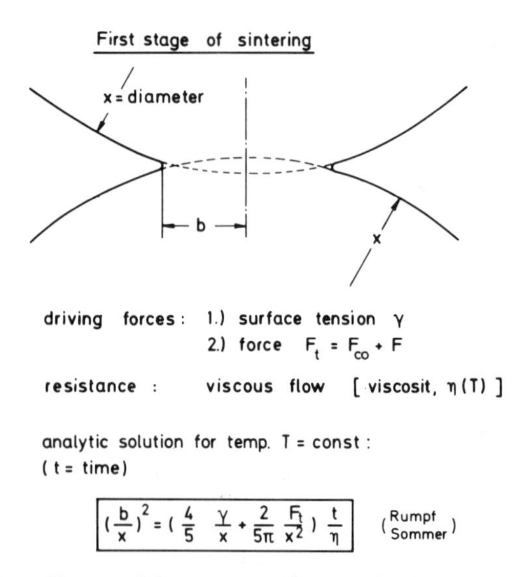

Figure 27.6. Sinter neck between two spheres at the first stage of sintering.

interested in the first stage of sintering, where adhesion forces are reinforced by sintering effects. If, for instance, a powder with a relatively low melting point is stored for a long time, the particles stick together and the flowing properties of the powder change in an often very unwelcome way.

In this first stage of sintering, the driving forces are surface tension γ and a force F_t. The force F_t can be the sum of a compression force F_{co} and the adhesion force F_a due to van der Waals (or liquid-bridge) adhesion around the sinter neck. F_a changes with the growth of the sinter neck, but this may be neglected in the very first stage of sintering.

The resistance is usually related to viscous flow or creep, which is expressed by the viscosity η. We are interested in the growth with time of the sinter neck radius b (Figure 27.6) between two spheres of diameter x. The temperature is constant, and the variables are b, x, γ, η, F_t, and t. From a dimensional analysis it is found that

$$\frac{b}{x} = f\left(\frac{\gamma \cdot t}{\eta \cdot x}, \quad \frac{F_t \cdot t}{\eta \cdot x^2}\right)$$

The analytical solution is based on a differential energy balance. In the time element dt, the dissipational energy due to viscous flow must be equal to the decrease of surface energy plus the work done by F_t. The exact calculation gives

$$\left(\frac{b}{x}\right)^2 = \left(\frac{4\gamma}{5x} + \frac{2}{5}\frac{F_t}{\pi x^2}\right)\frac{t}{\eta}$$

The factor $\frac{4}{5}$ differs from Frenkel's[6] factor, which is $\frac{3}{4}$. The difference is practically unimportant. However, Frenkel's derivation of the velocity field in the sphere and sinter neck is not exact in the sense of a rigorous theory.

If the force F_t is only van der Waals adhesion, we get

$$\frac{F_t}{x^2} = \frac{F_{vdW}}{x^2} = \frac{\hbar\bar{\omega}\ \mathrm{p}}{32\pi a_0^2}\frac{1}{x}$$

$$\left(\frac{b}{x}\right)^2 = \left(\frac{4}{5}\gamma + \frac{\hbar\bar{\omega}}{80\pi^2 a_0^2}\right)\frac{t}{x\eta}$$

The two terms in parentheses have a fixed relation, independent of x. For $\gamma = 500$ dynes/cm, $\hbar\bar{\omega} = 5$eV, and $a_0 = 4 \times 10^{-8}$ cm, the relation of the surface force term ($\frac{4}{5}\gamma$) to the adhesion force term ($\hbar\bar{\omega}/80\ \pi^2 a_0^2$) is 63:1. The van der Waals adhesion can therefore be neglected.

A compression force ($F_{co} \approx F_t$) has the same influence as the surface energy, if

$$\frac{F_{co}}{x^2} = 2\pi\frac{\gamma}{x}$$

For $\gamma = 500$ dynes/cm we get

$$\frac{\epsilon}{1-\epsilon}\cdot P_{\text{isost}} \approx \frac{F_{co}}{x^2} = \frac{3.14 \times 10^2}{x/\mu\text{m}}\ \frac{N}{\text{cm}^2}$$

This equation allows an assessment of whether surface energy or compression forces have a greater influence on the early stage of sintering. If, for instance, the particles are of 100 μm size and $\epsilon = 0.5$, the influence of F_c is greater than that of γ, if an isostatic pressure is applied such that $F_{co}/x_2 > 3.14\ N/\text{cm}^2$. For a calculation of the sinter neck radius, the viscosity η should be known, but this value is generally not obtainable.

BONDING MECHANISMS AND STRENGTH OF AGGLOMERATES

For moist agglomerates having less than 30% of the void spaces filled with liquid, a theoretical tensile strength can be calculated from the mean value of the number of contact points times the adhesion force component in the tensile strength direction. The average adhesion force depends on the average a/x value in the system. Experimental values of strength indicate a value of 0.05 for the average a/x value. This is shown in Figure 27.7 in the region of $0 \leq S \leq 0.3$.

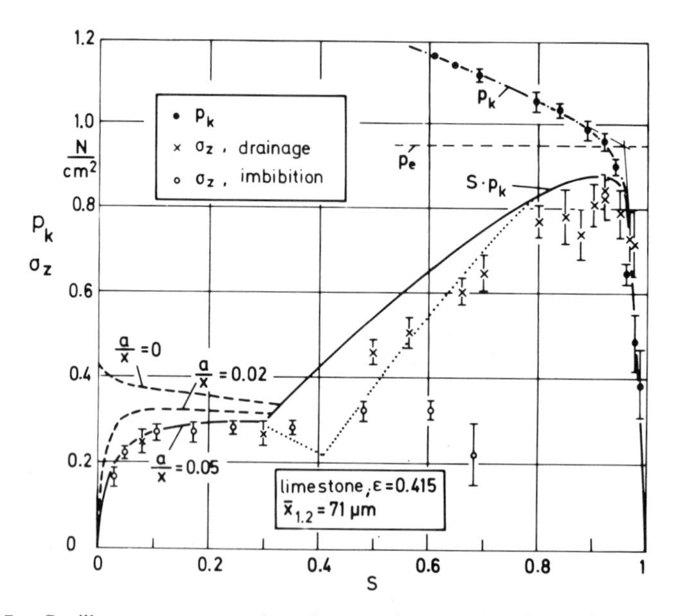

Figure 27.7. Capillary pressure p_k and tensile strength σ_z as a function of the liquid saturation S.

In reality, of course, the distance has a distribution. At some points there is real contact combined with elastic and probably plastic deformation. At other points the distance is still larger but small enough to transmit adhesion forces. Each contact is characterized by a force–elongation curve. If the whole agglomerate is strained, the total force, transmitted in a certain cross-sectional area, is the sum of the contact forces. At a certain total elongation a maximum total force T_{max} is reached. The breakage occurs in the cross-sectional area with the smallest T_{max}. This gives the tensile strength. In the simplified theory it is supposed that all contact forces in the normal direction are equal in the case of breakage. It has been shown experimentally that this simplified theory is applicable to large monosized spheres and the liquid-bridge bonding mechanism.[7]

If more than about 80% of the void space between the particles is filled with liquid (capillary state, liquid saturation $S > 0.8$), a capillary suction (pressure P_k) is formed in the liquid space and the agglomerate is held together by the outer pressure. The tensile strength σ_z equals the product P_k times S.

We have simultaneously measured capillary pressure in the agglomerate, saturation, and tensile strength with wet limestone powders. In Figure 27.7 capillary pressure (P_k) and tensile strength of agglomerates (σ_z) made from

71 μm limestone particles are plotted against saturation S. The measured values of tensile strength just reach up to the theoretical curve $(S) \cdot (P_k)$, calculated from the measured capillary pressure. Figure 27.8 also shows that the capillary liquid mechanism, $S > 0.8$, gives a maximum tensile strength, which is about three times as great as the strength in the range $0 > S > 0.3$ (the liquid-bridge state).

The measured tensile strength is different, depending on whether the saturation is realized by drainage or by imbibition. The capillary pressure after drainage is much higher than after imbibition at comparable S (Figure 27.7), because the porous system is a sequence of caverns and necks. With drainage the capillary pressure depends on the radii of curvature in the necks; with imbibition it depends on the radii of curvature in the caverns.

The tensile stress/strain behavior of wet agglomerates is shown in Figure 27.8 for limestone pellets made from 65 μm particles. The maximum stress and maximum strain both increase with increasing saturation. If we compare the curves for drainage with $S = 0.65$ and those for imbibition with $S = 0.68$, we see that the somewhat smaller maximum stress in the case of imbibition is reached at a much greater strain. When, after imbibition, the agglomerate is strained, the liquid situation is changed into drainage. So approximately the same stress is reached, combined with much greater strain. This can quantitatively be derived from capillary-pressure hysteresis curves.

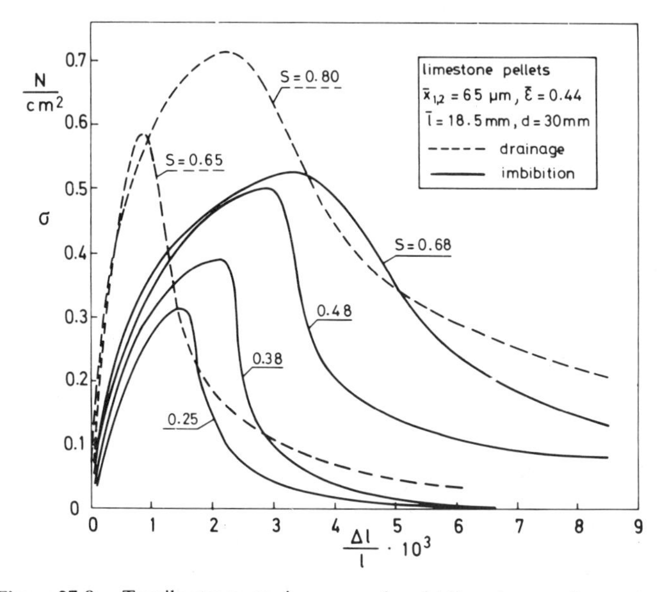

Figure 27.8. Tensile stress–strain curves of moist limestone agglomerates.

In general, tensile stress, shear stress, and strain experiments are necessary to understand and predict the mechanics of agglomerates.[8-9]

OTHER INFLUENCES ON ADHESION FORCES

Influence of Surface Roughness on the Adhesion Forces

In reality adhesion forces can be considerably changed by surface roughness. The influence of surface roughness is maximum with van der Waals forces, since they depend sensitively on the microgeometry in the contact region, while electrostatic adhesion of ideal insulators with opposite surplus charges is independent of the contact geometry and distance as long as $a/x \ll 1$.

The quantitative relations are demonstrated in Figure 27.9 for a sphere with radius R supporting a half-spherical roughness peak of radius r, which is in contact with the half space. The calculated adhesion force is plotted against the radius r of the roughness peak. The van der Waals forces are calculated for $R = 0.5$, 5, and 50 μm. Each of the three curves has a sharp minimum that lies some orders of magnitude below the maximum value for the smooth sphere with $r = 0$ or $r = R$. At the minimum both the small and the large sphere contribute to the adhesion force. On the left side of the minimum, with smaller r, the attraction due to the large sphere increases,

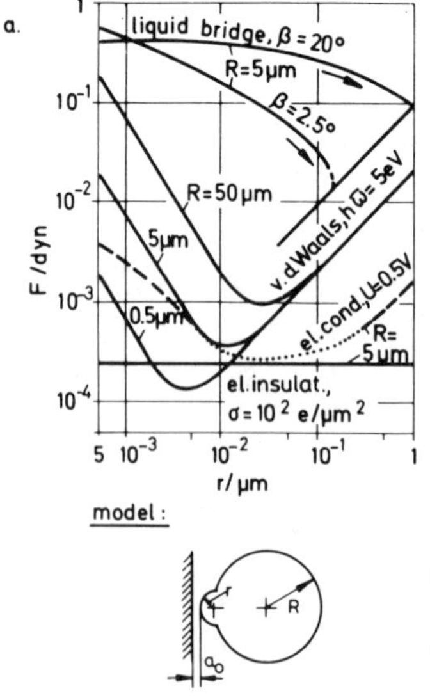

Figure 27.9. Adhesion force F of liquid bridge, van der Waals, and electrostatic interaction between a sphere and a half space as a function of the radius r of a roughness peak.

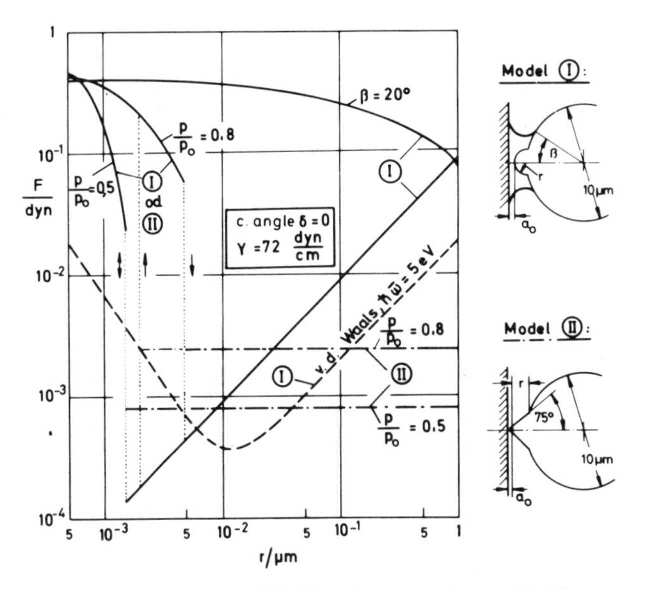

Figure 27.10. Adhesion force F of liquid bridges between sphere and half space as a function of the radius r of a roughness peak. Parameter is the relative vapor pressure p/p_0.

since the distance r is decreasing. On the right side of the minimum, the large sphere has no more influence and the van der Waals adhesion force increases in proportion to r. We see that the van der Waals adhesion can drop to very low values, below electrostatic forces. The same lowering of van der Waals adhesion is produced if, instead of roughness peaks, fine particles of 10^{-2} to 10^{-1} μm are placed between the spheres. In practice this effect is utilized to prevent agglomeration of fine dry powders by adding a very fine silica powder of 10^{-2} to 10^{-1} μm particle size.

Liquid bridges are much less sensitive to roughness peaks if the bridge angle β is not too small. If β is very small, the liquid-bridge forces are also very sensitive to roughness peaks as shown in Figure 27.10 in the case of very small liquid bridges due to capillary condensation.

Again the adhesion force is plotted against roughness peak radius r. The radius of the large sphere is $R = 5$ μm. The liquid-bridge force for $\beta = 20°$ and the van der Waals force have been taken from Figure 27.9. The volume of the liquid bridge formed by capillary condensation depends on the relative vapor pressure p/p_0 and is calculated according to Kelvin's law. At a very small radius r, the liquid bridge is formed between the large sphere and the plate. At a certain size r of the roughness peak, the bridge between the large sphere and the plate bursts and only a small bridge between the small sphere and the plate remains. The adhesion force drops abruptly and rises again with increasing r. Here we see again a great influence of roughness peaks. We recognize also that in dry systems with relative humidities above 50% the liquid-bridge adhesion due to capillary condensation may be larger than van der Waals adhesion with small roughness peaks. Complete wetting has been assumed, that is, a contact angle $\delta = 0$ and the surface tension of water $\gamma = 72$ dynes/cm. In reality the two adhesion forces superimpose, whereby the van der Waals forces can be somewhat smaller because of the water volume.

As a consequence of these calculations, we conclude that adhesion is mainly due to liquid-bridge forces and van der Waals forces if there are no solid bridges or viscous binders. Only under special roughness geometries, where van der Waals forces have a mimimum, electrical adhesion forces may reach the same order of magnitude.

If the particles are not in contact and the distance a increases, liquid bridges collapse at distances $a/x < 1$, depending on bridge volume. Van der Waals attraction forces drop practically to zero for $a > 1\ \mu$m. Thus at distances larger than 1 μm, that is, $a/x > 1$, only electrostatic attraction or repulsory forces are still effective. They play an important role in agglomeration kinetics of particles suspended in a gas by forcing particles together, and in diffusion, turbulence, and convection of particles.[2]

Influence of Elastic Deformation

The existence of adhesion forces causes bodies to undergo elastic deformation at their contact points. The elastic deformation occurs to the extent where the elastic repulsion force balances the attractive force. This elastic deformation causes the effective adhesion force to increase because the force required to separate the contacting surfaces must overcome the sum of the attractive and repulsive forces. Dahneke[10] has analyzed the influence of elastic deformation on adhesion for a few specific cases and derived approximate relationships for the separation-force enhancement. New calculations, however, show that the results given by Dahneke should be reduced by a factor of 8. Hence it follows that the influence of elastic deformation on particle adhesion can be neglected in most cases.

Influence of Plastic Deformation

If plastic deformation occurs at the point of contact between two bodies, a permanent surface of contact is established. The van der Waals force of adhesion is greater for two plane surfaces in contact than for point contacts. The additional van der Waals attractive force due to plastic deformation is equal to $f_{\mathrm{pl}} \cdot P''_{\mathrm{vdW}}$, where P''_{vdW} is the force of attraction per unit area of contact between two flat surfaces, and f_{pl} is the surface area of contact. The quantity P''_{vdW} is given by[1]

$$P''_{vdW} = \frac{\hbar\bar{\omega}}{8\pi^2 a_0^3}$$

Since the additional adhesion force from plastic deformation depends on the area of permanently established contact (due to plastic deformation), this area must be determined. An eleastic deformation of a particle produces a contact circle at the contact point. According to the theory of Hertz, the normal stress σ_1 within the contact circle varies from zero at the circumference to a maximum value p_0 in the center according to the formula

$$\left(\frac{\sigma_1}{p_0}\right)^2 = 1 - \frac{c}{C_0}$$

where C_0 = radius of the contact circle
c = distance from the center
σ_1 = stress at position c

If p_0 exceeds a certain limit, which we denote as p_{pl}^{H}, plastic deformation occurs in the center.

The stress at which plastic deformation begins with small particles and roughness peaks is much higher than the macroscopic yield strength. Faulhaber,[11] using a micro-compression test apparatus, found a value for p_{pl}^{H} of around 10^6 N/cm² for 10 μm quartz glass spheres.

As the compression force increases, p_0 becomes greater than p_{pl}^{H} and the area f_{pl} of the plastically deformed zone increases. The area f_{pl} is given by

$$f_{pl} = \frac{F_t}{p_{pl}^{H}} \cdot \frac{1 - (p_{pl}^{H}/p_0)^2}{1 - \frac{1}{3}(p_{pl}^{H}/p_0)^2}$$

where F_t is the total compressive force. A pseudo-yield strength p_{pl} can be defined as F_t/f_{pl}. It follows from the above equation that

$$p_{pl} = p_{pl}^{H} \frac{1 - \frac{1}{3}(p_{pl}^{H}/p_0)^2}{1 - (p_{pl}^{H}/p_0)^2} \qquad (1)$$

If the plastic part of deformation is much greater than the elastic one, $p_0/p_{pl}^{H} \gg 1$ and hence $p_{pl} \approx p_{pl}^{H}$. For small glass spheres and predominant plastic deformation, $p_{pl} \approx p_{pl}^{H} \approx 10^6$ N/cm³. Similar values can be expected for crystalline materials, since in very small volumes the yield strength is much greater than the macroscopical yield strength owing to dislocation movement. In small volumes below 1 μm extension, there is very little chance for dislocation existence and dislocation movement. The yield strength in this case approximates the molecular shear strength, while the macroscopic yield strength of metals is on the order of some 10^4 N/cm^2.

If in addition to any adhesion force F_0 an outer compression force F_{co} is applied, the total plastic deformation is

$$f_{pl} = \frac{F_t}{p_{pl}} = \frac{F_0 + F_{co}}{p_{pl}}$$

In contrast to the case with elastic deformation (due to F_{co}), this plastically deformed area remains if the compression force is removed. It causes an additional van der Waals attraction force $f_{pl} \cdot P''_{vdW}$.

In the case of separation, the separation force must overcome the total adhesion force

$$\begin{aligned} F &= F_0 + f_{pl} P''_{vdW} \\ &= F_0 \left(1 + \frac{P''_{vdW}}{p_{pl}}\right) + F_{co} \frac{P''_{vdW}}{p_{pl}} \end{aligned}$$

With $\hbar\bar{\omega} = 1$ to 2 eV as an estimated value for limestone and $a_0 = 4 \times 10^{-8}$ cm we get

$$P''_{vdW} = (0.3\text{–}0.6)\ 10^4\ \text{N/cm}^2$$

$$\frac{P''_{vdW}}{p_{pl}} \leq 10^{-2} \qquad \text{for hard materials}$$

$$F \approx F_0 + F_{co} \frac{P''_{vdW}}{p_{pl}} \tag{2}$$

Figure 27.11 shows some of the first results of measurements executed by Schutz of our Institute with 60 μm limestone particles. A set of particles were laid on a plane-polished steel substrate that could be fixed in a centrifuge in two opposite situations. In one situation the particles were pressed against the substrate by the centrifugal force, which then equaled the compression force F_{co}. After that, the substrate was fixed in the opposite situation and the particles were separated from the substrate by the centrifugal force, which then equaled the adhesion force F.

Each curve represents one measured distribution function of adhesion forces. The speed of the centrifuge, and therefore the separating centrifugal force, is increased stepwise. This force is plotted as the abscissa. At each step the percentage of particles that have left the substrate is measured (ordinate). They are all particles with a smaller adhesion force than the force indicated at the abscissa; the compression force F_{co}. The curve with

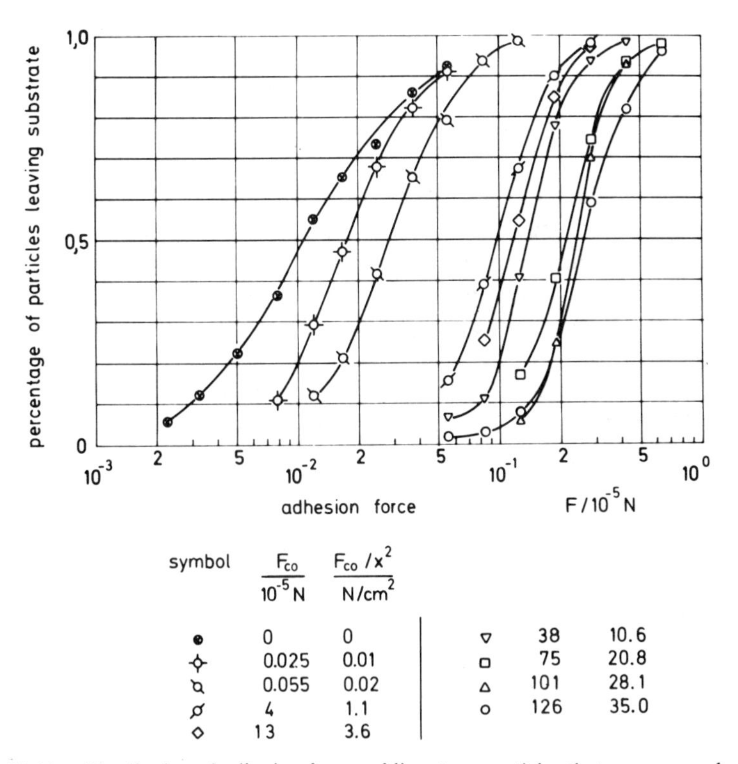

Figure 27.11. Distribution of adhesion forces of limestone particles that were pressed against a substrate by a centrifugal force.

$F_{co} = 0$ is flat. The particles have been spread very loosely on the substrate. In this case a one-point contact is very likely. Since the limestone particles are irregularly shaped, a great variety of contact geometries is possible. Therefore, the standard deviation of the measured adhesion force distribution curve is great. If the particles are pressed against the substrate by an increasing compression force, the distribution curves get steeper and shift to higher adhesion forces. The change in the inclination of the curves indicates that the particles are changing their position, probably from one-point to more-point contacts. The contact geometry becomes more homogeneous. Above $F_{co} = 4 \times 10^{-5}$ N the curves remain parallel. It can be assumed that with higher compression forces the particles do not change their position any more. $F_{co} = 3.8$ dynes is about 400 times the median value of F for $F_{co} = 0$.

Such compression forces occur in practice when particles are compressed in a packing. If the packing is stochastically homogeneous with porosity ϵ

and the compression is isostatic, the relation between isostatic pressure P_{isot} and the compression force F_{co} at the contact points is

$$P_{isot} = \frac{1 - \epsilon}{\epsilon} \cdot \frac{F_{co}}{x^2}$$

In a bulk powder with $\epsilon \simeq 0.5$, we get $P_{isot} \simeq F_{co}/x^2$.

The values F_{co}/x^2 are given in Figure 27.11. Pressures between 1 and 10 N/cm^2 are very common in bulk powders.

In Figure 27.12 the median values F_{50} of the distribution curves in Figure 27.11 are plotted against the compression force F_{co}. At $F_{co} = 0$ the adhesion force is $F_{50}(0) = 10^{-7}$ N. It increases very steeply with relatively small compression forces until the threefold adhesion force $F_{50} = 1.2 \times 10^{-6}$ N is reached. This indicates that in this first region the contact position changes from one-point to more-point contact. Above $F_{50} \simeq 10\ F_{50}(0)$ the adhesion force increases linearly with F_{co}, with

$$\frac{dF_{50}}{dF_{co}} = 0.8 \times 10^{-3}$$

If the increase of adhesion is due to plastic deformation, we derive from

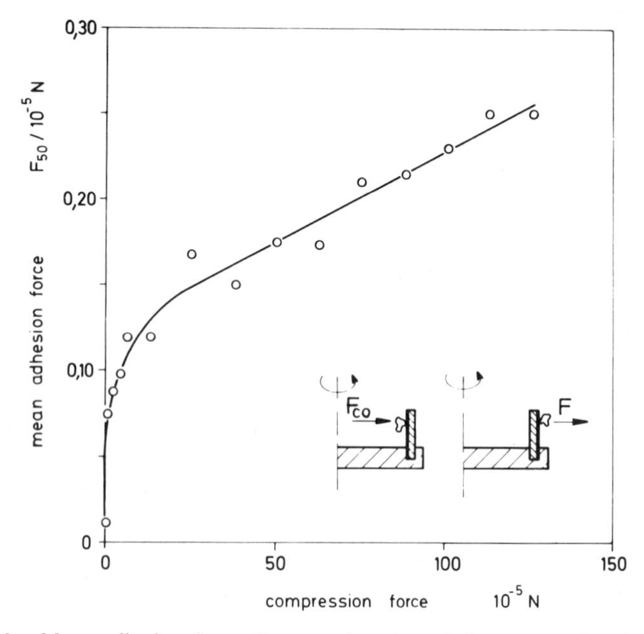

Figure 27.12. Mean adhesion force F_{50} as a function of the compression force F_{co} for limestone particles.

Equation 1

$$\frac{dF_{50}}{dF_{co}} = \frac{P''_{vdW}}{p_{pl}} = 0.8 \times 10^{-3}$$

with $P''_{vdW} = (0.3\text{–}0.6) \times 10^4\ \mathrm{N/cm^2}$ we get

$$p_{pl} = (3.8\text{–}7.5) \times 10^6\ \mathrm{N/cm^2}$$

This value is in satisfactory agreement with Equation 1 for $p^H_{pl} = 10^6$ $\mathrm{N/cm^2}$, measured with quartz–glass spheres.[11] $p_{pl} > p^H_{pl}$ indicates an elastic deformation superimposed on the plastic deformation at the contact edges of the particles. This remaining elastic stress diminishes the effect of the van der Waals adhesion pressure in the contact area. From Equation 1 $p^H_{pl}/p_o =$ 0.9 to 0.95 can be estimated.

In tableting and briquetting F_{co}/x^2 reaches values on the order of some $10^3\ \mathrm{N/cm^2}$ instead of 126 $\mathrm{N/cm^2}$ as in the experiments referred to above. It is not likely that the linear function $F_{50}(F_{co})$ can be extrapolated to much higher values of F_{co} for, in addition to the elastic stress, grinding will take place at the contact edges or in the whole particle. It will be necessary to determine experimentally how adhesion forces change with high compression forces.

In general, the calculation of adhesion forces and the relevant influences indicates the tendencies. But they are restricted to geometrical models that deviate considerably from reality if the particles have an irregular shape. All measurements of adhesion forces of irregularly shaped particles show a broad distribution of adhesion forces. They often are log-normal distributed and range over one to two orders of magnitude.

SUMMARY

Agglomeration is due to different adhesion mechanisms. We have a most comprehensive knowledge of the liquid capillary mechanism in moist agglomerates with above 80% saturation. The liquid-bridge adhesion mechanism has also been very well studied theoretically. Van der Waals adhesion can be calculated for spherical particles but depends very sensitively on contact geometry and surface roughnesses. Electrostatic adhesion forces are normally smaller. An additional compression force increases particle adhesion by changing the particle position from one-point to more-point contact and, after that, by plastic deformation if the compression force is higher than about 400 times the original adhesion force. A new formula for the first stage of sintering is presented that can be used to assess the influence of a compression force in comparison with surface tension.

ACKNOWLEDGMENT

The authors are indebted to Dr. Sommer for his substantial contribution to the sintering calculations, W. Schutz for his measurements of adhesion forces, W. Muhr for his calculations of electrostatic forces, and K. H. Sartor and K. Schaber for their calculations of liquid bridges. This work was kindly supported by the Deutsche Forschungsgemeinschaft, Bonn-Bad Godesberg.

REFERENCES

1. H. Krupp, "Particle Adhesion, Theory and Experiment," *Adv. Colloid Interface Sci.*, **1** (2), 111–239 (1967).
2. K. Borho, "Agglomeration und Wandansatz bei Gas-Feststoff-Stromungen aufgrund elektrostatischer Aufladungen," *Chem.-Ing.-Tech.*, **45**, 387–391 (1973).
3. H. Schubert, "Untersuchungen zur Ermittlung von Kapillardruck und Zugfestigkeit von feuchten Haufwerken aus kornigen Stoffen," Thesis University Karlsruch, 1972.
4. H. Schubert, "Haftung zwischen Feststoffteilchen aufgrund von Flussigkeitsbrucken," *Chem.-Ing.-Tech.*, **46**, 333–334 (1974).
5. W. Pietsch, "The Strength of Agglomerates Bound by Salt Bridges," *Can. J. Chem. Eng.*, **47**, 403–409 (1969).
6. J. Frenkel, "Viscous Flow of Crystalline Bodies under the Action of Surface Tension," *J. Phys.*, **9**, 385–391 (1945).
7. H. Schubert, "Tensile Strength of Agglomerates," *Powder Tech.*, **11**, 107–119 (1975).
8. H. Rumpf, "Die Wissenschaft des Agglomerierens," *Chem.-Ing.-Tech.*, **46**, 1–11 (1974).
9. H. Schubert, W. Herrmann, and H. Rumpf, "Deformation Behavior of Agglomerates Under Tensile Stress," *Powder Tech.*, **11**, 121–131 (1975).
10. B. Dahneke, "The Influence of Flattening on the Adhesion of Particles," *J. Colloid Interface Sci.*, **40**, 1–13 (1972).
11. F. R. Faulhaber, Thesis University Karlsruhe 1966.

28

Strength and Microstructures of Dried Clay Mixtures

W. O. Williamson

This account, based on published and unpublished information, reviews the relations between the transverse strength (modulus of rupture) of a dried fine-grained clay, with or without nonplastic additions, and the corresponding microstructures. The nonplastic additions were coarse or fine quartz, tremolitic talc, chrysotile asbestos, glass fragments, glass spheres, and Madagascar and Mexican graphite. All except the glass spheres had been comminuted industrially.

SOME FACTORS AFFECTING DRY STRENGTH

The bonds broken during dry strength tests are not commonly intracrystalline but are between clay-mineral platelets or their aggregates, or between these and nonplastic grains. Such grains, except those of graphite for instance, are not usually fractured during tests. The nature of the bonds is discussed elsewhere.[1,2] Evidence is presented that the bonds, although of several types, are often ionic. The ions involved may remain incompletely dehydrated, even in test pieces dried at 105 to 110°C.

Transverse strength tends to increase with the number of bonds per unit bulk volume and with the extent of the surface over which an individual bond produces effective adhesion. Strong bonds need not imply strong test pieces, as they help to maintain open porous structures, which are weak because they contain abundant flaws.[1,2]

Transverse strength is influenced not only by bonds, but by the effectiveness of flaws in concentrating the applied stresses and thus breaking the bonds. A connection between this effectiveness and the porosity of the test piece may be anticipated. However, ions originally present in the interstitial water not only affect the porosity and pore characteristics of the dried test piece, but also contribute to the bonds. Thus bonds and the microstructures within which they exist are not independent variables.[1,2]

Dried clay-containing test pieces, with or without nonplastic additions, often have apparent porosities of 20 to 40%, and their strength commonly increases with decreasing porosity. Frequently, however, no obvious quantitative relation is discernible between strength and porosity.[1-3] Indeed, within limits, strength and porosity can increase together.[1,2,4] The absence of a simple or universal relationship is not surprising, as percentage of apparent porosity represents merely the total void space, in a constant volume of material, accessible to the fluid used in determining it. It neglects pore size, shape, and size distribution. For instance, a pore may have an overall shape that makes it unlikely that it will operate as a dangerous flaw, while the really dangerous flaws might be slotlike spaces between clay–mineral platelets, or packets of platelets, in its walls.[1,2] Moreover, the stepped edges of packets in contact with the basal planes of platelets in adjacent packets could constitute further regions of dangerous flaws, although the void spaces in these regions might contribute little to the total porosity.[1,2]

Most clay-mineral crystals are flat and thin. The same is true, in greater or lesser degree, for Madagascar graphite flakes, cleavage fragments of tremolite, and grains of ground quartz, glass, and Mexican graphite. Tremolitic talc and especially asbestos contain fibrous particles.

During extrusion these shape factors promote particle orientation and, as a result, pore orientation and interconnection. Dry strength is thus anisotropic.[1,2] Extruded test pieces are broken across, not along, planes or directions of particle and pore orientation, and thus higher, not lower, strength values are ascertained. These strength values may become higher still if the interstitial liquid in the plastic mixture contained dispersants, because these promote more intense particle parallelism and closer particle packing. The dispersants may have been deliberately added or may have resulted from reactions between the interstitial liquid and solid particles of feldspar, nepheline syenite, glass,[5] and so forth, which release alkali metal ions.

The interstitial solutions in clay–water systems are commonly somewhat acidic in the absence of extraneous sources of alkalies. However, interaction between the clay and the water occurs progressively and involves not only ion exchange, but also gradual chemical decomposition of the solid phases. The resulting changes in the nature and concentration of the ions produce corresponding changes in the degree of flocculation of the clay minerals and, ultimately, in the microstructure and the associated strength of the dried clay. Such changes are among the factors that cause dry strength to vary with the time and temperature of contact between the clay and water before and during the desiccation of the test pieces.[2,6] Thus the dry strength of even a monomineralic, homoionic, monodispersed clay may be expected to change with the time and temperature of aging in the plastic state,[2] although no reference to relevant experiments has been found.

Extrusion produces not only particle parallelism resulting from laminar plastic flow, but also slip bands, caused by compression within which the clay-mineral platelets lie at large angles to the surface, which is under tension during cross-breaking (Figure 28.1). In some circumstances these slip bands weaken the test piece. Planes or directions of particle orientation that are favorable for higher strength are not in these slip bands but instead are in the matrix. Those in the bands are not as favorably oriented and can cause fracture.

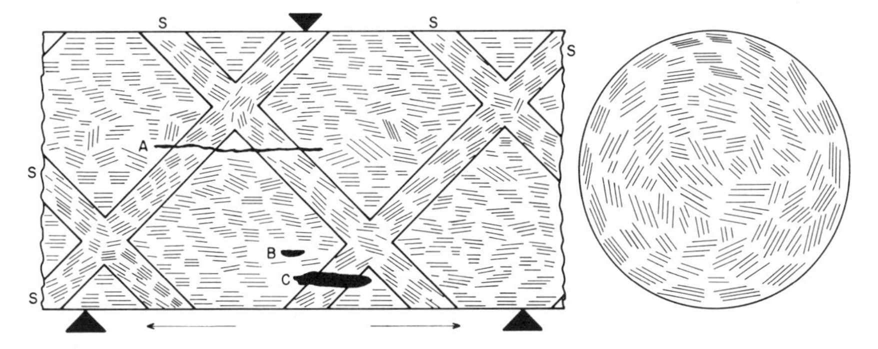

Figure 28.1. A schematic section containing the axes of an extruded clay cylinder (with corresponding transverse section). The cylinder is being broken by three-point loading, which produces tensile stresses in the directions of the arrows. Traces of the clay platelets are represented by lines.

METHODS AND MATERIALS

The preparative, extrusion, testing, microscopical, and fractographic techniques were essentially those already described,[3,6] but in addition some specimens were examined by scanning electron microscopy. The plastic bodies had water contents slightly below those at their sticky points. Data on the materials used, summarized here, are given more fully elsewhere. Similarly, the results of strength determinations appear in simplified form, although their validity was checked statistically in many instances.

The clay was largely kaolinite[3,6] with some micaceous minerals. Its particle-size distribution included 50 wt % < 1.0 μm esd (effective sieve diameter). An original water content of 39.6 wt % corresponded to a dry modulus of rupture of 775 psi.

The coarse and fine quartz particles were somewhat flat angular chips. A weight distribution[3,6] of the coarse quartz, on U.S. Standard Screens, was +140/Tr., −140 + 200/2.3%, −200 + 230/4.1%, −230 × 270/33.7%, −270/59.9%.

The two different samples[7] of finer quartz were quartz A (specific surface 2 m^2g^{-1}) and quartz B (specific surface 0.5 m^2g^{-1}). They had an average particle size of 1.1 and 4.2 μm, respectively.

The tremolitic talc used also contained considerable talc and antigorite, as indicated by X-ray diffraction.[8] Polarizing microscopy revealed many cleavage fragments of tremolite, with some fibrous mineral constituents. The particle-size distribution of the total material included 81, 43 and 12 wt % < 49, 15 and 1.3, μm esd, respectively.

The chrysotile asbestos had a wide range of fiber lengths, but 95% of its fibers were less than 13.5 mm long. The fibers were bundles of fibrils, each about 200 Å across, and were not very open structured.[4]

The crushed glass was composed of rather flat, sharp chips that, on U.S. Standard Screens, gave approximately +100/0, −100 + 200/75, −200/25%. The glass spheres had a similar size distribution.[5]

Weight distributions, on U.S. Standard Screens, of the Madagascar (flaky) and Mexican (nonflaky) graphites appear in Table 28.1.

MICROSTRUCTURES OF EXTRUDED CYLINDERS

The cylinders contained packets of mutually parallel clay-mineral platelets. The plane of mutual parallelism within each packet, which is the basal plane of the clay-mineral crystals, was angularly related to the corresponding planes in neighboring packets, although the angular differences were

Table 28.1 Screen analysis of Madagascar (flaky)[9] and Mexican (non-flaky)[10] graphites

Source	Percentage Weights Retained on U.S. Standard Screens								
	20	30	40	50	60	70	80	−70	−80
Madagascar	0.2	1.1	16.5	68.2	81.0	92.6	ND	6.8	
Mexico	ND	ND	8.3	46.5	79.5	88.8	97.9		2.3

often small. As an approximation the packets can be regarded as lying in a series of cylindrical shells, coaxial with extruded cylinder, with the widest shell having the diameter of the cylinder. The cylinder underwent a large reduction in diameter when forced through the circular orifice of the die. Thus there was no obvious tendency for the packets to lie on paraboloidal surfaces, convexly curved in the direction of flow, although this tendency was very apparent in certain cylinders extruded under other conditions.[11] More specifically, the clay-mineral packets tended to be parallel to the axis of the cylinder and, towards the exterior, parallel to the curved surface (Figure 28.1).

The added nonplastic grains were mostly much coarser than the clay-mineral crystals. Except for the glass spheres, the nonplastic grains tended to be flat, and the preferred orientation of their greatest cross-sectional areas resembled that of the major surfaces (basal planes) in the clay-mineral packets. Elongated or fibrous grains, abundant in tremolitic talc or asbestos, tended to parallelism with the axis of the cylinder, especially in its outer layer (Figure 28.1).

Slip bands were very variable in thickness and often less than 12 μm wide. They crossed the planar and linear parallel structures just described. Ideally, they coincided with the curved surfaces of two sets of cones coaxial with the extruded cylinder. Within each set the cones followed each other successively and each cone protuded into the one ahead. One set pointed in the direction of extrusion and the other in the opposite direction.[11] Thus, in a section containing the axis of the cylinder, the traces of the slip bands intersected at about 90° and met those of the curved surface at approximately 45°. The clay-mineral packets in the slip bands lay at small angles to their wall (Figure 28.1).

Electron microscopy did not show conclusively whether the slip bands terminated at the actual curved surface of the cylinder or beneath the very thin, highly oriented layer of clay-mineral platelets sometimes found at this surface. At the magnification of the optical microscope, the slip bands appeared to reach the surface itself.

THE INFLUENCE OF SLIP BANDS ON BREAKING BEHAVIOR

Whether or not the slip bands ended at or beneath the curved surfaces, fractography demonstrated that test pieces frequently failed along them.[3] Such bands contained slotlike pores between clay-mineral platelets or between their packets, which lay across the directions of the applied stresses and were thus likely to be sites for the initiation and propagation of the cracks that lead to failure. The measured transverse strength is probably affected, not only by the mere presence of slip bands, but by their periodicity and width and by variations of the microstructure within them. Among factors that might influence these characteristics are the mineralogy, particle-size distribution and water-content of the clay, the ratio between the diameters of the extruder reservoir and orifice, and the rate of extrusion. Relevant published information is minimal, although there is evidence that more rapid extrusion reduces the distance between the successive slip bands, while, at similar extrusion rates, the distance increases in coarser-grained bodies.[12]

Coarse quartz,[3] tremolitic talc,[8] asbestos,[4] and crushed glass,[5] when up to 25, 20, 25, and 25 wt % respectively, was added, produced corresponding increases in modulus of rupture of 4, 47, 27, and 9%, respectively. Reduction in the water content of the clay–quartz mixture from 39.6 to 29.3 wt % changed the strength increase from 4 to 16%. This was the only mixture for which the effects of varying the water content, as well as the proportion of nonplastic grains, were studied in detail.[3,6]

Strengthening resulted from the simultaneous operation of two mechanisms, while, for glass, a third was involved concurrently. Added nonplastic grains tended to reduce the number of slip bands per unit length of extruded cylinder, so that fewer dangerous flaws occurred on the surfaces under tension and the test piece appeared stronger. It was noticed that the addition of most nonplastics increased the pressure required to initiate extrusion.[6,9,10] This, presumably, implied that the plastic mass had become more resistant to the compressional stresses necessary for the formation of slip bands. Added glass, however, lowered the pressure required for extrusion[5] because it produced an interstitial alkaline solution that dispersed the clay and allowed it to flow more readily. As might be expected, slip bands remained relatively abundant unless the mixtures were very rich in glass (50 wt %).

Many nonplastic particles were large enough to project into or across those slip bands that had been able to develop in their presence. These produced strengthening by impeding the propagation of cracks. For continued growth, cracks had to circumvent obstructing particles that, because of the orientation caused by laminar plastic flow during extrusion, tended to lie with their greatest cross-sectional areas or, if prismatic or fibrous, their lengths across the slip bands and the paths of the cracks therein (Fig. 28.1). Cracks arising from flaws in the matrix of the slip bands would encounter similarly oriented nonplastic grains athwart their paths. However, it is not yet known whether nonplastic grains impede the nucleation and the continued growth of cracks.

Increasing additions of coarse quartz grains changed the fractography of test pieces from angular and faceted because of failure along intersecting systems of slip bands, to relatively smooth because the slip bands had been reduced in number and continuity.[3] These changes were obvious to the unaided eye. Similar changes were observable in test pieces of some of the mixtures containing nonplastics other than quartz. However, mixtures containing asbestos[4] were notable for irregularly fractured surfaces, often without marked faceting, from which fibers not only protruded, but sometimes linked the two halves of the broken test piece. Here part of the total fracture mechanism involved rupture at clay–fiber boundaries and the partial or complete withdrawal of fibers from their clay sheaths.

STRENGTH INCREASES PRODUCED BY CRUSHED GLASS

There was evidence that crushed glass caused strength increases, not only by reducing the number of slip bands and by inhibiting the ready propagation of cracks along them, but also by producing an interstitial alkaline solution that dispersed the clay.[5] Thus the intermittent shaking of 50 wt % of glass in water, during 6 hours at room temperature, produced a supernatant solution of pH 10. When the clay was tempered with this solution, instead of with distilled water, its transverse strength increased by 9%. Although the interstitial solution in the clay–glass mixtures might not have attained so high a pH, it may be reasonably expected to disperse the clay and thus strengthen it. Dispersion was in fact suggested by the progressive decrease in the percentage of apparent porosities of the dried test pieces that accompanied the addition of increasing weights of glass to the clay.

At some time during the preparation, ageing, and drying of the mixture, the glass fragments developed mantles of clay-mineral crystals having their basal planes approximately parallel to the surface of the substrate. This was found by examining thin sections. Such occurrences resemble the ionotropic deposition of clay-mineral platelets that occurs, for instance, around sodium chloride crystals that are in contact with aqueous suspensions of montmorillonite.[13]

The clay mantles around glass fragments implied the presence of boundary regions that, in the dry state, might be conducive to the overall strength of the clay–glass mixture. That such boundary regions were relatively strong was suggested by the reported adhesion between clay and glass, which became more marked when the original clay slurries were alkaline.[14]

If clay mantles developed around the other nonplastic grains studied, they were not obvious in thin sections, although they are known to form around quartz grains in soils[15] and can be produced in the laboratory by percolating clay suspensions through columns of quartz grains.[16]

ADDITIONS CAUSING ONLY STRENGTH REDUCTIONS

Fine quartz A and B[7] and Madagascar[9] and Mexican[10] graphites resulted in 15, 3, 15 and 24% reductions in the modulus of rupture, respectively, when as little as 10 wt % of quartz or 5 wt % of graphite was added. If any increases in strength occurred, which seems improbable, they must have accompanied smaller additions than those investigated.

Numerous slip bands still developed in the presence of the percentages of added quartz or graphite cited. The quartz grains, especially those in quartz A, were commonly too small to lie across the slip bands and thus too small to increase the strength in the manner possible for coarse quartz. Thus it could be argued that the modulus of rupture need not fall below that of the clay without admixture because, in this clay, the number of slip bands that exert some control on the measured strength is not less than that in the clay containing the fine quartz. The observed strength reduction could result, however, from the very numerous clay–quartz boundaries (potential sites for dangerous flaws) exposed on the surfaces under tension during testing. These boundaries are numerous because of the small size and extensive specific surface of the quartz grains. In accord with these concepts, quartz A, which had about four times the specific surface of quartz B, produced five times the percentage decrease in modulus of rupture caused by the latter. A further possibility is that, although developing slip bands tend to remove small nonplastic particles from their paths,[12] some quartz grains become trapped within them and modify their microstructure in such a way that cracks can be more readily initiated and propagated.

Both strengthening and weakening of clays and bodies by added quartz has been reported but without discussion of the mechanisms involved.[17,18] Examination of the published data suggests that strengthening occurred when the quartz was markedly coarser than the associated clay-mineral crystals, as was found in the present study.

Many particles of Madagascar[9] and Mexican[10] graphite were large enough to lie across slip bands and thus to promote strengthening, yet they caused weakening even when only 5 wt % was present. Grains of the two types of graphite differed in both internal structure[19] and shape. Those of Madagascar graphite were flaky, while those of Mexican graphite were irregular but had a tendency to be simultaneously somewhat flat and elongated, but not flakelike.

The stresses accompanying extrusion often fractured Madagascar graphite into small blocks, many of which remained in mutual contact. Others, near the original surfaces of the flakes, were partially or completely separated by intrusive clay or protruded into their clay matrix.[2] The resulting clay–graphite boundary resembled the hooking adhints that promote adhesion between certain pairs of materials. However, because of the numerous flaws at or near the boundary and within the graphite grain itself, weakening, not strengthening, resulted.

Optical microscopy suggested that Mexican, unlike Madagascar, graphite was relatively free from gross flaws, including those produced during extrusion. However, its presence weakened the test pieces, presumably because dangerous flaws developed at its boundary with the clay.

The curved surfaces of test pieces, especially of those containing Madagascar graphite, showed graphite grains, partially fractured during contact with the die surfaces to produce trails of finer fragments. Often there existed a larger fragment ahead of the trails that was a remnant of the parent grain. This was a further indication that graphite-produces dangerous flaws.

The nature of clay–graphite boundaries is worthy of further study. Clay minerals are hydrophilic, while graphite particles are sometimes regarded as essentially hydrophobic, which suggests that dried clay would not adhere strongly to graphite. However, the edge faces of graphite crystals, or surfaces produced by transverse fracture, as well as defect regions on the basal plane, can adsorb water,[19] which indicates that at least part of the crystal surface is hydrophilic and thus might form an effective bond with clay. Moreover, natural graphite crystals contain fine-grained inorganic materials that could impart some hydrophilic character to their entire surface. A sample of Mexican graphite, for instance, left 13.3% of ash, and much of this was derived from kaolinite.[19] Nevertheless, the outer clay layers of dried extruded cylinders sometimes flaked away spontaneously from underlying graphite grains, suggesting poor adhesion between the clay and at least part of the grain surface.

STRENGTH LOSS CAUSED BY SMALL GLASS ADDITIONS

The addition of 10 wt % of crushed glass to the clay decreased the modulus of rupture by 5%, although additions of 25 and 50 wt % increased it by 9 and 8%, respectively.[5] During passage through the die, the glass fragments tore the clay surface and produced imperfections visible to the unaided eye. These could weaken the test piece and their effects can be overcome only when sufficient glass is present to strengthen the clay by the mechanisms previously discussed.

When 50 wt % of glass spheres, similar in size to the crushed glass fragments, was added to the clay, the surface imperfections just described were not obvious. However, the modulus of rupture was reduced by 25%. This weakening may be caused by the ability of a crack, propagating in a slip band or elsewhere, to circumvent a sphere more readily than an irregular glass chip, often lying with its greatest cross-sectional area across the path of the crack. The slower reaction of the spheres with interstitial water, because their specific surface was less than that of crushed glass, and the consequent less-effective dispersion of the clay, may be involved also. However, the different effects on clay strength, caused by the addition of irregular and of spherical glass particles, are worthy of detailed study.

STRENGTH REDUCTION CAUSED BY AN EXCESS OF NONPLASTICS

Mixtures containing sufficient coarse quartz, crushed glass, tremolitic talc, and asbestos declined in strength, although smaller percentages had produced strengths above that of the clay without admixture. In the mixtures studied, the clay content was always more than adequate to cover the surfaces of the nonplastic grains. Slip bands were rare or absent in such mixtures so that dangerous flaws must be associated with other types of microstructure, including those at or near the boundaries between the clay and the nonplastic grains. For instance, the adhesion at these boundaries may be less than the cohesion within the clay itself. Moreover, the nonplastics grains were commonly larger than the clay-mineral crystals in their matrix and, unlike this matrix, they did not shrink during drying. Hence they could influence the spatial distribution or orientation of the vector linear drying shrinkage in their vicinity. The linear shrinkage is unlikely to be scalar because of the flatness of most clay particles and their tendency to common orientation. Pores might thus develop in the dried clay, near the nonplastic grains, having shapes and sizes conducive to their operation as dangerous flaws.[20]

Clay, in the interstices between nonplastic grains, may develop drying cracks. These reduce transverse strength and seem to form especially if increasing percentages of relatively coarse nonplastic grains are added to clay and the total water content of the various mixtures is kept constant.[3] Because the clay is often more hydrophilic than the nonplastic grains and has a more extended specific surface, the water tends to concentrate increasingly on the diminishing fractions of clay when the nonplastic content of the mixture is raised. The clay becomes more and more hydrous and its drying shrinkage, and the corresponding tendency to crack, increases.

Increasing amounts of nonplastic grains may become progressively more difficult to mix effectively with clay and thus the grains segregate and form local porous regions that diminish the strength of test pieces in which they occur. Fine quartz, especially quartz A, behaved in this way and whitish clumps of grains were obvious on fracture surfaces.[7] The fracture surfaces of clay–asbestos mixtures showed a similar phenomenon because the more flexible of the fibrous components tended to roll up into tiny balls.[4]

CONCLUSIONS

The transverse strength of dried, extruded, clay-containing cylinders depends on interparticle or interaggregate bonds and on flaws in the microstructure. Slip bands, produced during extrusion, are one source of flaws, and fracture tends to occur along them. The propagation of cracks becomes more difficult and strength is increased if the nonplastic grains (coarse quartz, tremolite cleavage fragments, crushed glass) are broad enough, or long enough (fibers in tremolitic talc or asbestos), to lie across the slip bands. In addition to lying across slip bands, crushed glass reacts with interstitial water in the plastic mixture to produce an alkaline solution that disperses, and thus strengthens, its clay matrix. Nonplastic grains promote weakening if they are too small to lie across the slip bands (fine quartz). Grains of Madagascar and Mexican graphite, although large enough to lie across the slip bands, cause only strength reductions owing to weakness of the clay–graphite boundaries and the development, during extrusion, of flaws in the graphite grains.

Progressive additions of nonplastics reduce the number of slip bands, but any strengthening effect is finally annulled, and weakening supervenes, because of the increasing numbers of clay–nonplastic boundaries and thus of the dangerous flaws situated at or near these boundaries. In some circumstances segregation effects during mixing, or the development of drying cracks in the clay, accompany the addition of large quantities of nonplastics and contribute to the loss of strength.

ACKNOWLEDGMENT

The author is grateful to the students whose theses are cited in the references. Their experimental data, obtained under his supervision, contributed substantially to the present account. Dr. E. W. White gave the author invaluable help in scanning electron microscopy.

REFERENCES

1. W. O. Williamson, "Strength of Dried Clay-A Review," *Amer. Ceram. Soc. Bull.*, **50** (7), 620–625 (1971).
2. W. O. Williamson, "The Strength of Clays at Small Moisture Contents," *Miner. Sci. Eng.*, **6** (1), 3–8 (1974).
3. F. L. Kennard III and W. O. Williamson, "Transverse Strength of Ball Clay," *Amer. Ceram. Soc. Bull.*, **50** (9), 745–748 (1971).
4. F. A. Beightol, "Effects of Asbestos on the Dry Strength of Clay," B.S. Thesis, Ceramic Science Section, Department of Material Sciences, Pennsylvania State University, 1974.
5. J. Brun, "Dry Strength of Clay–Glass Mixtures," B.S. Thesis, Ceramic Science Section, Department of Material Sciences, Pennsylvania State University, 1973.
6. F. L. Kennard III, "Factors Affecting the Dry Strength of Clay," M.S. Thesis, Ceramic Science Section, Department of Material Sciences, Pennsylvania State University, 1970.
7. F. K. Koons, "The Effect of Fine Silica on the Strength of Dried Clay," B.S. Thesis, Ceramic Science Section, Department of Material Sciences, Pennsylvania State University, 1974.
8. D. W. Wirth, "The Effect of Talc on the Dry Strength of Clay," B.S. Thesis, Ceramic Science Section, Department of Material Sciences, Pennsylvania State University, 1972.
9. D. L. Amig, "Effects of Graphite on the Dry Strength of Clay," B.S. Thesis, Ceramic Science Section, Department of Material Sciences, Pennsylvania State University, 1973.
10. J. M. Ceriani, "Effects of Non-Flakey Graphite on the Dry Strength of Clay," B.S. Thesis, Ceramic Science Section, Department of Material Sciences, Pennsylvania State University, 1973.
11. J. H. Weymouth and W. O. Williamson, "Effects of Extrusion and Some Other Processes on the Microstructure of Clays," *Amer. J. Sci.*, **251** (2), 89–108 (1953).
12. F. Moore, "The Physics of Extrusion," *Claycraft*, **36** (2), 50–54 (1962).
13. W. T. Higdon, "Studies of Ionotropy—A Special Case of Gelation," J. Phys. Chem., **62** (7), 1277–1281 (1958).
14. S. Anderson, D. Tandon, L. B. Kohlenberger, and F. G. Blair, "Strength of Adhesion of Dried Clay Slurries to Window Glass as a Function of Slurry pH," *J. Amer. Ceram. Soc.*, **52** (9), 521 (1969).
15. D. I. Sideri, "On the Formation of Structures in Soil: II. Synthesis of Aggregates; on the Bonds uniting Clay with Sand and Clay with Humus," *Soil Sci.*, **42,** 461–481 (1936).
16. R. J. Hunter and A. E. Alexander, "Surface Properties and Flow Behavior of Kaolinite. Part III: Flow of Kaolinite Sols through a Silica Column," *J. Colloid Sci.*, **18** (9), 846–862 (1963).
17. E. Kieffer, "On Testing the Binding Power of Ceramic Raw Materials," *Ber. Deut. Keram. Ges.*, **12** (9), 477–479 (1931).
18. V. Hofmann and A. Rothe, "Plasticity and Dry Bending Strength of Kaolins and Clays with and without the Addition of Quartz," *Ber. Deut. Keram. Ges.*, **47** (5), 296–299 (1970).

19. S. M. Kemberling and P. L. Walker, Jr., "Compaction of Natural Graphite," *Tanso*, **2** (52), 1–7 (1968).
20. W. O. Williamson, "Microstructures of Plastic or Dried Clay-Bodies," *Proc. Int. Seminar Clay Miner. Ceram. Processes Prod., Milan,* **1974,** F. Veniale and C. Palmonari, eds., Cooperativa Libraria Universitaria Editrice, Bologna, pp. 47–57.

29

Extrusion Defects

G. C. Robinson

Extrusion is the process of shaping an object by pushing a material through an opening the size and shape of the cross section of the object. Extrusion equipment may be of two types, the piston extruder and the auger extruder. There are many variations within each type. The products of extrusion may be very small, such as tubes of under 0.06 in. diameter, or very large, such as 36 in. diameter pipes. The process is used to form crude natural clays or refined nonplastics, including oxides, graphite, and metal powders.

A variety of green-body defects can arise during the forming process. These include lamination, surface and edge tearing, bridge and core cracking, column splitting, segregation, and preferred orientation of particles. The causes and potential cures for these problems are discussed in this chapter.

MATERIAL FLOW DURING EXTRUSION

Flow Characteristics

The flow pattern of material through the conveying zone of the equipment and then through the forming zone is significant to extrusion quality. Flow is the result of a force applied to the extruded mass that causes its forward movement. Forcing a Newtonian liquid through an orifice in a cylinder results in a straight-line relationship between the applied force and the rate of flow. In contrast, a Bingham plastic substance requires an initial force to start flow. This behavior is characteristic of many bodies. The slope of the rate of flow versus pressure curve is a measure of the mobility of the substance. Departures from the linear flow–pressure relationships, characteristic of dilatant and thixotropic materials, are also encountered in extrusion of ceramic materials. To achieve extrusion control and minimize defects, such nonlinear behavior must be kept to a minimum.

Piston Extrusion

Flow patterns occurring during piston extrusion have been studied by Astbury et al.[1] Lines imprinted transversely to the direction of column flow showed that the center moves in advance of the exterior of the column. The amount of displacement increases with the taper of the die, as shown in Figure 29.1. The drag of the die tends to produce shear planes or cracks

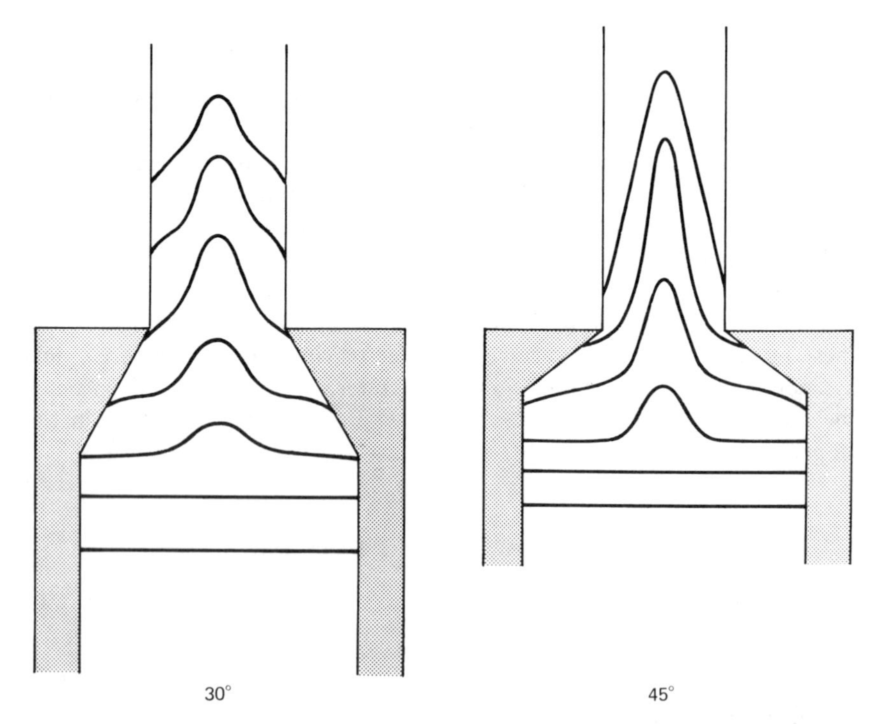

Figure 29.1. The influence of die taper on retardation of flow of exterior surfaces.

that extend from the surface of the column and cut across the flow lines into the interior of the column. One or both of these patterns may appear as cracks in the finished product.

Auger Extrusion

Auger extrusion will produce a pattern of flow similar to that produced by piston extrusion in the die section of the extruder. However, additional flow characteristics are developed by the auger.

Flow is not as efficient in the auger extruder. Forward movement depends on friction between the clay and the barrel lining to restrain the clay from turning with the auger. The clay must either slide on the surfaces of the auger or shear at intermediate surfaces between the auger surface and the interior barrel surface to move forward. The clay is discharged as a single rope (or multiple ropes for multiple wing augers) at the tip of the auger. The twisting of the auger causes the rope to be discharged as a coil, while the spacer and die tapers cause compression of the coil into a single column of extrudate. The hub of the auger produces the hole in the center of the coil. The coils may fracture in low-plasticity materials to produce a series of concentric and tapered cones, which are then forced together during travel through the die taper. Thus auger extrusion has the potential of producing a number of weakness planes that may develop into flaws in the finished product.

STRUCTURES PRODUCED BY EXTRUSION

Lamination

Lamination has been the curse of extrusion since inception. There is always laminar flow in extrusion, but correct design of the die and raw-material mixture can prevent laminar flow from developing into cracks or weaknesses. The development of laminar flow during auger extrusion is shown in Figure 29.2. Lumps of different-colored clays were alternately fed into an extruder and then the column was sectioned to show the flow patterns. The section cut perpendicular to the direction of extrusion shows the characteristic circular pattern, while longitudinal sections show twin peaks of forward displacement. There is a lag in the center of the column produced by the dead zone of the hub of the auger and the resulting hole in the coiled rope. There is also a lag at the exterior of the column produced by the drag of the die on the exterior surface of the extrudate. Failure of the column to be completely knitted together results in weak zones that may separate into

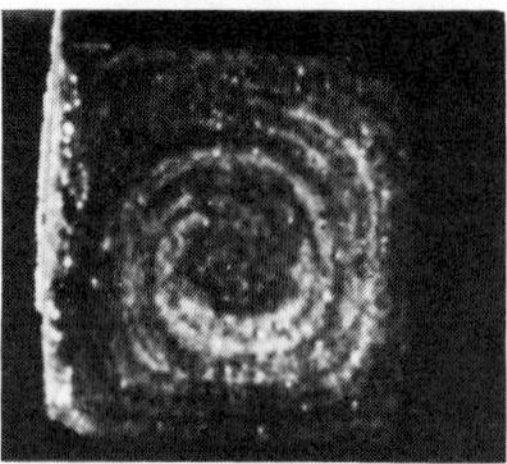

Figure 29.2. Laminar flow development from auger tip (top) to extruded column (bottom).

cracks as a result of forces developed during drying, firing, or even freezing if this uncommon treatment is applied.

Figure 29.3 shows lamination cracks appearing in the interior of an extruded object. These cracks, which are parallel to the flow patterns in both longitudinal and transverse sections, normally develop in materials that are difficult to dry. Elimination of lamination cracks (but not laminar flow) can be accomplished by improving the drying behavior of the material, changing the die design to minimize displacement, or achieving better knitting of the extrudate.

Figure 29.3. Laminations cracks in longitudinal section (left) and transverse section (right).

Surface and Edge Tearing

An extruded column may exhibit surface cracks. One type is predominant at the corners of a rectangular column. Small cracks enter at the edges and cut at an angle across the column. This cracking has been called "edge tearing," "dog teeth," or "feather edging" (Figure 29.4). The cracks arise from the greater friction at the corners of the die than on the faces. The presence of large particle sizes of nonplastics will increase the tendency for formation of this flaw.

Other cracks may appear over the entire surface. These cracks have a square to rectangular pattern and result from interfacial friction between

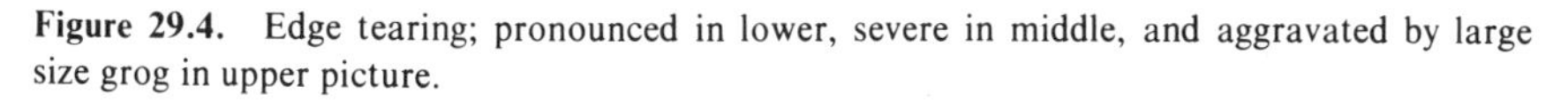

Figure 29.4. Edge tearing; pronounced in lower, severe in middle, and aggravated by large size grog in upper picture.

the die and the column. The clay has a low plastic strength and is incapable of surmounting the friction drag without cracking. This type of surface cracking and penetration of the cracks into the interior of the column is shown in Figure 29.5.

Particle Orientation

The process of extrusion causes particles of flaky habit to orient parallel to the faces of the die. This orientation causes a greater shrinkage in the direction perpendicular to extrusion than in the lengthwise direction. Thus the flaky particles are a source of a laminated structure in a direction transverse to the direction of extrusion.

The orientation may cause a differential expansion during the removal of the last 2% of water from an extruded material. For example, comparison of the transverse and linear expansion for a 40% clay, 60% sericite mixture with the expansion of a 40% clay, 60% sand mixture has shown that the micaceous flakes of sericite cause a larger expansion and a greater differential than the rounded grains of sand.

Particle and Water Segregation

Analyses performed by Astbury, et al[1] and Robinson, et al[2] have shown that there is a difference in composition between the surface and the interior of an extruded column. The skin of a die-slickened extrudate will show a higher percentage of fine particles and a higher water content than the interior of the column. The nonuniform particle size, water, and shrinkage in the column cross section can lead to cracks in the finished unit. Measurements were made at different points in the cross section of a conduit tile by Eighmie and Robinson.[3] The water content was found to vary from 14.6 to 13.4% and the volume drying shrinkage varied from 9.3 to 7.5% on different locations in one plane of the cross section.

Bridge and Core Cracking

The introduction of cores and their supporting bridgework can produce weak zones and cracks in the finished product. The selection of bridge and core configuration has been the subject of many articles but remains an art. The separation in the column caused by insertion of the bridge has been minimized by streamlining, serrating edges, and changing the separation distance from the die.

A study of bridgework design has been presented by Merry.[4] There is a choice of supporting the bridge with a single cross bar or the use of an H-

Figure 29.5. Surface cracks of low-plasticity materials.

shaped support. The best support configuration depends on variations in the sand content in the plant raw material. With high sand contents, the H-bridge configurations were desirable and prevented the development of cracks. In contrast, clays with low sand content required the use of a single cross-bar configuration. Bridgework not only causes separation of the extruded column, but also retards the flow in some parts of the cross section and may even direct and flow to other locations.

Column Splitting

The friction at the die interface working against the cohesion of the material can attempt to pull the column apart into a variety of patterns. A high-friction sandy material may pull apart in a continuous split as shown in Figure 29.6. A material of high cohesion and high strain together with high friction may split into four segments with a central twisting core as shown in Figure 29.7.

Enhancement of Extrusion Structure

A number of methods have been employed to make extrusion structure more apparent. When the plastic extrudate is frozen, expansion of the freezing water causes an opening of the laminar structure. Such a technique may be informative, but it should be remembered that a material that cracks under freezing conditons may not show such faults under normal drying and firing procedures.

Liquid penetrants and fluorescent-die penetrants have been used to examine extrusion structures. Cutting apart a fired product will frequently

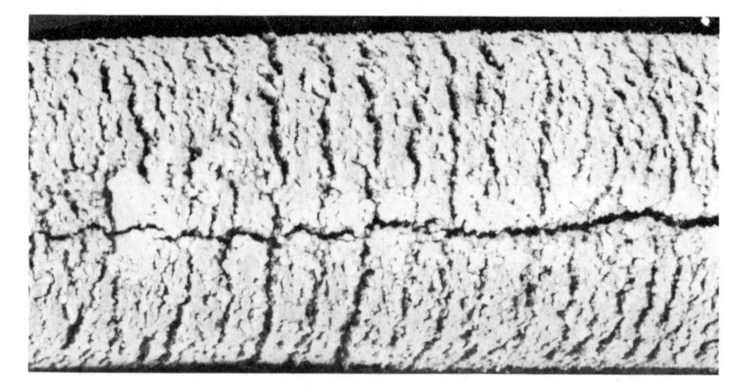

Figure 29.6. Column splitting in sandy material.

Figure 29.7. Splitting of column into four exterior segments and a central twisting core.

display internal lamination cracks. Examination of the fracture pattern occurring after strength determinations on plastic dry or fired products may reveal extrusion structures. The extrusion of different-colored materials or the application of patterns prior to extrusion may assist in examining the flow mechanisms occurring during extrusion.

RAW MATERIAL PROPERTIES DETERMINING EXTRUDABILITY

Adhesion

Adhesion of the extrudate to the metal parts of the extruder is significant in determining flow. This adhesion is responsible for the resistance against turning of clay with the auger, for the sliding of the clay on the auger, and for the drag between spacer lining and clay, and die lining and clay, which causes surface tearing and lamination.

The adhesion between substances can be evaluated by an experimental piston extruder operated without an orifice. The raw material for extrusion is packed into the cylinder by completely closing one end of the cylinder and pushing the material against the closure until a preselected pressure level is applied to the piston. The closure is removed and the cylinder is reversed with the piston introduced into the closure side of the cylinder. The slug of clay is pushed downward away from the closure side and the force required to cause movement of the clay slug through the cylinder is determined and used to indicate adhesion. The surface area of this clay slug can be measured and the adhesion is expressed as the force per unit area. The cylinder metal and its surface finish should duplicate the barrel lining, auger surface, or die surface being evaluated.

Adhesion can be changed by changing the structure and composition of the metal or the extrudate. This provides a variety of potential control measures to decrease adhesion: (1) polishing the metal, (2) chrome plating the metal, (3) heating the metal, (4) increasing the water in the extrudate, (5) changing from angular to rounded particles in the extrudate, (6) adding lubricants to the extrudate, and (7) lubricating the interface between metal and extrudate. Extrusion quality and cost are optimum when the adhesion against turning is high and all other adhesions are low.

Internal Friction

The raw material must slide on itself whenever the die opening is smaller than the barrel diameter or when the compression of the rope coils is greater than the void space within the coil. The resistance against internal flow can be evaluated by the same device as used for adhesion measurement. Internal friction is evaluated by placing a die restriction on the open side of the reversed cylinder and measuring the increase in force required to extrude through the opening. A series of progressively smaller openings can be used and a curve is obtained of the force required versus increasing opposition to flow. Another approach is to measure the rate of flow through an opening versus the pressure causing flow.

The internal friction indicates the energy required for extrusion and the extent of "forging" desirable in extruder design. Materials of high internal friction should be extruded through equipment in which there is little size reduction occurring between the die diameter and the auger diameter. The size reduction can be large and coring extensive with materials of low internal friction.

Cohesion

Cohesion is the plastic strength of the extrudate. The stress at the elastic and plastic yield points can be determined on the extruded column by conventional means. The value is significant to extrusion quality.

A material of low cohesion will surface tear when the adhesion to the die is greater than the cohesion. The lower the cohesion, the more likely is the appearance of surface defects. Very high cohesion will favor carryover of auger twisting into the extruded column but will lessen the forward displacement caused by spacer and die drag.

Cohesion can be increased by increasing the quantity of colloidal material in the mix, reducing the water, or adding organic plasticizers or deflocculants. Cohesion can be reduced by adding grog or other nonplastics. The smaller the particle size of a nonplastic addition, the more effective it is in reducing cohesion.

Plastic Strain

Plastic strain indicates how much you can bend or stretch the extruded column before it cracks. Plastic strain is symptomatic of lamination behavior. The higher the value of strain, the greater the forward displacement and the greater the carryover of auger twist. Any material with a strain greater than 0.10 is likely to show lamination faults. Also, such materials are difficult to dry.

High or low values of strain are undesirable. Low values of strain foretell surface tearing. The raw material cannot be formed against the dragging interfacial friction and instead tears apart or cracks. Intermediate values of strain give the best extrusion. Strain can be increased by adding plasticizers and reduced by adding small-size nonplastic particles.

The low limit of strain is influenced by the cohesion of the extrudate and, as a result, surface tearing is better predicted by the product of stress and strain than by strain alone. A product of less than 1.0 predicts surface tearing, while a product between 1 and 4 indicates good extrusion.

Particle Shape

Platelike particles contribute their own characteristic laminar structure. These particles will orient and produce laminae across the cross section of the column without necessarily displaying any displacement structure in the direction of column flow.

Mica, sericite, vermiculite, and graphite are illustrations of flaky substances that can cause troublesome lamination. The lamination can be minimized by reducing the particle size, reducing the orientation force by minimizing die taper, or extruding at a soft consistency. Agglomerating the flaky particles into nodules of sufficient strength to resist collapse during extrusion will also produce lamination.

Drying Behavior

It is difficult to answer the question of whether or not lamination cracks are the fault of extrusion or of drying. The flow patterns and weak zones are developed during extrusion, but usually there is no visible crack in the plastic extrudate. The cracks develop during the drying process as the result of forces applied by shrinkage gradients. Reducing the shrinkage gradients will reduce or eliminate the lamination crack. Slower drying schedules can reduce lamination cracking, and changing the body composition to improve drying behavior will eliminate lamination cracks. The addition of nonplastic particles in two size classifications will improve drying behavior. Some nonplastic particles should be added in the size range between 28 and 65 mesh to reduce drying shrinkage, while particles of nonplastics in the −200 mesh size should be added to increase the permeability of the body. This combination will make a crack-free unit.

Abrasiveness

The abrasiveness of the material determines the rate of wear of the metal parts of the extruder. This property helps determine maintenance cost of

Figure 29.8. Drying cracks caused by Large-particle-size grog.

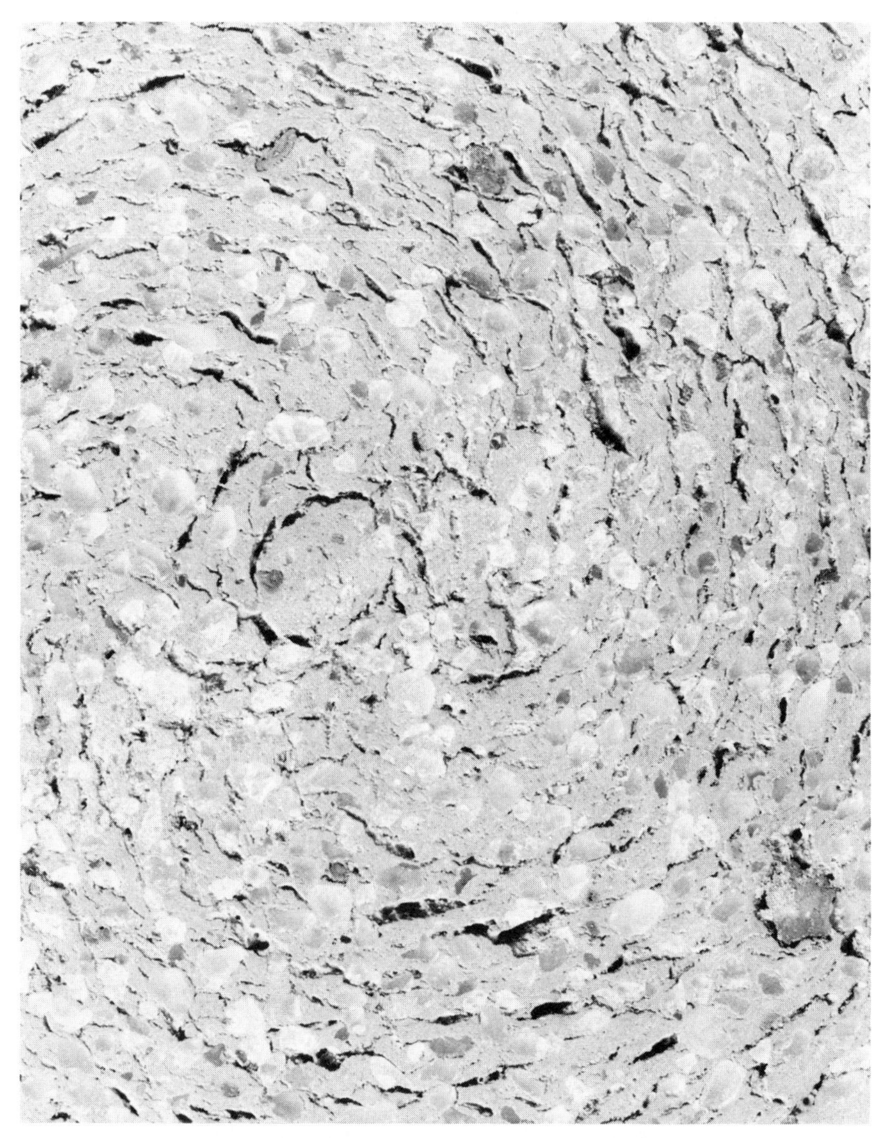

Figure 29.9. The tranverse lamination cracking with $-14 + 28$ mesh grog.

extrusion and selection of materials for the extruder. Thus ceramic cores will give greater life because of greater abrasion resistance than metals.

Particle-Size Distribution

Particle-size distribution is an important raw material variable. Nonplastic particles larger than 28 mesh encourage edge tearing, drying cracks (Figure 29.8), and transverse lamination cracking (Figure 29.9). Particle sizes of nonplastics between 28 and 100 mesh will break up large lamination cracks into many small cracks. A combination of these particles with −200 mesh nonplastics will eliminate the small lamination cracks. The −200 mesh non-plastics are more effective than an equal quantity of larger-size particles in reducing plasticity and cohesion and working toward elimination of surface cracking.

The ideal particle-size distribution would be a somewhat uniform gradation of sizes with 30% nonplastics. Natural clays with good extrusion characteristics will approximate the foregoing suggestions for particle size and quantity of a nonplastic.

MACHINE VARIABLES THAT DETERMINE EXTRUSION

The Auger

The auger has been the subject of extensive study. It influences the rate of production, the cost of production, and the quality of the extrudate. The characteristics of the auger are explained in detail elsewhere.[5-11] The significant design features of the auger are its pitch, diameter, hub diameter, taper, and number of wings.

Speed of Auger Rotation

The speed of auger rotation can influence quality and productivity. The work of Goodson and Dodd[11] showed that the extrusion rate is a function of both the rate of rotation of the auger and the quantity of contained water. At any given water content there appears to be an optimum rpm that gives maximum output. Furthermore, the optimum shifts with changing water content.

The speed of rotation influences the friction developed between the barrel lining and the clay. The best speed to use depends on the type of clay. Insufficient speed will permit slipping and turning with the auger, while correct speed will hold the clay against twist of the auger. Insufficient speed will also produce heating of the material within the brick machine.

Turndown Ratio

The significance of the ratio of die diameter to auger diameter has been mentioned in previous sections. This should be selected to suit the characteristics of the material being extruded. Furthermore, this ratio will affect the rate of column output.

The Die

Die configuration has been studied from the standpoint of its role in energy consumption, but there have been very few studies of the relation between die design and the product quality. The design of extrusion dies remains an art with very few sources of supply in this country. The following studies are being made in our laboratories to help establish the relation between die design and quality of extrusion.

The die may be defined as the entire section of the machine forward of the auger or it may be separated into two zones—the spacer zone between the auger tip and the entrance point to the shaper cap, which approximates the cross-sectional dimensions of the extruded column. In the latter case, as used herein, the shaper cap is considered as the die.

Tests were made with three different brick raw materials using dies of different taper, different entrance angle, different diameter, and different length. The force required for extrusion, volume shrinkage, water of plasticity, dry bulk density, and column consistency was determined for each extrusion. A mass of data was compiled for this investigation and the results were disappointing. Variables of taper, entrance angle, and die length had only a small influence on the properties measured. Perhaps the key result of this work was to point out that a 1% change in forming water, say from 17 to 18%, had a much greater influence on all the properties evaluated than did the die parameters.

However, changes in die taper, die length, and entrance angle have been shown to have a significant influence on edge tearing, lamination, and plastic strength. Increasing die taper reduces edge tearing and increases internal laminations. Increase in extrusion speed at a single die taper reduces the tendency for edge tearing.

Vacuum

Evacuation or reducing the pressure within the extruder can dramatically improve the extrusion behavior of some materials. As a consequence, de-airing or vacuum extrusion is the common practice. However, the effects of de-airing are not linear with respect to pressure. Little improvement is obtained with a pump operating to, say, 24 in. of mercury vacuum, whereas large improvements are observed by increasing the vacuum from 25 to 29 in.

Although de-airing is universally accepted, it is sometimes inappropriate to the particular material being extruded. Very plastic materials may give better performance with extrusion without evacuation instead of with de-airing.

Die Balance

The quality of the extrusion is influenced by the position of the die. The extent of separation between the tip of the auger and the die determines the tendency for one part of the column to deform forward with respect to the remainder of the column. A die position close to the auger tends to make the exterior of the column run faster than the interior. A lengthy separation between the auger tip and the die will cause the center to run fast. At a condition of balance, the column will move without a tendency for longitudinal displacements. The correct placement is a function of the material and also of the water in the material. A change in the quantity of water will change the correct die position. Incorrect placement of the die can produce column splitting or other defects in the extruded column.

Centering the die is also important to performance. An off-center die will encourage the column to twist in one direction or another and may introduce strains that will later appear as cracks. The die needs to be aligned with the center line of the auger and also with cores and bridgework. However, in some instances, cores are displaced to encourage flow of materials into sections of the die that are normally slow moving.

REFERENCES

1. N. F. Astbury, H. R., Hodkinson, and F. Moore, "The Problem of the Clay Column," "The Mechanics of Extrusion," and "The Physics of Extrusion," papers presented at the 29th Annual Meeting of the Institute of Clay Technology, Sept. 29, 1962, Harrogate, England; Chambers Engineering News Letter No. 13.
2. G. C. Robinson, R. H. Kizer, Jr., and J. F. Duncan, "Raw Material Parameters Determining Extrudability," *Am. Ceram. Soc. Bull.*, **47** (9), 822–832 (1968).
3. T. L. Eighmie, "The Influence of Die Parameters on the Extrusion of Structural Clay Products," Master of Science Thesis, Clemson University, 1970.
4. E. A. Hawk, "Handling the Difficult Extrusion Problem," *Amer. Ceram. Soc. Bull.*, **33** (11), 326–327 (1956).
5. J. George Seanor and W. P. Schweltzer, "Basic Theoretical Factors in Extrusion Augers," *Amer. Ceram. Soc. Bull.*, **41** (9), 560–563 (1962).

6. G. Caprix and A. Laratta, "Screw Extrusion of Bodies," *Trans. Br. Ceram. Soc.*, **64** 19–26 (1965).

7. A. J. Reed, "Auger Design," *Amer. Ceram. Soc. Bull.*, **41** (9), 549 (1962).

8. W. J. Johnson, "Evaluation of NCPRC Research Auger," *Amer. Ceram. Soc. Bull.*, **41** (9), 550–555 (1962).

9. H. H. Lund, S. A. Bortz, and A. J. Reed, "Auger Design for Clay Extrusion," *Amer. Ceram. Soc. Bull.*, **41** (9), 554–559.

10. J. R. Parks and M. J. Hill, "Design of Extrusion Augers and the Characteristic Equation of Ceramic Extrusion Machines," *J. Amer. Ceram. Soc.*, **42** (1), 1–6 (1959).

11. F. J. Goodson and A. E. Dodd, *Conf. Mixing Agitation Liquid Media* **July 1951,** Institute Chemical Engineers pp. 333–341, London.

PART FIVE

PROCESSES AND APPLICATIONS

Tape casting is one of the latest ceramic processes practiced on a large commercial scale. The process is a major departure from the more traditional methods of ceramic processing. Because many details of the process have not been widely published, a rather extensive treatment appeared justified and is the subject of Chapter 30.

Two applications, thermal control coatings and thick films, illustrate the importance of powder characterization in these processes. In the final chapter, brief comments are provided by various members of a panel discussion on future directions in processing research.

30

Tape Casting of Ceramics

R. E. Mistler
D. J. Shanefield
R. B. Runk

Tape casting is an important process for forming large-area, thin, flat ceramic parts. A "tape" is formed by the use of a scraping blade, or "doctor blade," which evenly coats a moving surface with a slip. The coating dries to form a green tape that can be cut into different shapes. The cut tape is fired to form a flat ceramic body that is essentially two-dimensional.

The chief advantage of the tape-casting process over other forming techniques is that it is the best method of forming flat articles with thicknesses in the 1 to 50 mils range. These articles are virtually impossible to dry press and most difficult, if not impossible, to extrude. If the plate is to be pierced with numerous holes, dry pressing would be even more difficult because of problems in uniform die fill.

BACKGROUND

Tape casting or doctor-blading is really a very ancient art that is more commonly known as knife coating in many other fields, including the paper, plastics, and paint industries. In the paint industry the technique is used to test the covering power of paint formulations.

The marriage of a more ancient art, slip casting, which is familiar to all ceramists, and doctor-blading took place within the past 25 years. Tape casting is more closely related to slip casting than to any other "traditional" ceramic-processing technique. Many of the tricks and problems in working with ceramic suspensions are common to both. The means for removing the liquid carrier or suspending agent during drying are different: tape casting involves evaporation, while slip casting utilizes absorption in a porous mold. Also the sizes and shapes of products, manufactured by the two processes differ considerably.

Use of tape casting to form ceramics apparently occurred first during World War II in the development of dielectric materials useful as substitutes for mica in high-quality capacitors. Glenn Howatt obtained a patent[1] describing such a process and founded a company that is now part of Gulton Industries. Howatt's patent is for "forming ceramic materials into flat plates, especially useful in the electrical and radio fields." This is still the principal field of application today, although the present end uses are in many cases far beyond anything imagined in 1945.

The Howatt patent teaches a process that is a combination of slip casting and doctor-blading, since the slurry is spread on absorbent bats where the liquid vehicle is removed.

The limitations of the tape-casting process are in thickness and lack of "three-dimensional" capability. Dry pressing and extrusion seem to have the production edge when thicknesses of $\frac{1}{8}$ in. or more are desired in plates. Tapes can be made this thick, but pressing seems to be less of a problem. As noted earlier, tapes are essentially two dimensional, and the technique is not practical for pieces that have lips, ledges, tapered holes, and blind holes.

At present the main ceramic applications of the tape-casting process are in the electronics industry. Capacitor dielectrics, piezoelectrics, and thick- and thin-film substrates are the major products. Other products include tape glazes and metallizing preparations.

There are numerous other references and patents relating to tape casting in the ceramics field. The most important of these are included in reference 2.

This chapter covers binder–solvent selection, raw material characterization, dispersion of the ceramic powder, precasting treatments, casting process and apparatus, including precision casting, drying of the tape, punching to shape, and multilayer casting. We draw mostly on our experience in casting high-alumina ceramic substrates for the discussion and illustrative examples. We have not been limited to this one material, however, and have experience in casting the following materials: lead zirconate–lead titanate (PZT), lead zirconate–lead titanate doped with lanthanum (PLZT), barium titanate, steatite, forsterite, porcelains, glass, manganese–nickel–cobalt thermistor materials, ferrites, bonded silicon carbide varistor materials, and calcium–aluminum silicate compositions.

BINDER–SOLVENT SELECTION

There are two principal types of solvents, water and organic liquids. In either case the binder should meet the following criteria: (1) form a tough flexible film when dried and present in low concentrations, well under 10 wt %; (2) volatilize to a gas leaving no residual carbon or ash during firing; (3) not be adversely affected by ambient conditions during storing: (4) be relatively inexpensive; and (5) be soluble in an inexpensive, volatile, non-flammable solvent in the case of organic solvent systems.

The binder, or plastic, is thus of the "film-forming" type. These film formers usually are long-chain polymers that possess characteristics of internal flexibility in contrast to polymers that have substantial three-dimensional linkages, which tend to be rather rigid. Just about every plastic imaginable has been used to make tapes, but some of the more popular are poly(vinyl acetate), poly(vinyl chloride), poly(vinyl chloride–poly(vinyl acetate) copolymers, polystyrene, poly(vinylidene chloride), poly(vinyl alcohol) (water system), polymethacrylates of many types, and cellulose nitrates.

Some of the factors that must be considered in the selection of binder are: (1) thickness of tape to be made; (2) casting surface–glass, plastic, metal; and (3) solvent type desired. The extremely volatile solvents, such as acetone, are most suitable for the thinnest films, and the lower volatility solvents, such as toluene, seem best for the thicker films. When trying a new material in the laboratory we generally used a poly(vinyl chloride–poly(vinyl acetate) copolymer with MEK as the solvent for casting up to a thickness of 10 mils and poly(vinyl butyral) with toluene or trichloroethylene as the solvent for thicker casts. Table 30.1 gives the typical binder–solvent–plasticizer systems for thick and thin casting trials.

Plasticizers are normally added to obtain sufficient flexibility of the film for easy handling. The plasticizer is often present in amounts greater than the binder itself. The list of plasticizers is much too long to include here, but binder manufacturers are in a position to recommend those suitable for use with their products. Milling of the slip (discussed in a later section) is best done before additions of the binder and plasticizer because of the very high viscosity that results after their addition. After binders and so forth are added, the mill is run until all are thoroughly dissolved and mixed. In some

Table 30.1. Typical binder–solvent–plasticizer systems for tape casting

	Thick (0.010 in.)	Thin (0.010 in.)
Binder	3.0 poly(vinyl butyral)[a]	15 vinyl chloride-acetate[b]
Solvent	35.0 toluene	85.0 MEK
Plasticizer	5.6 poly(ethylene glycol)	1.0 butyl benzyl phthalate

Note: Numbers given are parts by weight per 100 parts ceramic powder.
[a] Butvar Type B-76, Monsanto Company, St. Louis, Mo.
[b] VYNS, 90:10 vinyl chloride–vinyl acetate; copolymer supplied by Union Carbide Corp.

cases the binders are predissolved in a portion of the solvent system for faster mixing.

STARTING MATERIALS

For any ceramic process it is essential that the starting powders be well characterized. This is especially true for tape casting. The important parameters that should be monitored on all powder lots are average particle size and distribution, surface area, and trace-impurity level.[2]

At this point we draw upon the vast amount of information that was generated in developing a process for the manufacture of fine-grained alumina substrates by tape casting. The characterization and processing would be very similar for any ceramic material. Briefly, the process consists of nonaqueous wet ball milling, followed by tape casting on pastic film, punching, and firing. A flow diagram in shown in Figure 30.1.

Table 30.2 gives the starting materials. The major component is a dry-ball-milled Bayer-process alumina, either Alcoa-A-16 or Reynolds RC-172 DBM. This powder is essentially alpha alumina, with a medium particle size of 0.4 μ, and a B.E.T. surface area of about 11 m^2/g. As received, the alumina is agglomerated, and nearly spherical clusters of particles are visible at 300× magnification (Figure 30.2). Tap densities of 1.3 and 1.5 g/cc have been reported for this type of material. Figure 30.3 is a TEM of the alumina particles sprayed onto glass after ultrasonic dispersion in water. Shadowing was done with Pt-carbon evaporated at a 23° angle without rotation of the sample. The length of the shadow indicates that the particles are not the flat platelets often observed[3] in other types of alumina powder but are roughly as high as they are wide.

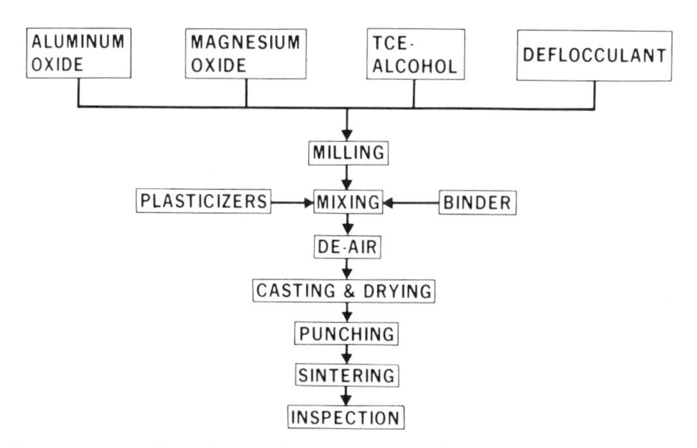

Figure 30.1. A flow diagram for an alumina-substrate tape-casting process.

Figure 30.2. SEM of as-received agglomerated alumina powder.

Table 30.2. Materials and milling procedure

Process Step	Material	Function	Parts by Weight
Mill for (24 hours) (first stage)	Alumina powder	Substrate material	100.0
	Magnesium oxide	Grain growth inhibitor	0.25
	Menhaden fish oil[a]	Deflocculant	1.7
	Trichloroethylene	Solvent	39.0
	Ethyl alcohol	Solvent	15.0
Add to above and mill for 24 hours (second stage)	Poly(vinyl butyral)[b]	Binder	4.0
	Poly(ethylene glycol)	Plasticizer	4.3
	Octyl phthalate	Plasticizer	3.6

[a] Haynie Type Z-3 (air treated), Jesse S. Young Company, New York, New York.
[b] Type B-98 (molecular weight approximately 32,000), Monsanto Company, St. Louis, Missouri.

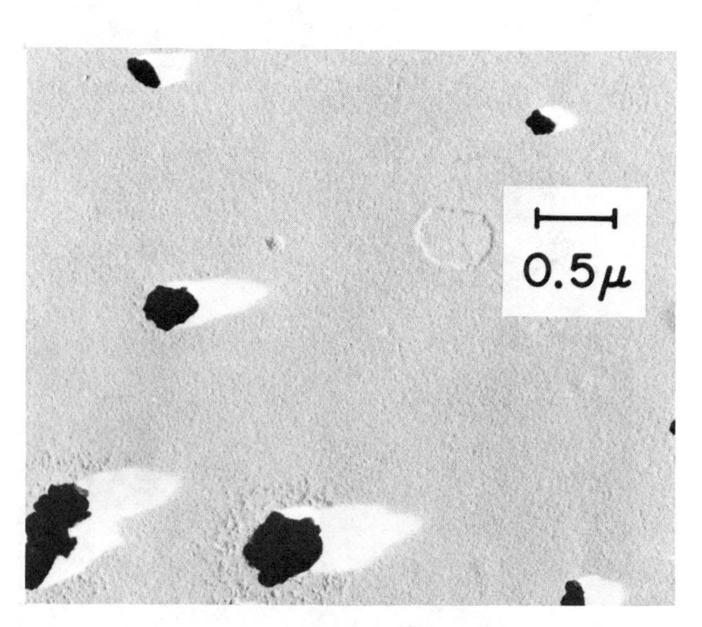

Figure 30.3. Transmission electron micrograph of alumina powder, spray dispersed and shadowed at 23° angle to the plane of the illustration. Length of shadow indicates that particles are not flat platelets.

The binder used in this process is poly(vinyl butyral), and the plasticizers are poly(ethylene glycol) and octyl phthalate. The solvent is an azeotropic mixture of trichloroethylene and ethyl alcohol. Two liquids are used to make up the solvent, because the solubility of a polymer is generally greater in optimized mixed-solvent systems than in any individual pure solvent.[4]

According to the supplier, the menhaden fish oil deflocculant contains polyunsaturated ester molecules. It has an iodine number of 100 and a saponification number of 185.[5] It consists mostly of glyceryl esters of fatty acids, and 76% of the acid groups are unsaturated types, such as oleate and linolenate.[5]

DISPERSION OF THE CERAMIC POWDER

Milling Procedure

As shown in Table 30.2, the milling is done in two stages, with and without binder and plasticizers. An 85% alumina porcelain is satisfactory for use in the mill lining and grinding media. If there is concern with mill impurity pickup, many other mill and grinding-media materials are available. The advantages gained must be weighed against the added cost, particularly if the process is to be used in large-scale production. The mill is half filled with the media, which are $^{13}/_{16}$ in. cylinders. The milling is done at 65% of critical speed. The charge is added in proportion to 3000 g of dry alumina powder added to a mill having an empty capacity of 2.3 gal. This charge can be scaled up or down depending on the size of the mill used.

Figure 30.4 shows the effects of variations in the milling procedure. Black circles denote 24 hour, second-stage milling. Low densities resulted from either insufficient milling time or excessive milling time in the presence of the binder. The effects of excess milling are presently unexplained. The effects of insufficient milling are apparent from the discussion below.

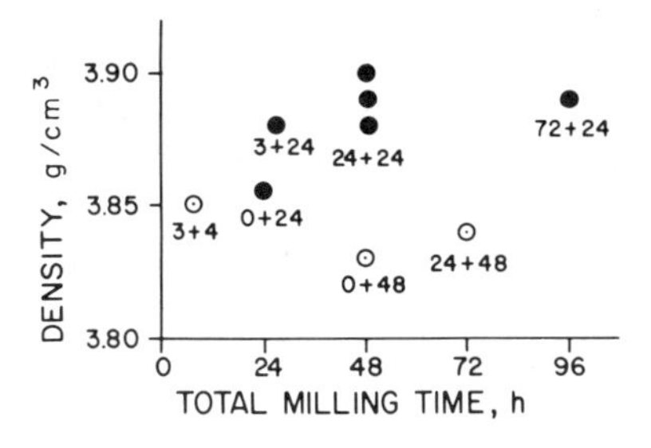

Figure 30.4. Densities of fired substrates versus milling time. First number denotes hours of ball milling in first stage (before adding binder and plasticizers); second number denotes hours of milling in second stage.

Impurity Pickup

During milling, Ca and Si from the mill lining and grinding media are added to the slip, as shown by emission spectrographic analysis of the milled and dried slip, expressed as element ppm by weight as follows:

	As-received	Milled 2 days
Ca (ppm)	100	300
Si (ppm)	300	1500

If the milling is done with high-purity lining and media (99+% Al_2O_3), the resulting tape does not sinter adequately under the firing conditions used, and the substrate has open porosity and a density of 3.80 g/cc or less. However, if talc is added to the higher-purity composition sufficient to provide the 1500 ppm of Si, the resulting tape sinters to 3.88 g/cc and contains essentially zero open porosity. Polymethylphenylsiloxane silicon resin (SR-82, General Electric Company, Waterford, New York) can be substituted for the talc, resulting in a fired density of 3.89 g/cc. The degree of dispersion of the impurities appears to be important, since equivalent additions of SiO_2 powder (Quso, Type 30, Philadelphia Quartz Company, Philadelphia, Pennsylvania) yield low-density substrates. Also, low density results when Mg is not included in the starting materials as a grain-growth inhibitor.

Deagglomeration

The term "weak agglomerate" is used to mean a group of particles that are weakly bonded together, which is consistent with the definition in Chapter 5. The term "solid agglomerate" is defined operationally in this paper as a group of particles that cannot be separated by 40 kHz ultrasonic treatment at 100 watts for 20 minutes in an aqueous dispersion.[6,7] The word "floc" is used as defined in Chapter 5.

As mentioned earlier, the as-received alumina powder is highly agglomerated. We have found that ultrasonic treatment at 40 kHz and above reduces the measured particle size of this material to a limiting value, at which the agglomerates originally present were broken down. Wet and dry sieve analyses[8] and other studies have differentiated between some levels of bond strength in these agglomerates.

It is generally true that a large surface area tends to favor agglomeration. (For an example of this correlation, see the data on nine different alumina powders in Figure 30.5 of reference 9.) The large area appears to offer more possibilities for interparticle contact and adhesion, thus leading to porous, poorly packed agglomeration and low green density. However, good sinterability requires both the high surface area and the good packing. It is a remarkable feature of the present process that a powder with a surface area of about 11 m^2/g, which is high for alpha alumina, can be effectively deagglomerated and deflocculated, resulting ultimately in good packing.

Direct evidence of the deagglomeration produced by milling was obtained by sedimentation particle-size analysis. Slip of the composition given in Table 30.2 was milled for 2 days, and the relative particle size was determined as follows: 7.56 g of slip was added to 40 cc of a 50 vol % solution of benzene in ethanol, but ultrasonic dispersion was *not* used before the sedimentation measurement. The measured median size was 0.4 μ, but the absolute accuracy of this value was not trustworthy, because the density and viscosity of the solution were assumed to be the averages of the benzene and alcohol values. For relative comparison, however, the particle-size distribution of unmilled slip was determined after it was diluted in benzene–alcohol in the same manner and ultrasonically dispered for 20 minutes. The particle sizes were within 15% of those of the milled slip, showing that 2 days of milling is equivalent to the ultrasonic dispersion in breaking down agglomerates.

Milled slip was also diluted in benzene–alcohol and, in addition, it was ultrasonically treated for 20 minutes. These particle sizes were also within 15% of those of the unmilled but ultrasonically dispersed slip, showing that the hard particles are not significantly reduced in size by the wet milling.

Particle-size determinations of A-16 alumina dispersed by other means are consistent with the above. The diameters of the particles that were separable by smearing and spraying using a TEM and the centrifugal action of the MSA particle-size analyzer[8] were generally in the same size range.

Surface-Area Effects

Unmilled slip of the composition shown in Table 30.2 consisted of A-16 alumina with a surface area of 10.2 m^2/g. This powder was dried and heated in air at 500°C for 2 hours to remove the organic constituents. The resulting material was white powder that adhered together only about as strongly as powder pressed between two fingers, and sintering was therefore not great. After a 300°C outgassing, the BET surface area was 10.8 m^2/g. Slip samples milled for 1, 2, and 7 days were treated in the same manner and yielded surface area that were 11.2, 12.6, and 11.8 m^2g, respectively indicating a slight increase in surface area with milling, but this small increase might be due to impurities from the grinding media.

Deflocculation

The Mechanism. If less than half the amount of menhaden oil indicated in Table 30.2 is used during milling, the slip viscosity becomes too high to allow grinding media motion when the mill is rotated. However, the presence of the full amount of deflocculant provides a paintlike fluidity with good milling action, in spite of the high solid-to-liquid ratio shown in Table 30.2, and in spite of the high surface area of the solid material present.

Two models that are commonly proposed[10] to explain the action of deflocculants are charge repulsion and steric hindrance. Each assumes that the deflocculant is adsorbed onto the solid particles.

Generally the steric hindrance model is useful in explaining the stabilization on nonaqueous dispersions.[11,12] However, traces of water were shown[13] to influence the zeta potential and agglomerate size of TiO_2 dispersed in nitrobenzene, and charge effects might therefore be important in some essentially nonaqueous systems. For example, charge repulsion has been proposed as the mechanism for the deflocculation of alumina in xylene.[14]

In comparisons of the effectiveness of oleic acid versus stearic acid as deflocculants for alumina in nonaqueous liquids, the presence of a carbon–carbon double bond in the deflocculant has been reported[14,15] to enhance the deflocculation effect. Also, infrared spectra of the adsorbed species have shown[16] that the double bonds in oleic acid strongly influence the effective area covered by each molecule adsorbed on TiO_2 dispersed in benzene.

The influence of the carbon–carbon chain length on deflocculation has also been reported[17] for alumina in various nonaqueous liquids. The chain-length influence on adsorption has been reported for alumina in water[18] and for Fe_2O_3 in heptane.[19] Generally an optimum length has been found for each of these effects.

The adsorption of various esters onto alumina has been shown[20] to be increased when the ester is less soluble in the solvent (toluene, chloroform, etc.).

To estimate the degree of adsorption, a solution of menhaden oil in the trichloroethylene–alcohol mixture was filtered through the alumina powder and evaporated to determine the unadsorbed residue. Additional blanks were run without the alumina and without the oil. The amounts of oil residues in the first two cases were indistinguishable to within the $\pm 5\%$ precision of the experiment. Therefore, there was little or no deflocculant adsorbed. In the third case the residue was less than one-tenth of the usual amount of oil in the composition, showing that there was not sufficient oil-like adsorbate (grinding aid, etc.) on the as-received alumina to dissolve and mask a significant adsorption of the menhaden oil. (These mixtures were not milled because of difficulty in separating the milled alumina from the small amount of solvent.)

If trichloroethylene is substituted for the alcohol in the slip composition, the viscosity becomes excessive. The menhaden oil is very soluble in this liquid but less soluble in the alcoholic mixture and insoluble in pure ethanol.

A hypothesis consistent with the above is that a certain minimum amount of menhaden oil is required in solution to drive the adsorption process by mass action. Increasing the solubility decreases this effect. However, less than 5% of the oil is actually adsorbed.

The zeta potential of the alumina in trichloroethylene–alcohol in the presence of the menhaden oil was estimated to be +74 mV, calculated from a measured[21] electrophoretic mobility of 0.5 μm cm/V second. However, when the menhaden oil was omitted, the mobility was the same. This indicates that the deflocculant probably does not operate through charge repulsion, and the steric-hindrance model is likely to describe the system. It should be noted that the mobilities were measured on dilute, unmilled suspensions. It is known[22] that the zero point of charge of alpha alumina can be altered by abrasion, but experimental difficulties prevented the use of concentrated, milled suspensions. Also, traces of water were probably present. However, if a strong charge-repulsion effect existed in this system, it would probably be detected here, even without milling.

The pH of water in contact with the standard milled slip is 8.0 $\pm$ 0.5.

Chemical Characterization. Using A-16 with a surface area of approximately 11 m^2/g, a series of compounds was substituted for fish oil in otherwise standard mill batches. The 2.3 gal mill jar was used in all cases and at least one standard mill using menhaden oil was made from each bag of A-16 alumina. The results are reported in Table 30.3. A measure of the viscosity during the first day of milling was estimated qualitatively by the loudness of the ball sound.

All the slips marked "thixo." in the table were pseudoplastic and thixotropic and appeared highly viscous when not being agitated. However, the viscosity of the slip when nearly at rest is evidently not an important consideration, as shown by the fact that these thixotropic slips were sufficiently fluid when agitated and were easily cast.

Mill batch 236 adhered to the cellulose acetate carrier film after casting and drying, and the surfaces of the forcibly peeled-off tape were rough. Therefore, the amine has a deleterious effect on the overall system, although it deflocculates. Batch 251 was prepared using 120 g of the B-98 Butvar binder in place of the fish oil deflocculant in the first stage of milling. Fluidity was achieved during the first hour of milling and was maintained

Table 30.3. Substitute deflocculants

Mill Batch Number	Material Replacing Menhaden Oil, 55 g	Chemical Functional Groups Present	Results			
			Molecular Weight	Slip Viscosity	Tape	Fired Substrate
211	No Deflocculant	None	—	Solid		
185A	Nujol	(Possible C=C and high MW)		Solid		
252	Propylamine	Base	59	Viscous		
236	Octadecylamine	Base	270	Thixo.	One half cracked	Rough surface
—	Trichloroacetic acid	Acid	163	Solid		
191	Oleic acid	Acid, C=C	282	Solid		
209	Octadiene	C=C	110	Solid		
249	Glycerine	OH	92	Solid		
189	Phthalates and polyethylene glycol (plasticizers)	OH, COO	278, ect.	Solid		
182, 185	Octyl and other phthalates (plasticizer)	COO	278, etc.	Solid		
251	Poly(vinyl butyral) (binder)	COO, OH, (C=C)	32,000	Fluid	Sticks to cellulose acetate	

207	Glyceryl tristearate	OOO	891	Solid		
210	Ethyl oleate	COO, C=C	310	Solid		
237	Glyceryl monooleate	COO, C=C	357	Thixo.	One half cracked	
239	Ditto, but 1413 cc alcohol, no TCE	COO, C=C	357	Solid		
204, 208	Glyceryl trioleate	COO, C=C	885	Thixo.	Good	Good
250	Glyceryl trioleate plus octadecylamine	COO, C=C, Base	885, 270	Thixo.		
177, 253	Corn Oil	COO, C=C; see Table 30.4	885; etc.	Fluid	Good	Good
296	Untreated menhaden oil (Haynie LCP)	COO, C=C (NH_2); see Table 30.4	885, etc.	Fluid, slightly viscous		
215	Extra menhaden oil (110 g total)	COO, C=C (NH_2); see Table 30.4	885, etc.	Fluid, slightly viscous		

during the remaining 23 hours. According to the supplier,[23] this material contains about 19% hydroxylated species and about 2% acetate ester. It might also contain residual, unpolymerized vinyl groups and, therefore, double bonds.

The odor of the menhaden oil indicates that an amine base might be present. Therefore, the composition of mill batch 250 was designed to duplicate the functional groups present in the menhaden oil (amines and double bonds, etc.), but with the use of only synthetic, characterizable materials. However, the viscosity was equally low in batches 204 and 208 using glyceryl trioleate alone as a deflocculant. The glyceryl trioleate results demonstrate that the goal of a synthetic deflocculant is achievable, at least experimentally.

Inspection of Table 30.3 indicates that a combination of a carbon–carbon double bond, an ester group, and a molecular weight of at least 357 was necessary in the series of deflocculants that were studied here. For example, the ethyl oleate molecule contains a double bond and an ester group, but its molecular weight is too low, and the deflocculation was unsatisfactory. Glyceryl monooleate, with the same functional groups and also a higher molecular weight gave sufficiently fluid slip for casting.

Glyceryl tristearate has ester groups and a high molecular weight but no double bonds, and it was unsatisfactory. For comparison again, glyceryl monooleate has double bonds in addition and was satisfactory.

Nujol is a refined mineral oil commonly used for dispersing powder samples for infrared analysis. It is used for this purpose because its infrared spectrum does not usually show COO group absorptions or components other than carbon–carbon single and double bonds. Its viscosity indicates the probable presence of components with molecular weights on the order of a few hundred. This material, usually lacking COO groups, did not give effective deflocculation.

Corn oil is similar to menhaden oil in that it contains glyceryl esters of unsaturated fatty acids, as shown in Table 30.4, and therefore all three attributes mentioned above are present. This oil gave excellent results as a substitute for menhaden oil. The slip viscosity appeared slightly low, indicating excellent deflocculation.

Other complex materials, such as poly(vinyl butyral), evidently do deflocculate and might be worth futher study. Also, glyceryl monooleate might be improvable, with the use of extra plasticizer to decrease the cracking during drying. It is soluble in alcohol but requires the presence of trichloroethylene (TCE) for effective deflocculation (see batch 239 results).

A wider range of materials appears to deflocculate the slip when the alumina surface areas are $\leq 10.5\ m^2/g$, but we have not thoroughly evaluated these compositions.

Table 30.4. Analysis of various oils

			Weight Percent of Fatty Acid Groups[b]			
			Saturated		Unsaturated	
Material	Iodine Number[a]	Saponification Number[a]	Stearic	Other	Oleic	Other
Untreated menhaden[5]	170	191	1	23	0	76 (30% linolenic)
Air-treated menhaden[46]	100	185	—	—	—	—
Untreated corn[5]	123	192	3	12	50	35
Glyceryl trioleate (calculated)	86	190	0	0	100	0
Glyceryl tristearate (calculated)	0	190	100	0	0	0

[a] Defined in reference 5, p. 793. The iodine number is a measure of unsaturation, and the saponification number is a measure of ester content.
[b] Percentage of total fatty acid groups only, not of total oil. Natural oils are reported in reference 5, p. 776 to be mostly glyceryl esters of fatty acids.

PRECASTING TREATMENTS AND CONTROLS

After the slip is milled it is transferred from the mill to storage vessels or chambers for de-airing. A vacuum is pulled on the slip using a roughing pump to a level below the boiling point, usually until agitation due to escaping air ceases. Some times the slip is actually permitted to boil for a few minutes at this point. One must be careful of excessive solvent loss if boiling is carried on for a long time.

For the alumina process, the slip is pumped into the casting machine through a series of two filters, one with 37 μm openings and one with 10 μm openings. A peristaltic-type finger pump is used with Tygon tubing. The filters, which are nylon, remove any large pieces of unground alumina or undissolved binder that may be present. Both of these can cause defects in the cast tape or fired product.

As in any slip-casting operation, the control of viscosity and specific gravity are important. Early in our development work it was standard practice to check these parameters on every batch of slip that was produced. Later we developed confidence in the process reproducibility and the "feel" of the operator and eliminated all but occasional checks on these parameters. The normal operating viscosity of the slip in Table 30.2 was in the range of 2500 to 3500 cp* and the specific gravity was 1.9 to 2.0 g/cc.

CASTING PROCESS

The Casting Machine

There are a great many variations of this simple device. Basically, however, one has some type of container, one side of which can be moved vertically to form an opening or gate through which the slip escapes. The gate height is controlled by micrometer screw adjustments, often with large dial indicators for convenient viewing. The assembly is customarily fixed in position with a carrier or substrate material drawn under the gate. The substrate is most often a metal belt or a plastic film made of Mylar, polyethylene, or the like, or in some cases individual glass plates. Many factors influence the thickness of the wet cast film at a given gate opening, for example, the hydrodynamic head of the slip reservoir behind the blade, the viscosity of the slip, and the speed with which the film is cast. All these are susceptible to variation, and suitable controls must be instituted for uniform production.

The casting machine used in the alumina substrate work is shown in Figures 30.5 and 30.6. It is an enclosed chamber 4.75 in. × 12 in. × 25 ft. The floor of the machine is a smooth aluminum metal plate on which a strip of cellulose acetate film moves. The film is supplied continuously from a spool and enters the slip chamber through a slot in the end of the chamber. Electrostatic brushes can be mounted at this point to remove any dust particles from the carrier film.

The doctor blade at the exit side of the chamber is adjusted by means of micrometer screws. To produce the desired final substrate thickness of 0.026 in., about a 0.063 in. gap between the blade and the carrier film is used. The layer of slip passes through the gap with its thickness controlled by the micrometer settings. The thickness of the freshly cast layer of slip (just beyond the doctor blade) is continuously monitored by a gamma-ray back-scatter or X-ray transmission instrument, and the gap between the doctor blade and carrier film is adjusted accordingly. This can be done automatically or manually.

The normal casting speed utilized in this work was 6 in./minute. This rate is dependent on the length of the machine, the drying conditions, and whether the machine is to be operated in a continuous or batch mode.

* Brookfield Viscometer: #4 spindle, 20 rpm.

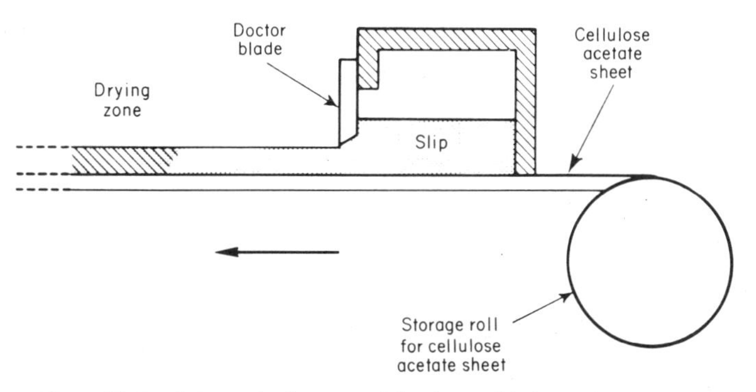

Figure 30.5. Schematic diagram of the doctor-blade casting machine.

Continuously variable speed controls were located at the payoff end of the machine so that a single operator could process the tape.

The Carrier Material

As mentioned above, many materials have been used as substrates or carrier films in the tape-casting process. One usually tries to use the material that

Figure 30.6. Laboratory model of the doctor-blade casting machine.

does the job for the lowest cost. In most cases the binder solvent system plays an important role in the selection of a carrier. A continuous stainless steel or chrome-plated steel belt may be the best carrier if the tape is to be stripped continuously at the end of the machine. The fine-grained alumina substrate process used a unique carrier material, cellulose triacetate (Kodacel TA401, Eastman Chemical Products, Inc., Kingsport, Tennessee). According to the supplier, this material, which is 0.015 in. thick, contains $\simeq$15 wt % of diethyl phthalate plasticizer (bp 290°C), an organic liquid that is similar to one of the plasticizers given in Table 30.2. TGA and DTA studies of the film showed 16% weight loss in N_2 between 100 and 320°C and an endothermic peak at 280°C. After casting and drying, the cellulose acetate is 4.7% thicker and is an average of 3.3% narrower than as-received, probably because of absorption of solvent and/or exchange reactions among the plasticizers.

In our experiments thus far, substitute carrier-film materials that did not become narrowed did not yield satisfactory results, and the dry tape cracked and/or stuck to the film excessively or insufficiently. The cast material must stick to the carrier film sufficiently to prevent extreme curling during drying, but it must not stick so much as to prevent the film from being peeled away from the dry tape prior to firing.

The movie-grade cellulose triacetate also provided an extremely smooth and uniform casting surface, which we believe contributed to the smooth-surfaced, low-defect alumina substrates produced. Thicknesses down to 0.005 in. were used in the standard process with equivalent results.

DRYING

In the substrate process, the layer of slip dries slowly while it is carried through the machine on the moving plastic film. Filtered air is blown through the machine in a direction opposite to that of the moving slip. The air flow rate is 5 ft/minute. This arrangement provides dry air at the exit end in contact with the almost dry tape. At the other end in contact with the freshly cast wet slip is air that has passed through the entire casting machine and is saturated with solvent vapor. The difference in solvent content between the slip and the air is slight at any point in the machine. This controls the rate of solvent evaporation, allowing the remaining solvent in the slip layer to redistribute itself with only a small vertical gradient of concentration. This small gradient of concentration tends to minimize curling and cracking of the slip as it dries. Heat increases the drying rate, but its use is limited, because the boiling point of the solvent, 71°C, must not be exceeded if controlled, bubble-free drying of the slip is to be achieved.

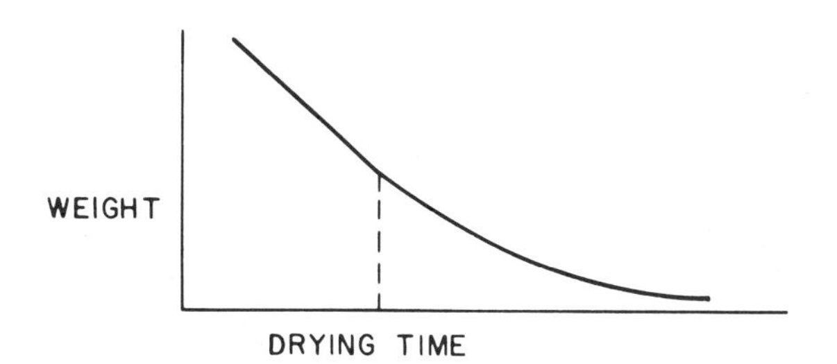

Figure 30.7. Schematic of weight versus time in a two-stage drying process.

The drying of slurries in general has been studied by other workers. A two-stage process is commonly observed with such materials,[24] the first stage of drying proceeding at constant rate, and the second stage at a gradually decreasing rate. These are shown in Figure 30.7. In either stage of drying, the solvent must move through three consecutive steps of transport: (1) solvent flows vertically through the slip to the surface, (2) solvent then evaporates at the surface, and (3) solvent vapor is swept away at the surface. The overall rate is determined by whichever is the slowest step.

In the first stage the slip is still fluid and solvent is easily transported through it (step 1) by liquid diffusion or capillary action. Step 2 is the slow step, being limited by the inflow of heat required to supply the latent heat of vaporization. In those cases involving very little air flow, step 3 can also be slow, causing a pileup of nearly saturated vapor at the surface and a consequent decrease in the evaporation rate. Therefore, the rate of total drying in this stage can be influenced by either temperature (step 2) or air flow (step 3). Neither of these is likely to be time dependent, and so a constant rate is usually observed for still-fluid slip.[24]

In the second stage the slip has become solid, and the solvent no longer flows by capillary action or fast diffusion. The solid is gel-like, and step 1 is limited by the slow diffusion of solvent through the gel. This would be the slowest step, and like many diffusion processes, it might be expected to follow the same sort of equations as radioactive decay, having a rate that decreases with time.

A solution[25] to the well-known Fick's law diffusion equations, set up for the case of an initially uniform solvent distribution and evaporation from one surface of a slab, is given by

$$W = W_0 \left\{ e^{-Dt(\pi/2d)^2} + \frac{1}{9\exp[-9Dt(\pi/2d)^2]} + \frac{1}{25\exp[-25Dt(\pi/2d)^2]} + \cdots \right\} \qquad (1)$$

where W = weight
W_0 = initial weight
D = diffusivity
t = time
d = thickness of slip.

At relatively long times, only the first term of the series is significant. A plot of log W versus t should then give a straight line. It is apparent from the equation that an increase in the diffusivity would have a strong effect on the drying rate.

Diffusivities in solids tend to vary with temperature according to the Arrhenius activation energy equation:

$$D = D_0 \exp \frac{-E}{RT} \tag{2}$$

A special apparatus was used to determine the drying rates of the slip at various temperatures and air flows. This is shown in Figure 30.8. Nearly laminar flow was achieved by filling the rectifier chamber with 10,000 parallel drinking straws, thus improving the reproducibility by minimizing random eddy currents. Slip was cast onto a glass sheet on the pan of the scale using a movable doctor blade. The weight of the slip was noted periodically. A thermocouple was imbedded into the slip sample.

Figure 30.9 shows a plot of experimental data obtained with the laminar flow apparatus. Two stages of drying are indicated (run 1), as illustrated by the constant rate (linear) portion of the curve, and the changing rate

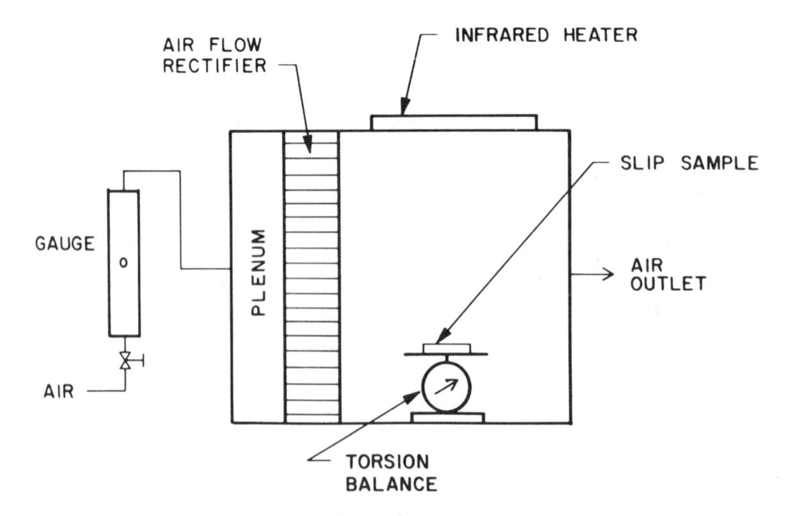

Figure 30.8. Experimental laminar flow-drying chamber.

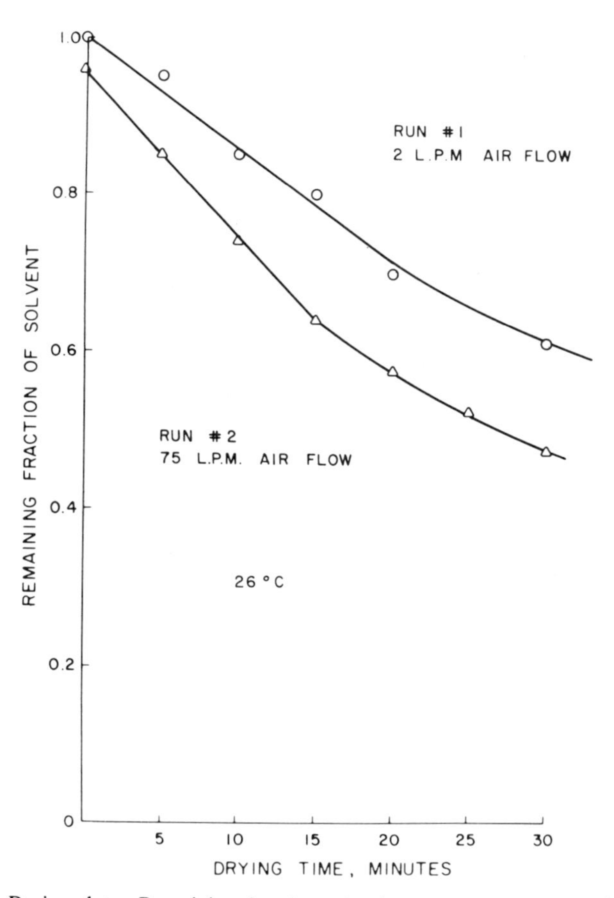

Figure 30.9. Drying data. Remaining fraction of solvent versus drying time at constant temperature.

(curved) portion. The experiment was repeated, but this time there was an air flow of 75 l/minute. These data indicate that the air flow has an effect on the first stage, as predicted by the model.

Figure 30.10 shows the same data replotted on semilog paper. The straight lines during the second stages of drying confirm the predicted exponential relationship, assuming that only the first term of the series is significant. Note that air flow does not affect the second stage slope, showing that diffusion through the gel is probably the slowest step, not evaporation or vapor removal. Also shown in Figure 30.10 are data taken at a slip temperature of 60°C. The indication is a speedup of drying during both stages. This is consistent with the idea of a latent heat effect during the first stage and an activation energy effect during the second stage. The diffusion through the solid was speeded to the limiting point where it was no longer the slowest step.

Five drying experiments were run at different temperatures, with the weighings being made only after the solidification of the slip (second-stage drying). A log W/W_0 versus time plot for each experiment gave values of $\pi^2 D/4d^2$ as the slopes. To obtain the Arrhenius activation energy, these five slopes were plotted against the reciprocal absolute temperature in Figure 30.11. The slope of this plot is $-E/R$ in the Arrhenius equation, and the activation energy E was thus obtained. The experimental value is 22 Kcal/mole, which is a reasonable value for such polymer–solvent systems.[26]

Several means of increasing the drying speed during actual casting runs were demonstrated by this experiment. Increased air flow during the first stage of drying is one method, and heat applied during either stage is another.

In the casting machine shown in Figure 30.6, it was found experimentally that air flow above 110 ft^3/hour (5 ft/minute) resulted in skin formation

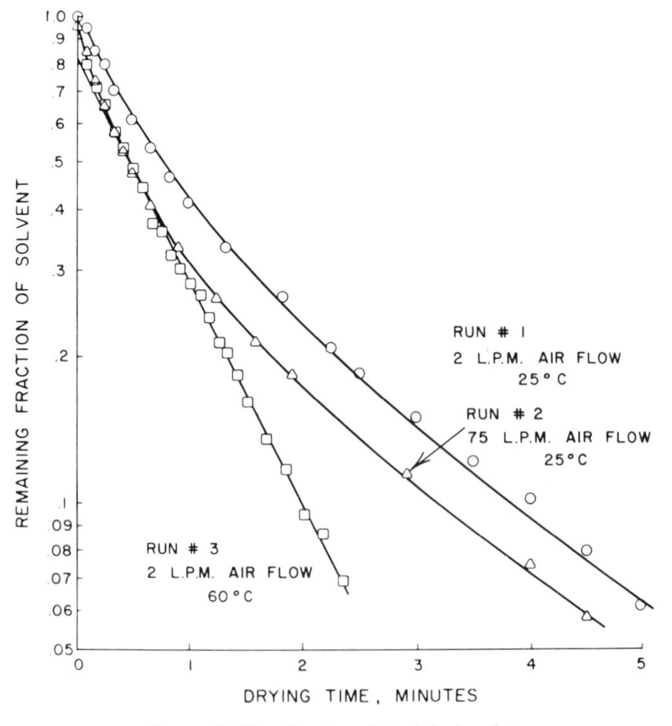

Figure 30.10. Semilog plot of drying data.

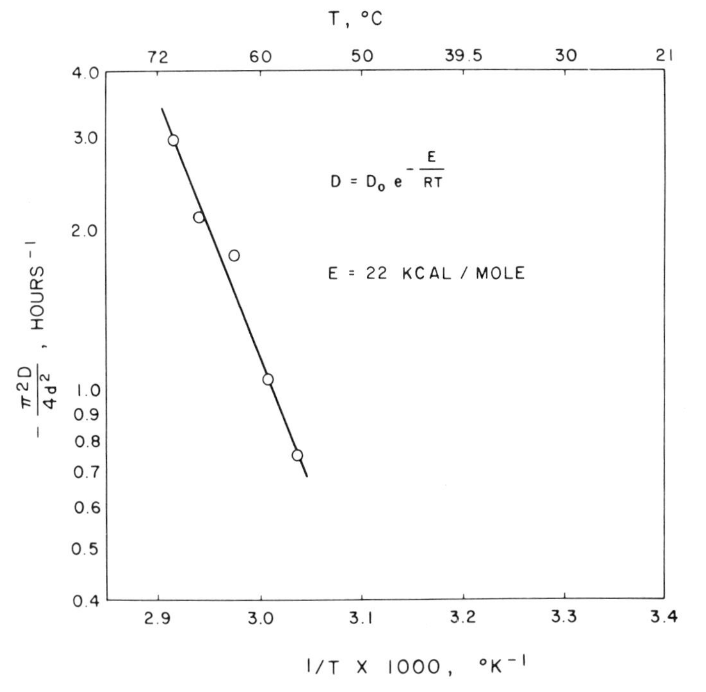

Figure 30.11. Arrhenius plot for diffusion-controlled stage of drying.

during the first stage of drying. Evidently the solvent is removed from the surface layer too fast for diffusion through the liquid to replenish the layer, even though this diffusion is fairly rapid. Once skin forms the first-stage drying is slowed, since all diffusion must then proceed through the solid skin. The formation of skin also has other detrimental effects on the quality of the final product. Therefore, the forced drying by increasing the air flow is limited to this level.

On our machine heat was applied to the drying slip by mounting nichrome heating elements on the glass top of the machine. The elements were positioned to bring the slip temperature up rapidly during the first stage of drying. A temperature of 50 to 80°C was maintained in the drying slip, with the highest temperature being at the driest end of the tape. Temperatures were measured by embedding a thermocouple in the traveling slip. The profiles of two typical forced-drying runs are shown in Figure 30.12.

This type of forced drying provided satisfactory results in that tapes were sufficiently dry to strip from the carrier at the end of a 25 ft casting machine.

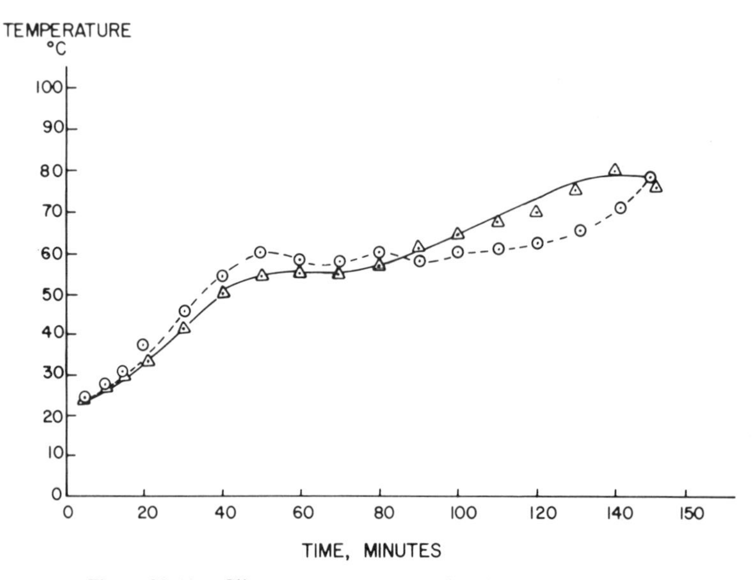

Figure 30.12. Slip temperature versus time in casting machine.

PRECISION CASTING

For some applications precise dimensional control of the final part and, therefore, of the as-cast tape is essential. This was the case in the development of a process for producing lead zirconate titanate tapes 1 to 12 mils thick with thickness tolerances of better than ± 0.3 mils over large areas.[27] Briefly, the primary requirement was to improve existing tape-casting technology by providing a precise gap between the blade and the casting surface. Further requirements were to provide a casting surface of a moving carrier tape of controlled thickness that would glide under the blade and over a supporting, laterally level bed with minimum friction. To provide these features we constructed a precision casting machine that had the following features:

1. A well-constructed, stable table base.
2. A curved bed, 1 ft wide and 14 ft long with its maximum height under the doctor blade.
3. A thin carrier tape with close thickness tolerances.
4. A doctor blade with an edge milled to a fraction of a mil in straightness.
5. A doctor-blade configuration to minimize the effects of hydrodynamic and suface tensile forces at the blade.

The major features of this machine are the curved bed of smooth tempered glass bent into the arc of a circle with a radius of 430 ft, and a two-bladed doctor blade. The curvature was provided to assure close contact between the thin carrier tape and the bed. The maximum height of the circle was located under the doctor blade so that the surface at this point would be horizontal in the longitudinal and the lateral direction, to prevent any tendency of the slip to flow under gravitational forces. The doctor-blade construction is shown in Figure 30.13. This construction with two blades was chosen to control the hydrodynamic and surface tensile forces as the slip passes under the blades. In Figure 30.13 the left blade is the casting blade. Because of hydrodynamic forces it is necessary to maintain a constant height of the casting pool at the right of the casting blade to maintain a constant height of the cast to the left of the blade. It is also beneficial to minimize the differential heights between the casting pool and the cast on the respective sides of the casting blade. The simple two-blade design satisfied these two conditions. By adjusting the relative heights of the two blades and the speed of the carrier tape, a constant low-level casting-pool height may be maintained for all but the initial and final few seconds of casting.

The other parameters that must be controlled to optimize casting thickness control include slip viscosity (function of percent solvent and temperature), casting speed, doctor blade setting, and drying shrinkage of the tape.

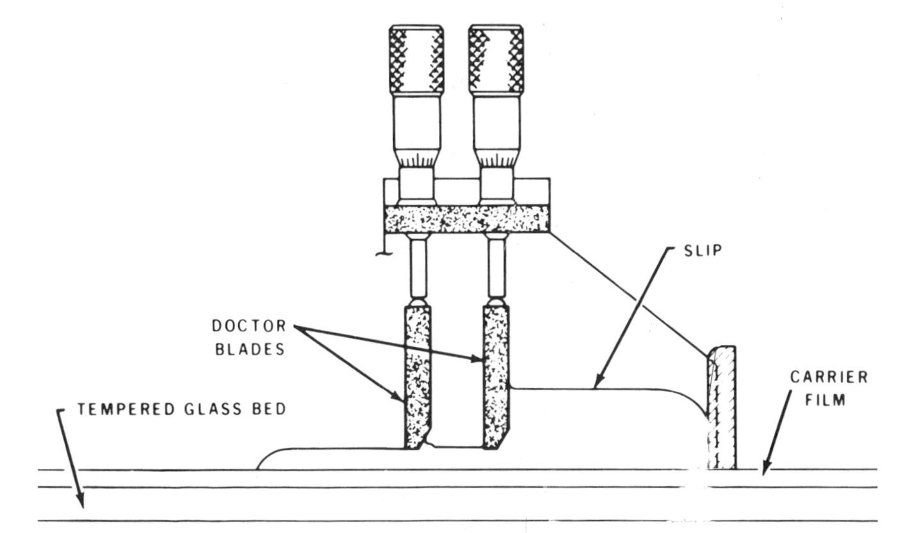

Figure 30.13. Side view of precision tape-casting doctor-blade showing two-blade construction and optimized flow conditions.

With the machine described above and by close control of the casting parameters mentioned, tapes of lead zirconate titanate have been reproducibly cast 8 in. wide and 10 ft long, with thicknesses from 1 to 12 mils and with thickness tolerances better than $\pm 4\%$ (e.g., 5.0 ± 0.2 mils).

TAPE CHARACTERIZATION

The density of unfired tape (often called the "green" density) is an important parameter for characterizing a tape-casting process. In experimental modifications of the tape, the green-density measurement can be useful in detecting poor packing of the powder or in detecting excessive polymeric binder content.

In the determination of bulk green density[28] the volume measurement can be made with a flat-faced micrometer caliper, provided about 15 thickness measurements are taken and averaged (for a substrate area of about 20 in.^{2}). The same volume can be obtained by extrapolating mercury porosimeter data to zero pressure. Typical bulk densities for the alumina tapes produced in our laboratory were 2.5 to 2.6 g/cc. Taking the organic materials into account, these bulk densities correspond to an alumina volume percent of 56, which is a densely packed green body, considering that no pressing is involved and the alumina particles are small.

The volume for calculating the apparent density[28] of the tape can be measured by a mercury porosimeter at 20,000 psi or an air pycnometer at atmospheric pressure. In either case the measured apparent green density of the alumina substrate tape ranged from 2.9 to 3.0 g/cc.

Another technique used for measuring the density of the tape was the Archimedes' method[29] using water. A difficulty was encountered in that water immediately began to be absorbed into the tape as the immersed weighings were begun, causing changing weight values and preventing bulk-density determinations by this means. Another approach[29] to bulk density determination is to thoroughly soak the tape sample with water and then weigh it while it is soaked but not immersed. This was not possible because the water immediately began evaporating, again causing continuously changing values. Also the tape expanded when immersed in all the liquids that we tried.

A method for extrapolating the water absorption back to zero immersion time was developed. The method rests on the assumption that the more water absorbed into the tape pores, the slower the further absorption. The amount absorbed should asymptotically approach a maximum value. One would then expect the following relationships to be observed.

$$\frac{dW}{dt} = k\left(\frac{\text{amount of unfilled open}}{\text{porosity remaining}}\right) = k(W_m - W) \qquad (3)$$

where W = weight of tape plus absorbed water at time t,
W_m = maximum value of weight reached by the tape at long times
k = empirical constant.

To simplify the calculations substitution can be made: Replace dW/dt with an equal quantity, $-d(W_m - W)/dt$. Then

$$\frac{d(W_m - W)}{W_m - W} = -kdt \tag{4}$$

Integration gives

$$\log (W_m - W) = -kt \tag{5}$$

This allows a linear plot to be obtained when $W_m - W$ is plotted versus t on semilog paper. Extrapolating back to zero time, one can obtain the initial weight of tape immersed in a buoyant fluid such as water before any swelling has taken place. From this and the unimmersed (dry) weight, one can calculate the bulk tape density using the Archimedes' method. The experimental plot did yield a straight line, and the calculated density of the alumina tape was 2.63 g/cc.

An estimate of the apparent density can also be obtained by a variation of this method. W_m minus the initial (extrapolated) weight is the amount of water absorbed. The volume of this is roughly the volume of the open porosity, subject to a possible error due to swelling. This volume can be combined with the other data to calculate[29] the apparent density. It was found to be 2.91 g/cc, which is in agreement with the other techniques tried.

A comparison of the bulk and apparent densities given above indicates that the tape samples contain some kind of open porosity. It should be noted that the tape must be broken into small pieces for use in the pycnometers or the porosimeter, and many cross-sections are thus exposed. Therefore, it is possible that the major surfaces of the tape are sealed by polymer and the pores are only accessible through the cross-sections.

Visible-light examination with a variety of lighting techniques at magnifications from 1 to 800× does not show evidence of porosity on any of the surfaces. However, SEM pictures of the surface at 10,000× show submicron-size pits. It could not be determined whether the bottoms of these pits were sealed with polymer. Also, microtomed cross sections appeared to show porosity, under the SEM, although these could have been an artifact of the microtoming operation.

Assuming that the tape is not significantly compressible, the mercury porosimeter pressure versus volume curves can be used to calculate pore size. A curve of this type on the alumina tape indicates that 50% of the pores are $< 0.09\ \mu$ in diameter. However, the tape is compressible, and this method cannot distinguish between porosity and compressibility while the pressure is applied.

The porosimeter operators reported some degree of permanent penetration of the tape by the mercury, proving that there is some open porosity. However, the weight gain due to Hg absorption cannot be used as a quantitative separation of porosity from compressibility because some mercury flows back out of elastic porous materials when the applied pressure is removed. The weight gain is only a qualitative test, but nevertheless it is a definite indication of unoccupied space in the original tape.

It should be noted that it is difficult to distinguish between open porosity versus a molecular level of penetration of the water and air into the polymer structure itself. The polymer might be especially permeable to these fluids if it were in a stretched state, having shrunk between the alumina particles during drying. The mercury might be a better test for macroporosity, but at 20,000 psi the mercury could conceivably force its way into the stretched polymer, thus creating discrete pores. Therefore, the original "pore" size might be as large as 0.09 μ or as small as a molecule.

As an increasing amount of polymeric binder is added to the tape composition, more of the interstitial space between the alumina particles is probably filled with binder. It has been found that an excess of binder above a certain level (4 g of binder for each 100 g of the alumina) causes maldistribution effects in the tape and curling or cracking during firing. The apparent density measured with an air pycnometer should be sensitive to the amount of organic chemical material (binder, etc.) present in the interstitial space, and this is probably why it was found to be useful in diagnostically assessing new formulations. A plot of bulk and apparent density versus percentage organics is shown in Figure 30.14. It should be noted that, while the apparent densities are sensitive to the amounts of organic materials, the bulk densities are less sensitive.

Attempts to correlate curling and cracking with the apparent-density measurements have only been made with data obtained by the use of the air pycnometer and not with other types of measurements. It is therefore unknown whether all the apparent-density methods are equally sensitive to excess binder or whether there are significant differences in fluid permeability. At any rate, the other apparent-density methods are less convenient than the air pycnometers.

It is well known[30] that the sinterability of a powder compact is dependent on the degree to which the powder particles are packed together. In the tape process, poor deflocculation might be expected to cause poor packing of the alumina particles and thus low fired density. The degree of alumina packing can indirectly be measured by determining the bulk density, provided the amount of other material (binder, etc.) is known.

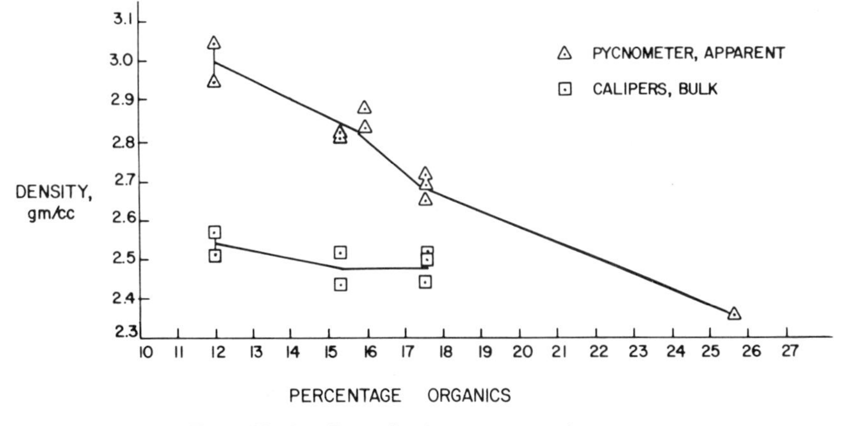

Figure 30.14. Green density versus organic content.

Bulk densities and the resulting fired densities are reported in Table 30.5 for various experimental tape compositions in which poor deflocculation and poor alumina packing probably occur. For comparison, data are also shown for standard tapes in which poor packing was not observed to occur. (It should be noted that good results have been reported when the other aluminas in Table 30.5 were used in other processes with increased firing temperatures.)

The packing factor in the tape can be calculated as follows: In 1 cc of standard tape the total weight is about 2.53 g (bulk density). Of this, 12.0% is organic material. Weight-loss measurements during 750°C binder

Table 30.5. Substitute aluminas

Alumina Powder Used	BET Surface Area (m^2/g)	Bulk Green Density (Caliper), (g/cc)	Bulk Fired Density (Archimedes), (g/cc)
Standard[a]	11.3–12.4	2.51–2.57	3.88–3.90
Linde A[b]	8.3	2.09	3.8, Porous
Calcined $Al_2(SO_4)_3$[c]	9.1	2.01	3.8, Porous
Kappa-theta[d]	16.6	1.97	3.8, Porous

[a] A-16, Aluminum Company of America, Pittsburgh, Pennsylvania.
[b] Linde A, Union Carbide Company, East Chicago, Indiana.
[c] Experimental, Western Electric Company, Princeton, New Jersey.
[d] Experimental, E. I. duPont De Nemours and Company, Wilmington, Delaware.

removal have shown that the tapes exposed to ambient air for several months contain less than 1% residual solvent. Spectrographic analysis has shown that inorganic impurities comprise about 1% or less of the fired substrate composition. Neglecting residual solvent and inorganic impurities, the alumina is the other 88% of the 2.53 g, or 2.23 g. The bulk density of the alumina itself in this configuration is therefore 2.23 g/cc. This is 56% of the density of sapphire (3.986 g/cc as measured by Wachtman[31] on a variety of synthetic sapphire samples).

As calculated from the micrometer bulk density, it is interesting to compare the 56% packing factor of the tape A-16 alumina with the corresponding value for dry-pressed A-16.

After pressing at 5000 psi, with a small amount of stearic acid as a lubricant, micrometer green densities measured in this laboratory are typically 2.18 g/cc (single action press), and Alcoa reported[32] 2.24 g/cc (double action press). These correspond to packing factors of 55 to 56%. Therefore, the packing in the tape-cast material is excellent, considering the fact that no external pressure is applied during its formation.

In 1 cc of standard tape, the weights of alumina and organic materials were as reported above. Knowing the densities of the sapphire (3.986 g/cc, reference 31) and the mixed organic materials (1.06 g/cc), the volumes of these materials are easily calculated. Alumina (2.226 g) has a volume of 0.558 cc, while the binder (0.304 g) has a volume of 0.287 cc. Thus the total weight is 2.530 g and the total volume is 0.845 cc. The apparent density of the total solids from these calculations is 2.99 g/cc. This is indistinguishable from the range of apparent-density values reported previously. Therefore,

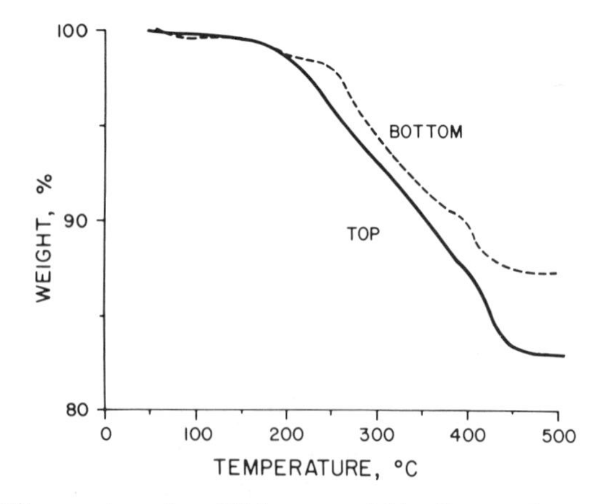

Figure 30.15. Nitrogen atmosphere TGA curves of thin slices cut from top and bottom sections of tape.

there is essentially no closed porosity, at least as measurable by these methods. These calculations are for 1 cc of tape, of which 0.845 cc is occupied by the total solids. Therefore, the open porosity is $\simeq$15% by these methods.

Figure 30.15 shows binder burnout TGA curves for thin sections cut from the top and bottom of the tape. The difference between the curves is related to slight warping during firing, but it is satisfactorily small. This also is an excellent characterization test to determine tape uniformity. If segregation of the solids and organics took place as a result of settling, there would be a large difference in binder content from side to side. The uniformity observed is a good measure of the degree of deflocculation, even during the drying stage. This, we believe, combined with a distribution of particle sizes leads to the excellent packing and relatively high bulk density achieved.

PUNCHING

The as-cast sheets of ceramic tape are usually cut to the desired sizes and shapes with a simple punch and die set actuated by a rotary punch press. This operation is called blanking. The pieces are punched oversize to allow for firing shrinkage.

In the case of the smooth-surfaced alumina substrate tapes, special care was taken to avoid scratching the relatively soft tape during handling and punching. Dust, punching scraps, and other foreign particles must not be pressed into the tape by the punch face or the ejector. A specially designed punch with a recessed central area that only contacted the tape at the very edges of the parts was used. In addition, careful vacuum-cleaning procedures between punching operations were instituted.

If holes are required in the finished ceramic, they can often be punched simultaneously with the blanking operation, especially if the holes are larger than 0.020 in. in diameter. For larger holes the operation is almost trivial and no serious problems are involved. But for many applications in the electronics industry, arrays of very small holes are desirable. Typical examples are alumina substrates for thin-film or thick-film microcircuits, in which electrically conductive metallization is applied to both sides, and holes must be provided for current to pass from one side to the other. The smaller the hole diameters, the more densely the components can be packed, and therefore the higher the scale of hybrid integration for designing a complex integrated circuit.

Punching is an inexpensive method for producing holes in tape-cast ceramics, because hundreds of holes can be produced in a single stroke of the punch into the die.

Alternative methods are available, such as laser drilling[33] and diamond drilling the fired ceramic. If holes are to be located within 1 mil of nominal, these are the preferred techniques. Piazza and Steele[34] have pointed out that multiple punching of unfired tapes can be utilized with reasonable yields if the tolerances on hole location are 8 mils or greater.

In determining whether punching will be less expensive for a given hole size, an important consideration is the expected life of the punch and die set. For holes of diameters greater than 0.020 in., punches made of hard materials, such as tungsten carbide, can be expected to last through several hundred thousand punch operations. A new punch and die set ordinarily costs several thousand dollars, and this method is therefore suitably inexpensive for mass production. However, when hole diameters approach 0.010 in., the punches are fragile, and laser drilling becomes cost competitive in spite of its necessarily time-consuming step-and-repeat mode of operation.

If the improved techniques of punching smaller-diameter holes reported below are used, the punching of the tape before firing can be restored as a good candidate for the less-expensive method.

Figure 30.16 shows a schematic cross section of a punch and die, in the middle of the operational sequence. As shown in the diagram, the punch has pushed a cylindrical slug of tape into the die cavity. The next step is for the punch to retract upwards, while the metal "stripper plate" keeps the tape from moving. The plate then rises also, the tape is removed (to be inspected and fired), and the next piece of tape to be punched is placed over the die.

Experiments in our laboratory, later confirmed at other facilities, have shown that 0.010 to 0.012 in. diameter punches tend to break or wear excessively during the first 20,000 operations unless the following design criteria are followed:

1. The larger-diameter shank of the punch should be 8 to 10 times the diameter of the punch itself, to provide rigidity against excessive flexing.
2. The punch tip should move down to a point below the bottom of the upper die hole (Figure 30.16) to prevent the tape slugs from remaining in the die after retraction and jamming the later operations by becoming compressively packed.
3. A partial vacuum should be applied to the larger-diameter die hole to aid in slug removal. This vacuum can conveniently be provided by a Bernoulli effect air jet in each die hole, emanating from a simple gallery of holes drilled at an angle down through the die block.
4. The punch-to-die clearance on each side is optimally about 0.0015 in. for alumina substrate tape. In other words, a punch diameter of 0.012 in. requires a die diameter of 0.015 in. Too small a clearance (0.0005 in. on

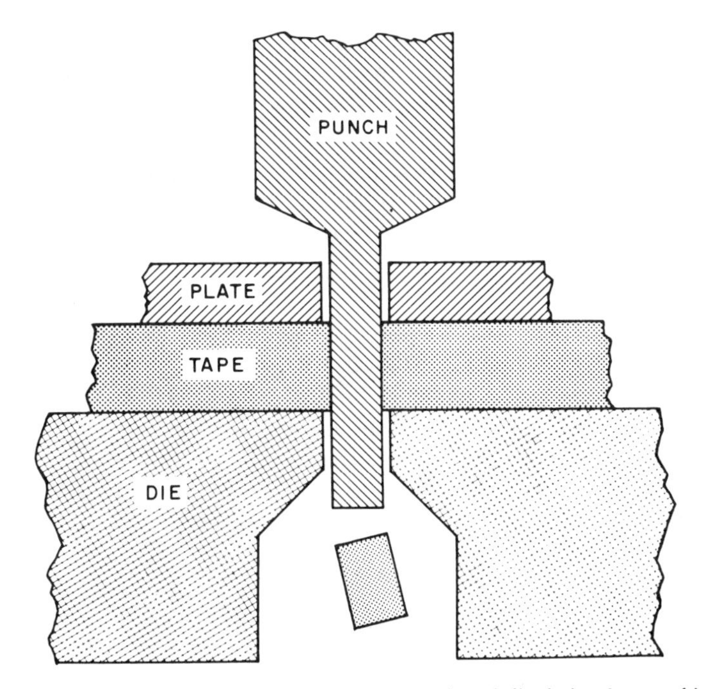

Figure 30.16. Schematic diagram of an optimized punch and die design for punching small holes in green ceramic tape.

each side) results in punch breakage, and too large a clearance (0.003 in.) results in irregular holes in the tape.

Because of irregularities in the tape, narrow punches occasionally flex sideways far enough to strike the die. Therefore, brittle punch materials, such as carbides, should be mated with relatively softer die materials, such as tool steel hardened to about Rockwell 60. After many thousands of operations with this combination of materials, dies were observed to have several "bites" taken out of them, but the tungsten carbide punches (available from Schneider and Marquard Company, Newton, New Jersey) did not break or visibly wear.

Another punch material having good wear properties for use with abrasive ceramic tape is boride-coated steel. (Borofuse, available from Materials Development Corporation, Medford, Mass.) This has the advantage of not requiring electrospark-discharge machining for punch fabrication, since boriding can be done by the punch maker after some of the other fabrication procedures are completed.

It should be mentioned that ordinary tool-steel punches are not sufficiently hard for large-scale production usage, and their tips become rounded after about 10,000 operations with alumina substrate tape. The holes in the tape produced by round tips are irregular and unpredictably small, especially after firing.

With the use of carefully designed punches and dies, hundreds of small holes having excellent cylindrical shapes can be punched simultaneously. No visible wear is typically observed on 0.012 in. tungsten carbide punches after 20,000 operations.

MULTILAYER STRUCTURES

A specialized technology commonly used in electronic ceramics, especially for multilayer capacitors, piezoelectric devices, transistor packages, hybrid circuit modules, and small printed circuit boards, is multilayer processing. Since the processing before sintering is an important part of this technology we cover it briefly in this section. Lamination and screened multilayer formation are two major techniques used to produce the multiple-layer structures.

Lamination

Schwartz and Wilcox[35] describe the lamination process as comprising the following steps: (1) tape casting thermoplastic sheets, (2) punching holes, (3) screen printing the metal electrode patterns, (4) laminating the sheets together, (5) punching the final shapes, and (6) sintering the monolithic structure.

Since we have described the processing steps before sintering with the exception of screen printing and laminating, we will concentrate on these latter aspects here. Metallization of the green ceramic sheets is accomplished by screen printing. Special vacuum chucks can be used to hold the tape in place and to provide suction for simultaneous metallization of through-holes. Close control of the metal-paste properties (viscosity and density) will aid reproducibility. The metal paste, which is screen-printed, must sinter compatibly with the ceramic material. Palladium, platinum, molybdenum, and tungsten have been used with various ceramic materials. There are numerous patents and papers[36-41] describing the different types of metallization used in multilayers.

The actual lamination process consists of applying pressure to several stacked green sheets at a suitable temperature to produce a monolithic structure. The temperatures used have ranged from room temperature to several hundred degrees Fahrenheit. Temperature and pressure depend to a large degree on the type and amount of binder. A typical range of pressures is from 200 to 20,000 psi. Care should be taken to insure equal application of pressure over the entire surface of the part. Sufficient time must be allowed to permit the entire part to reach thermal equilibrium. Another technique described in the literature[42] is continuous roll-to-roll lamination performed by passing two or more sheets of material between a set of nip rollers at pressures ranging from 100 to 1000 psi. This is done at room temperature at a speed of 20 ft/minute.

Screened Multilayer Structures

Recently a process has been reported[43] that eliminates the need for costly punching tools and allows low-cost, automatic processing of multilayered structures. It is known as the Screened Multilayer Ceramic (SMC). In the SMC process conductive layers and dielectric layers are alternately screen printed onto flexible unfired ceramic tape. For the conductive patterns, moly-manganese or tungsten are used, and a special screenable alumina paste is used for the dielectric layers when the base tape is alumina. The screened multilayer structure is cosintered at about 1600°C in a reducing atmosphere. The process avoids costs of alignment and punching, as well as the costs and time associated with tooling that are involved in lamination processes.

MULTILAYER CASTING

A multiple-layer casting technique has also been described in the recent literature.[44,45] The technique is not used to produce ceramic packages with buried conductors. In the process described, it was used to cast three layers of material simultaneously on top of one another. The purpose was to produce a gradient of grain sizes from the surface layers to the internal layers, with the ultimate goal of producing a high-strength material. A schematic of the triple doctor blade is shown in Figure 30.17. Details of the

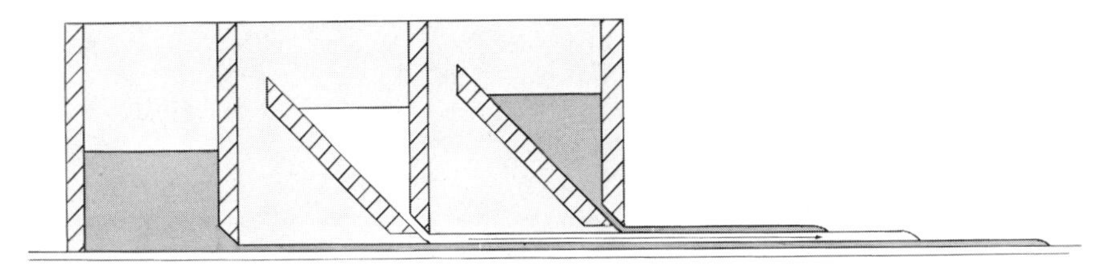

Figure 30.17. Schematic drawing of a doctor-blade arrangement for multiple-layer casting.

casting process are included in reference 44. This technique produced alumina substrates with flexural strengths of >100,000 psi.

SUMMARY

In this paper we have attempted to convey some of the many "tricks of the trade" a ceramist would use to develop a tape-casting process. Many of the examples we have used were taken from our own work in developing a process for producing high-quality fine-grained alumina substrates. Some areas have been emphasized more than others, mainly because of a lack of detailed literature in those areas. All steps before sintering were covered, however, in an attempt to create a coherent picture of the tape-casting process.

ACKNOWLEDGMENTS

Thanks are due to many Western Electric employees at the Engineering Research Center, both past and present, for aid in developing the technology described in this paper.

REFERENCES

1. G. N. Howatt, "Method of Producing High Dielectric, High Insulation Ceramic Plates," U.S. Patent 2,582,993, (Jan. 22, 1952).
2. (a) H. W. Stetson and W. J. Gyurk, "Alumina Substrates," U.S. Patent 3,698,923 (Oct. 17, 1972); (b) D. J. Shanefield and R. E. Mistler, "Manufacture of Fine-Grained Alumina Substrates for Thin Films," *West. Electr. Eng.*, **15** (2), 26–31 (1971); (c) G. N. Howatt, R. G. Breckenridge, and J. M. Brownlow, "Fabrication of Thin Ceramic Sheets for Capacitors," *J. Amer. Ceram. Soc.*, **30** (8), 237–242 (1947); (d) J. L. Park, Jr., " *Manufacture of Ceramics,*" U.S. Patent 2,966,719 (Jan. 3, 1961); (e) J. J. Thompson, "Forming Thin Ceramics," *Amer. Ceram. Soc. Bull.*, **42** (9), 480–481 (1963); (f) D. J. Shanefield and R. E. Mistler, "Fine Grained Alumina Substrates: I, the Manufacturing Process," *Amer. Ceram. Soc. Bull.*, **53** (5), 416–420 (1974); (g) H. W. Stetson and W. J. Gyurk, "Use of Menhaden Oil of Deflocculate Dry Ground Alumina in Manufacture of Substrates," U.S. Patent 3,780,150 (Dec. 18, 1973); (h) J. F. Argyle, G. O. Medowski, D. W. Ports, and R. D. Sutch "Fine-Grain Alumina Bodies," U.S. Patent 3,819,785 (June 25, 1974).
3. J. E. Davis, V. G. Carithers, and D. R. Watson, "Practical Use of Particle Size Distribution to Predict Compacting and Sintering Properties of Calcined Aluminas," *Amer. Ceram. Soc. Bull.*, **50** (11), 906–912 (1971).
4. H. Burrell, "Solubility Parameters," *Interchem. Rev.*, **14** (1), 3–16 (1955).

5. F. A. Norris, *Encyclopedia of Chemical Technology,* Vol. 8, R. E. Kirk and D. F. Othmer, eds., Interscience, New York, 1965, pp. 776–793.
6. G. D. Parfitt, *Dispersion of Powders in Liquids,* American Elsevier, New York, 1969, pp. 82, 116, 160–169, 296.
7. Terminology used as defined in ASTM D 1356 and E20, Part 33, American Society for Testing and Materials, Philadelphia, 1970, p. 10.
8. D. W., Jr. Johnson, D. J. Nitti, and L. Berrin, "High Purity Reactive Alumina Powders: II, Particle Size and Agglomeration Study," *Amer. Ceram. Soc. Bull.* **51** (12), 896–900 (1972).
9. J. J. Burke, ed., *Ultrafine-Grain Ceramics,* Syracuse University Press, Syracuse, 1970, p. 146.
10. G. D. Parfitt, *Dispersion of Powders in Liquids,* American Elsevier, New York, 1969, p. 315.
11. V. T. Crowl and M. A. Malati, "Adsorption of Polymers and the Stability of Pigment Dispersions," *Discuss. Faraday Soc.,* **42,** pp. 301–312 (1966).
12. D. H. Napper and A. Netschey, "Steric Stabilization of Colloidal Particles," *J. Colloid Interface Sci.,* **37** (3), 528–535 (1971).
13. F. J. Micale, Y. K. Lui, and A. C. Zettlemoyer, "Mechanism of Deaggregation and Stability of Rutile Dispersion in Organic Liquids," *Discuss. Faraday Soc.,* **42** pp. 238–247 (1966).
14. H. Koelmans and J. T. G. Overbeck, "Stability and Electrophoretic Deposition of Suspension in Nonaqueous Media," *Discuss. Faraday Soc.,* **18,** pp. 52–63 (1954).
15. A. E. Lewis, "Polar-Screen Theory for the Deflocculation of Suspensions," *J. Amer. Ceram. Soc.,* **44** (5), 233–239 (1961).
16. A. F. Sherwood and S. M. Rybicka, "Surface Properties of Titanium Dioxide Pigments," *J. Oil Colour Chem. Assoc.,* **49** (8), 648–649 (1966).
17. L. A., Romo, "Effect of C_3, C_4, and C_5 Alcohols and Water on the Stability of Dispersions with Alumina and Aluminum Hydroxide," *Discuss. Faraday Soc.,* **42,** pp. 238–247 (1966).
18. S. G. Dick, D. W. Fuerstenau, and T. W. Healy, "Adsorption of Alkylbenzene Sulfonate Surfactants at the Alumina–Water Interface," *J. Colloid Interface Sci.,* **37** (3), 595–602 (1971).
19. T. Allen and R. M. Patel, "Adsorption of Long-Chain Fatty Acids on Finely Divided Solids Using a Flow Microcalorimeter," *J. Coloid Interface Sci.,* **35** (4), 647–655 (1971).
20. R. R. Stromberg, A. R. Quasius, S. D. Toner, and M. S. Parker, "Adsorption of Polyesters on Glass, Silica, and Alumina," *J. Res. Natl. Bur. Stand.,* **62** (2), 71–77 (1959).
21. Zeta-Meter, Inc., New York, unpublished work.
22. D. J. O'Connor, P. G. Johansen, and A. S. Buchanan, Electrokinetic Properties and Surface Reactions of Corundum," *Trans. Faraday Soc.,* **52** (5), 229–236 (1956).
23. "Butvar," Technical Bulletin 6070, Monsanto Company, St. Louis, Missouri, p. 6.
24. A. F. Greaves-Walker, *Drying Ceramic Products,* Industrial Publications, Chicago, 1968, p. 19.
25. T. K. Sherwood, "The Drying of Solids—I," *Ind. Eng. Chem.,* **21** (1), 12–16 (1929).
26. L. Slifkin, *Encyclopedia of Physics,* R. M. Besancon, ed., Reinhold, New York, 1966, p. 171.

27. R. B. Runk and M. J. Andrejco, "A Precision Tape Casting Machine for Fabricating Thin, Organically Suspended Ceramic Tapes," *Amer. Ceram. Soc. Bull.*, **54** (2) 199–200 (1975).
28. Terminology used as defined in ASTM E 12-70. American Society for Testing and Materials, Philadelphia, 1970.
29. "Bulk Density, Apparent Porosity, and Apparent Specific Gravity of Fired Porous Whiteware Products," ASTM C373-56, Vol. 13, 1970, p. 326.
30. W. D. Kingery, *Introduction to Ceramics,* Wiley New York, 1960, p. 40.
31. J. B. Jr. Wachtman, National Bureau of Standards, private communication.
32. Alcoa Reactive Aluminas," Aluminum Company of America, Pittsburgh, Pennsylvania, 1967, p. 3.
33. J. Longfellow, "High Speed Drilling in Alumina Substrates with a CO_2 Laser," *Amer. Ceram. Soc. Bull.*, **50** (3), 251–253 (1971).
34. J. R. Piazza and T. G. Steele, "Positional Deviations of Preformed Holes in Substrates," *Amer. Ceram. Soc. Bull.*, **51** (6), 516–518 (1972).
35. B. Schwartz and D. L. Wilcox, "Laminated Ceramics," *Ceram. Age,* **83** (6), 40–44 (1967).
36. H. W. Stetson, "Method of Making Multilayer Circuits," U.S. Patent 3,189,978 (June 22, 1965).
37. B. M. Hargis, "Ceramic Metallic Composite Substrate," U.S. Patent 3,549,784 (Dec. 22, 1970).
38. M. Bennett, W. E. Boyd, and J. E. Nobile, "Fabrication of Multilevel Ceramic, Microelectronic Structures," U.S. Patent 3,518-756 (July 7, 1970).
39. G. N. Howatt, "Method of Eliminating Voids in Ceramic Bodies," U.S. Patent 3,635,759 (Jan. 18, 1972).
40. R. A. Gardner and R. W. Nufer, "Properties of Multilayer Ceramic Green Sheets," *Solid State Technol.*, **17** (5), 38–43 (1974).
41. D. L. Wilcox, "Ceramics for Packaging," *Solid State Technol.*, **14** (2), 55–60 (1971).
42. J. Ettre, and G. R. Castels, "Pressure-Fusible Tapes for Multilayer Structures," *Amer. Ceram. Soc. Bull.*, **51** (5), 482–485 (1972).
43. T. Ihochi, K. Otsuke, and H. Maejima, "Screened Multilayer Ceramics and the Automatic Fabrication Technology," *Proc. Semiconductor/IC Processing Prod. Conf. New York,* **1972,** p. 71.
44. R. E. Mistler, "High Strength Alumina Substrates Produced by a Multiple-Layer Casting Technique," *Amer. Ceram. Soc. Bull.*, **52** (11), 850–854 (1973).
45. R. E. Mistler, "Strengthening Alumina Substrates by Incorporating Grain Growth Inhibitor in Surface and Promoter in Interior," U.S. Patent 3,652,378 (March 28, 1972).
46. "Marine Oils" (booklet), Haynie Products, Inc., Baltimore, Maryland.

31

Zinc Orthotitanate Powders for Thermal-Control Coatings

Y. Harada
D. W. Gates

The objective of the present study was to develop a process for producing a pigment (finely divided powder) for use in spacecraft thermal-control coatings. The desired pigment was Zn_2TiO_4, a binary oxide that normally requires temperatures of 900°C or greater for its formation from ZnO and TiO_2. Thus the studies were directed toward maintenance of the powder in a discrete particle state under the relatively rigorous temperature regimes that would oridinarily encourage sintering of particles (conglomeration). Although a sintered material can be comminuted to a pigment particle size by grinding, the potential introduction of contaminants in such a process makes its elimination desirable.

A process was developed in which Zn_2TiO_4 powder of pigment size, that is, submicron to 5 μ particles, could be obtained at temperatures up to 1200°C. This was accomplished by increasing the ZnO and TiO_2 reaction rate at lower temperatures through the use of intimately mixed, finely divided oxalate precursors obtained by a mixed precipitation. Final heat treatment for complete reaction was conducted a very rapidly so as to minimize particle sintering.

TECHNICAL BACKGROUND

The temperature control of statellites and spacecraft is one of the most challenging technical problems confronting spacecraft designers and materials engineers. The ultimate objective of thermal design is to ensure that the spacecraft operates within a prescribed temperature range defined by the temperature limitations of the vehicle's materials and components.

One of the primary passive methods used to achieve thermal control has been the use of coatings with high reflectance or low solar absorptance (α_s) that are resistant to discoloration in a vacuum–ultraviolet environment. The inorganic pigment that has been found to exhibit high reflectance and stability is ZnO.[1] Two coatings that have been used extensively on various spacecrafts are Z-93 and S-13G. Both incorporate ZnO as a pigment, the former using potassium silicate as a binder and the latter using a methyl silicone.[1]

However, ZnO is a strong absorber in the ultraviolet region, limiting its ability to reflect solar energy. Therefore, more recent work[2] has been involved with the development of another pigment, zinc orthotitanate (Zn_2TiO_4), which exhibits higher reflectance than ZnO in the near ultraviolet as shown in Figure 31.1.

Synthesis of Zn_2TiO_4 at elevated temperatures by a solid–solid reaction has been reported by several investigators. Bartram and Slepetys[3] described its preparation from anatase titania and zinc oxide by reaction at 800 to 1000°C for 3 hours. Reaction times of 48 hours at 800°C to obtain Zn_2TiO_4 from a zinc oxide–titanic acid reaction has been reported.[4] A phase diagram (Figure 31.2) for the ZnO–TiO_2 system proposed by Dulin and Rase[4] shows the existence of a 1Zn to 1Ti compound, $ZnTiO_3$, as well as the orthotitanate. Both investigations showed the presence of secondary phases, such as $ZnTiO_3$, ZnO, and/or TiO_2, in their Zn_2TiO_4 products, indicating the need for accurate stoichiometry control to obtain a pure product. The presence of secondary phases has been found to be deleterious to the optical stability of Zn_2TiO_4.

The use of decomposable salt precursors to enhance reaction or sinterability of oxides has been reported by several investigators.[5-7] An example of this enhancement is that of a high-purity $BaTiO_3$ produced from coprecipitated $BaTiO(C_2O_4)_4 \cdot 4H_2O$.[8]

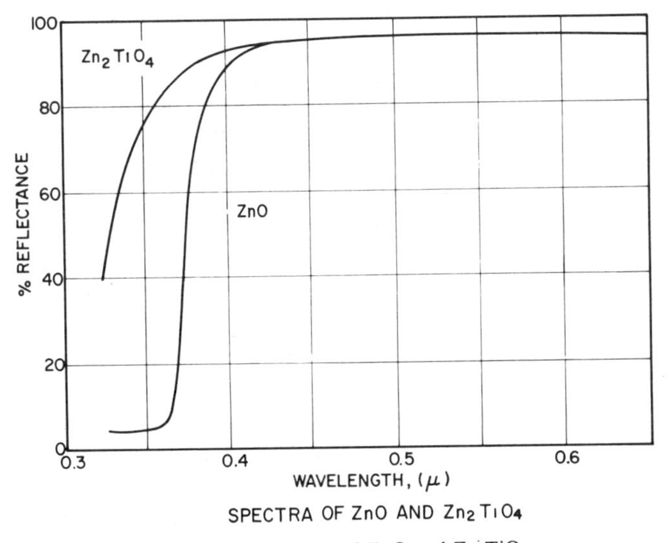

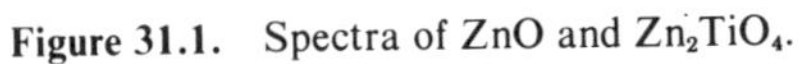

Figure 31.1. Spectra of ZnO and Zn_2TiO_4.

ZnO–TiO_2

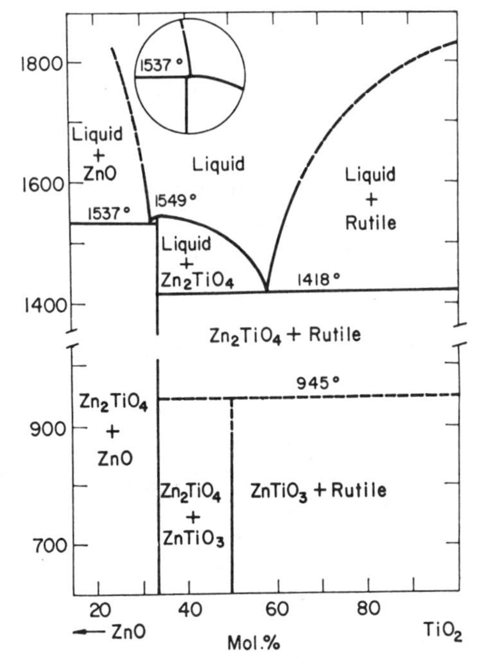

Figure 31.2. ZnO–TiO_2 system From Dulin and Rase.[4]

The present study was thus directed toward processing of the zinc- and titanium-source precursors so that rapid conversion to Zn_2TiO_4 under modest time/temperature requirements could be achieved. At the same time it was necessary to modify ultimate firing conditions to minimize sintering effects that are encouraged by such powder processing prior to firing. This paper describes the studies that were conducted in the areas of precipitation, decomposition, and firing, leading to the desired pigment material.

EXPERIMENTAL PROCEDURE

The zinc and titanium sources were reagent grade $ZnCl_2$ and purified $TiCl_4$, and the oxalic acid was reagent grade. Zinc chloride and oxalic acid solutions were prepared by conventional dissolution of the crystals in distilled water at 40°C. The $TiCl_4$ solution was prepared by dropwise addition of the $TiCl_4$ into distilled water, which was cooled in an ice bath to prevent formation of hydrolyzed titanium.

Precipitation of mixed oxalates was conducted by addition of a mixed-chloride solution (having a 2.05:1.00 Zn/Ti ratio) to oxalic acid at 40°C. This was followed by heating of the system with stirring to 90°C, where it was held from 1 to 4 hours, depending on the batch size.

Filtration was performed in a büchner-funnel system under slight vacuum. The precipitate was washed thoroughly with hot water until there was no evidence of acid in the filtrate.

Calcination and firing were performed in standard atmospheric Globar furnaces. Powders were contained in fused silica boats for firings up to 1300°C and in platinum crucibles at 1400°C.

RESULTS AND DISCUSSION

The solid-state process used to synthesize Zn_2TiO_4 in earlier work[2] is shown in Figure 31.3. The disadvantages of this oxide–oxide reaction are: (1) extensive grinding of the individual oxides and of the mixture prior to firing is necessary, (2) an extended period of 18 hours is required for complete reaction to form Zn_2TiO_4, and (3) additional grinding is needed to obtain a particle size of about 1 to 5 μ. The various comminution steps in this process present the danger of contamination from the milling media; the resulting impurities are a source of reflectance degradation and also can lower the stability to an ultraviolet–vacuum environment.

The precipitation process (Figure 31.4) was developed to minimize or eliminate such grinding and to decrease the processing times, both of which

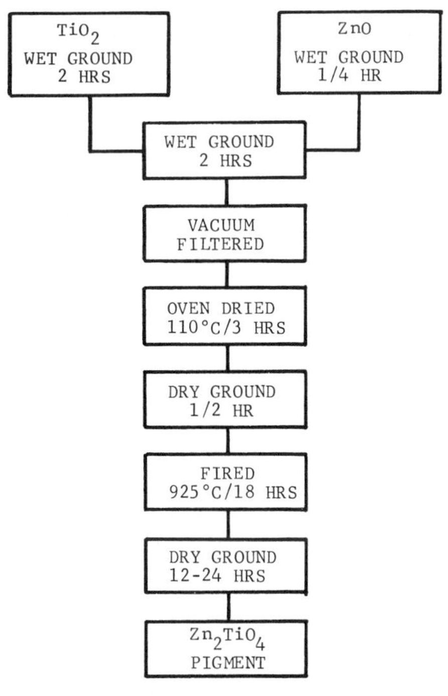

Figure 31.3. Solid-state process for forming Zn_2TiO_4.

would be desirable for larger-scale pigment production. The studies leading to the process shown in Figure 31.4 are discussed in the following sections.

Precipitation Studies

Complete precipitation of the mixed zinc and titanium oxalates in the aqueous medium was found to require a 2 hour hold at 90°C under continued stirring. X-Ray powder pattern examination showed the precipitate to be a mixture of $ZnC_2O_4 \cdot 2H_2O$ and a second phase whose lines suggest orthorombic crystallinity. The pattern for this titanium phase could not be found in the literature. This material, which is termed titanium oxalate in this paper, is presently being characterized.

Scanning electron microscope (SEM) photographs of the individual oxalates and the mixed precipitate are shown in Figures 31.5 through 31.7. The difference in particle size between $ZnC_2O_4 \cdot 2H_2O$ (2 to 10 μ) and the titanium phase precipitate (submicron to 5 μ) lead to a heterogeneous mixture in the mixed precipitate (Figure 31.7). Examination of different batches has shown uniform distribution of the two phases in each precipitation.

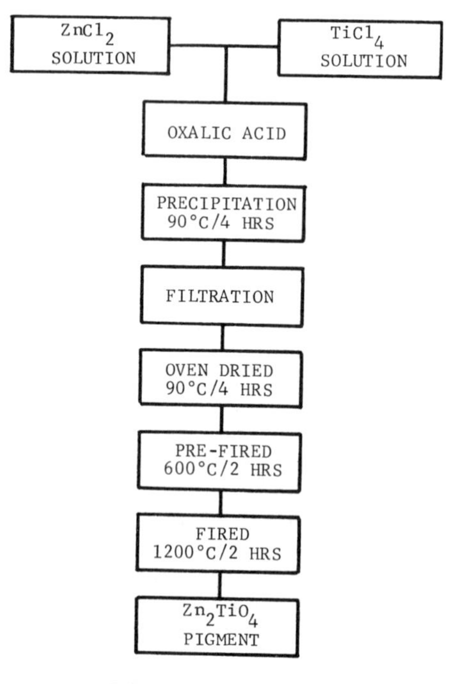

Figure 31.4. Precipitation process for forming Zn_2TiO_4.

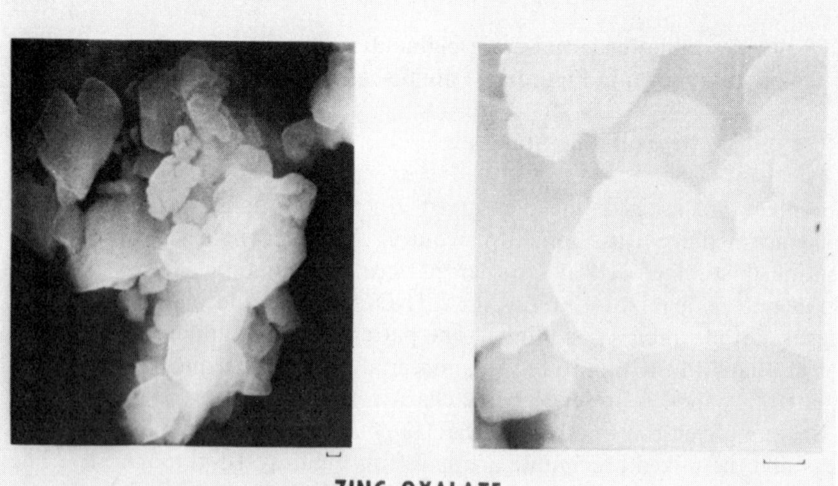

Figure 31.5. SEM microstructural view of zinc oxalate.

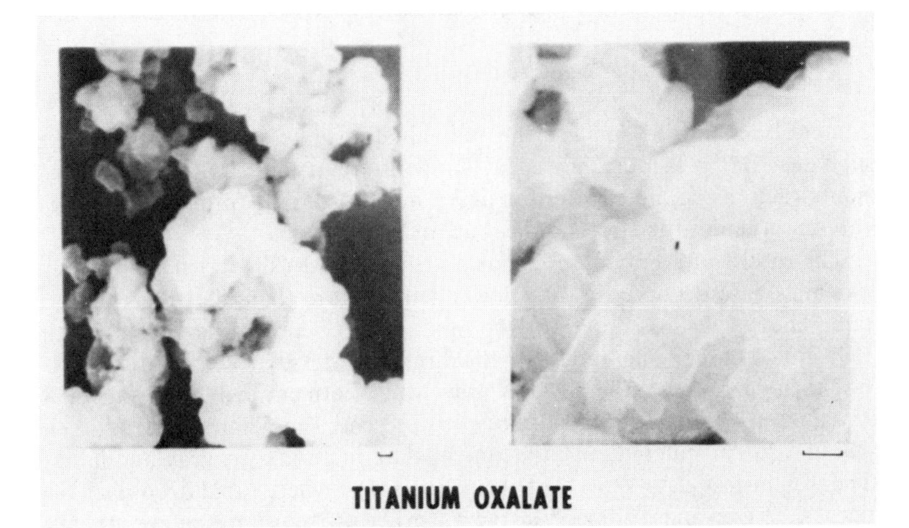

Figure 31.6. SEM microstructural view of titanium oxalate.

Figure 31.7. SEM microstructural view of mixed oxalates.

Decomposition Studies

Materials examined include the individual oxalates and the mixed precipitates. Samples were heated at temperatures from 120 to 700°C for 2 hours each, using direct insertion and removal from the furnace at temperature to minimize the effects of heat up and cool down.

The results of a gravimetric analysis (Figure 31.8) show that the titanium oxalate phase decomposes at a lower temperature than does the zinc oxalate. The weight-loss curve for the mixture lies between the curves for the individual components, reflecting the different rates of decomposition.

The appearance of the powders after heat treatment and the results of X-ray analysis using powder patterns are presented in Table 31.1. After the 400 and 500°C calcinations, the zinc oxalate material was gray and lumpy. The weight loss data show that this is the range where rapid decomposition occurs. D'Eye and Sellman[9] in their work with thorium oxalate attribute this coloring to the presence of carbon from the dissociation of carbon monoxide.

The titanium oxalate materials showed a lesser departure from white and were all free flowing. The mixed oxalate samples calcined at the various temperatures were also quite fine in particle size and did not display the aggregation shown by zinc oxalate samples. Samples calcined at 600 and

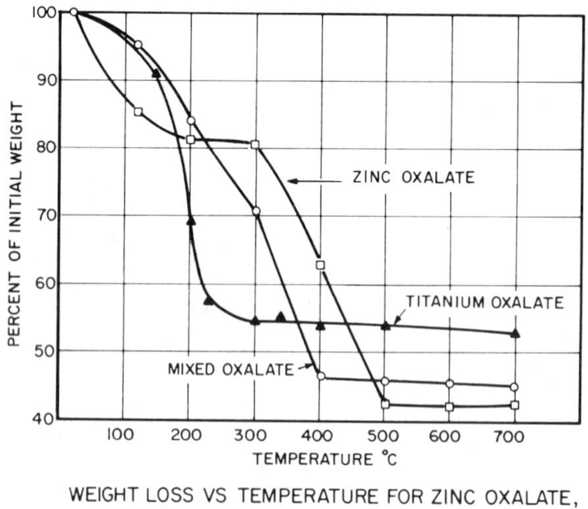

Figure 31.8. Weight loss versus temperature for zinc oxalate, titanium oxalate, and mixed oxalates.

Table 31.1. Appearance and phases present for zinc oxalate, titanium oxalate, and mixed oxalates as a function of temperature

	Appearance[a]			Phases Present[b]		
Temperature (°C)	Zinc Oxalate	Titanium Oxalate	Coprecipitate Oxalates	Zinc Oxalate	Titanium Oxalate	Coprecipitate Oxalates
120	White		White		TiOX	
200	White	Light yellow	White		TiOX	ZnOx + TiOx + X
300	White	Light yellow	Light yellow	ZnOx + X	TiO_2	X
400	Gray, Lumpy	Light yellow	Yellow	ZnO + X	TiO_2	ZnO
500	Light gray, Lumpy	Light yellow	Yellow	ZnO	TiO_2	ZnO + TiO_2
600	Tan	White	White	ZnO	TiO_2	Zn_2TiO_4 + ZnO
700	Yellow	White	White	ZnO	TiO_2	Zn_2TiO_4 + ZnO

Note: All samples calcined for 2 hours.

[a] All powders free flowing unless otherwise indicated.

[b] ZnOx, zinc oxalate; TiOx, titanium oxalate; X, apparently an intermediate phase in the decomposition of zinc oxalate.

700°C were white, although the lower-temperature samples showed some coloring.

X-Ray analysis of the various products revealed complete pyrolysis for Zn_2TiO_4 at 500°C, for titanium oxalate at 300°C, and for the mixed oxalates at 500°C. Formation of Zn_2TiO_4 was apparently initiated at 600°C. The existence of free ZnO but no TiO_2 in the X-ray pattern may be due to an extremely fine state of subdivision for the latter.

A SEM photograph of the 600°C material produced from mixed oxalates appears in Figure 31.9. A large range in particle size is evident for this partially reacted material, and the sharp edges of particle surfaces reflect the morphology of the as-precipitated material.

Heat-Treatment Studies

Complete conversion to Zn_2TiO_4 was conducted at temperatures from 900 to 1400°C. Firing times of 2 hours were used up to 1300°C; the 1400°C soak time was 5 minutes. To retain the material in unsintered form, samples were fired as loose powders using what may be termed "flash calcination." This involved direct insertion of the powder in a fused silica container into a furnace at the prescribed firing temperature. Approximately 15 minutes was needed for large batches (1000 g) to attain temperature. Samples were removed in a similar abrupt manner after the prescribed firing time.

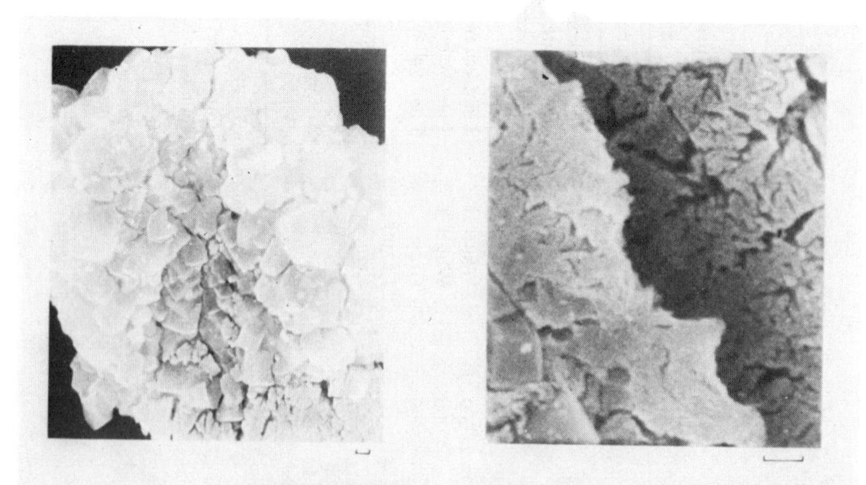

Figure 31.9. SEM microstructural view of mixed oxalates calcined at 600°C/2 hours.

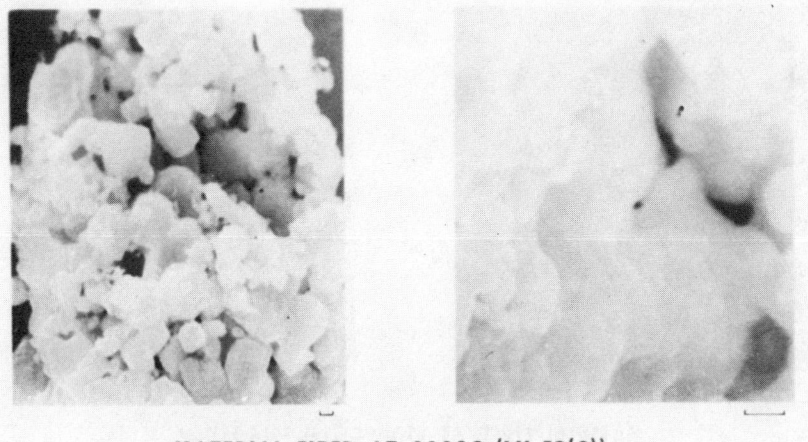

Figure 31.10. SEM microstructural view of mixed oxalates calcined at 900°C/2 hours.

All materials were precalcined at 600°C for 2 hours prior to the high-temperature (>900°C) treatments to eliminate the volatiles formed during decomposition. Flash calcination of the mixed precipitates without this pretreatment resulted in excessive blowing and loss of powder.

X-Ray evaluations revealed an essentially pure Zn_2TiO_4 after 2 hour firings at various temperatures from 900 to 1400°C. As designed for this study by the use of a 2.05 Zn to 1.00 Ti ratio, a slight excess of ZnO was also detected. Acetic acid leaching to form zinc acetate resulted in removal of the free ZnO. Gravimetric analysis showed the amounts to be from 2 to 4 wt % for various batches.

SEM studies showed that the 900°C material had a particle size range from submicron to about 2 μ (Figure 31.10). The particles exhibited rounded surfaces unlike the as-precipitated or 600°C materials. Material heated at 1200°C (Figure 31.11) showed an increase in size, exhibiting a range from about 2 to 5 μ, with no evidence of submicron particles. Sintering of particles was apparent from the necking behavior shown for the 1200°C material. Material heat treated at 1300°C/2 hours and 1400°C/5 minutes exhibited a similar morphology.

The final product at all heat-treatment conditions was a material that, although caked, could be reduced to a powder by manual crushing. Most importantly, the physical state of the Zn_2TiO_4 powder was amenable to direct incorporation into a silicone vehicle for use as a paint.

Thus the use of mixed-oxalate precursors to enhance the ZnO–TiO_2 reac-

Figure 31.11. SEM microstructural view of mixed oxalates calcined at 1200°C/2 hours.

tion at lower temperatures, coupled with a rapid-heat-treatment technique, flash calcination, provides a more rapid method for obtaining a Zn_2TiO_4 powder suitable for use as a pigment. Elimination of grinding of the powder, necessary in the solid–solid state process, yields a purer product less susceptible to color degradation in a space environment.

ACKNOWLEDGMENT

This work has been sponsored by the George C. Marshall Space Flight Center, National Aeronautics and Space Administration, under Contract No. NAS8-26791. The authors wish to express their appreciation to W. R. Logan for conducting the experimental studies and to J. E. Gilligan for his valuable contributions.

REFERENCES

1. G. A. Zerlaut, Y. Harada, and E. H. Tompkins, "Ultraviolet Irradiation in Vacuum of White Spacecraft Coatings," *Symposium on Thermal Radiation of Solids*, S. Katzoff, ed., NASA SP-55, Washington, D.C., 1965.
2. G. A. Zerlaut, J. E. Gilligan, and N. A. Ashford, "Space Radiation Environmental Effects in Reactively Encapsulated Zinc Orthotitanates and Their Paints," AIAA 6th Thermophysics Conference, April 1971, Paper No. 71–449.

3. S. R. Bartram and R. A. Slepetys, "Compound Formation and Crystal Structure in the System ZnO–TiO_2," *J. Amer. Ceram. Soc.*, **44** (10), 493–499 (1961).
4. S. R. Dulin and D. E. Rase, "Phase Equilibria in the System ZnO-TiO_2," *J. Amer. Ceram. Soc.*, **43** (3), 125–131 (1960).
5. D. T. Livey, B. M. Wanklyn, M. Hewitt, and P. Murray, Properties of MgO Powders Prepared by Decomposition of $Mg(OH)_2$," *Trans. Br. Ceram. Soc.*, **56**, (5), 217–236 (1957).
6. Y. Harada, Y. Baskin, and J. H. Handwerk, "Calcination and Sintering Study of Thoria," *J. Amer. Ceram. Soc.*, **45**, (6), 253–257 (1962).
7. R. E. Jaeger and T. J. Miller, "Preparation of Ceramic Powders by Liquid Drying," *Amer. Ceram. Soc. Bull.*, **53** (12) 855–859 (1974).
8. W. S. Clabaugh, E. M. Swiggard, and R. Gilchrist, "Preparation of Barium Titanyl Oxalate Tetrahydrate for Conversion to Barium Titanate of High Purity," *J. Res. Natl. Bur. Stand.*, **56** (5), 289–291 (1956).
9. R. W. M. D'Eye, and P. G. Sellman, "The Thermal Decomposition of Thorium Oxalate," *J. Inorg. Nucl. Chem.*, **1** (1/2), 143–148 (1955).

32

Characterization of Powders for Thick Films and Capacitors

K. K. Verma

A. Roberts

In thick-film applications, powders are utilized to prepare resistor and conductor pastes. These pastes or inks are screen printed on substrates. Resistors include printed patterns of resistive paste systems using metals, metal oxides, and glass-frit mixtures. Conductors are printed patterns of conductive-paste systems consisting of metals, metal mixtures and alloys, and glass-powder mixtures. Some special applications, such as dielectric crossover and encapsulating layers, quite often are used on thick-film printed patterns but are covered in this chapter.

The capacitor industry uses powders for preparing a dielectric body and conducting electrodes. The latter includes internal electrodes between dielectric layers in multilayer capacitors, external or exposed electrodes for "single-plate" capacitors, and end termination of multilayer capacitors.

Pastes for use in thick-film and capacitor applications are prepared by grinding, blending, and dispersing powders in a liquid-vehicle system. Powder characterization is an important part of achieving the desired paste rheology, firing behavior, and final properties. This chapter discusses the rheology of pastes, the effects of particle characteristics on rheology, and how these factors affect performance.

PERFORMANCE CHARACTERISTICS

For a given choice of materials and process parameters (such as drying and firing), the final performance of a capacitor or thick-film component depends on a proper choice of physical characteristics of the powders and on the powder-preparation method used in the manufacture of dielectric bodies and electrode and resistor inks. The parameters on which performance testing is based are given below.

For capacitors the parameters include: (1) micro- and macrostructure of the dielectric body, (2) leaching and dewetting of electrode metallization during soldering, (3) bond strength and solderability, (4) screenability and area coverage of the paste system, (5) capacitance and dielectric constant, (6) dissipation factor, (7) insulation resistance, (8) temperature coefficient of capacitance, and (9) shorts and voltage failure.

For thick films the parameters include: (1) macrostructure of the film, (2) leaching and dewetting, (3) adhesion with the substrate and bondability, (4) screenability and area coverage, (5) resistivity, (6) noise, (7) electrostatic discharge, and (8) temperature coefficient of resistance.

PASTE SYSTEMS

The screenability of thick-film pastes depends on the rheology of the system, which in turn depends on the nature of the powders, vehicle system, surface active agents, and other modifiers present in the paste. The particle-size distribution of the powders plays a considerable role in controlling rheology and, therefore, viscosity of the paste.

Figure 32.1 shows the commonly found flow behavior of paste systems. A theoretically ideal paste would show a Newtonian behavior, in which case the flow response (shear rate) of the paste is linear with applied stress. This type of behavior may only be expected from low-viscosity liquid systems, and common thick-film pastes very seldom act this way. They do not usually respond to the applied stress linearly. This nonlinear or non-Newtonian behavior is represented by curves *3* through *5* in Figure 32.1. Good screenable pastes fall into this category and may be classified as plastic or pseudoplastic systems. They usually have high viscosity, considerable interaction among organic vehicle systems and/or dispersed

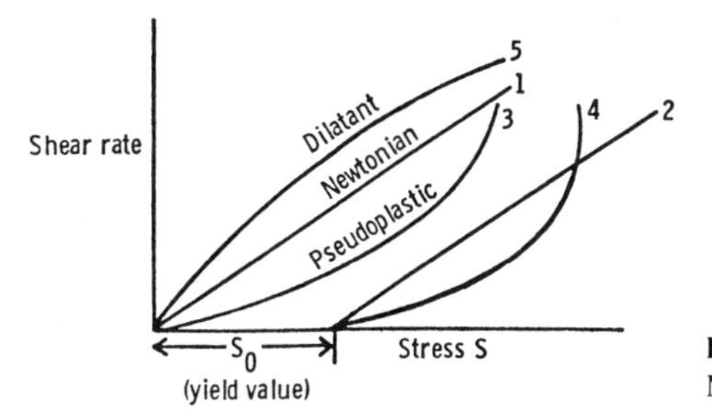

Figure 32.1. Typical flow curves. From Miller.[1]

powders in the paste, and some gel structure. Another common form of screenable pastes showing non-Newtonian behavior exhibits a phenomenon known as thixotropy. This behavior, represented in Figure 32.2, shows characteristic hysteresis believed to be due to the breaking up of the secondary bonds that form a gel structure, with increasing stress. The gel structure, however, can reform if the material is allowed to stand. This characteristic of the paste system can be utilized to an advantage in screenable pastes, since such behavior can, to an extent, eliminate secondary flow of screened pastes.

EFFECTS OF PARTICLE SIZE, SHAPE, AND DISPERSION ON A PASTE SYSTEM

Increasing the percent solids in a paste system normally increases its viscosity. This relationship, however, is affected by a tendency of the fine particles to form flocs rather than an increased amount of dispersed solids in the paste. Particle shape and size asymmetry will promote non-Newtonian behavior in the pastes. For example, at reasonably high concentrations, nonspherical particles are far more effective than spherical particles in producing non-Newtonian behavior. Such paste systems would generally show high yield values.

Powder particles that tend to flocculate at even low concentration destroy Newtonian behavior of suspensions. The strong surface forces of fine parti-

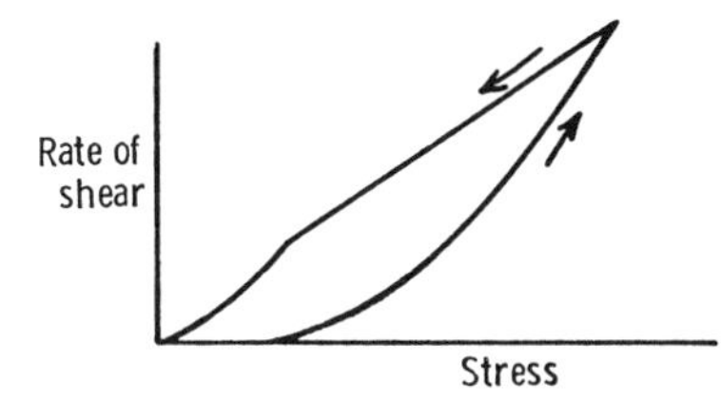

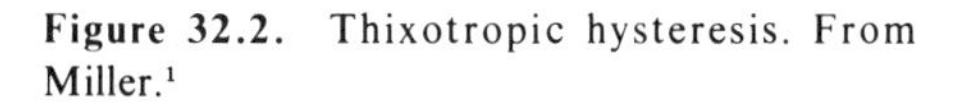

Figure 32.2. Thixotropic hysteresis. From Miller.[1]

cles bring the particles together, causing the paste system to gel. The smaller the particles, the larger the surface area and, therefore, the stronger the surface forces. In a paste system proper wetting of the particles by the vehicle is necessary for good flow characteristics. Highly surface active powder particles also require more vehicle to cover the surface, thus reducing the effective amount of vehicle available for flow; consequently, such systems exhibit plastic behavior and high viscosity.

A real screening paste system rarely consists of only one type of powder and vehicle. A practical system may consist of the following: (1) a mixture of powders, (2) more than one type of solvent and/or vehicle, (3) modifiers such as gelation materials, (4) surface-active agents, and (5) ingredients to control sensitivity of a paste to environment, temperature, drying, and firing.

EFFECTS OF POWDER PREPARATION AND POWDER CHARACTERISTICS ON PERFORMANCE

Experience has shown that in many instances more than one performance characteristic of a capacitor or thick-film component can be affected simultaneously by a choice of powder characteristic, such as a certain particle-size distribution or shape, and a processing parameter, such as longer time for mixing and dispersing of powder particles. An example is the poor and good macrostructures of two multilayer capacitor chips shown in Figure 32.3. A poor structure is represented by the discontinuity of the electrodes, and the voids and cracks in the dielectric body. These voids and discontinuities have been caused mainly by poor dispersion of organic binder in the dielectric body, which burns off during firing and leaves the poor structure. The essentially void-free structure obtained in the lower part of the figure by homogenizing dispersion of the binder system in a colloid mill simultaneously improved the capacitance, dissipation factor, insulation resistance, and temperature and voltage response of the capacitor chip.

Final quality and performance of a capacitor depends on both macro- and microstructure. Microstructure refers to the size and structure of grains in the dielectric body. A grain size of $\sim 1\ \mu$ is considered optimum for the most favorable combination of capacitance, loss, and dissipation factor. A homogenous mixture and controlled particle-size distribution are necessary to achieve the above requirements. With an excessive amount of fines, a homogeneous and better sintered body may be obtained, but extra fines may cause increased sintering and excessive grain growth. However, this can be restricted by small additions of grain-growth inhibitors. Again, increased amounts of blending and mixing are required for its homogeneous disper-

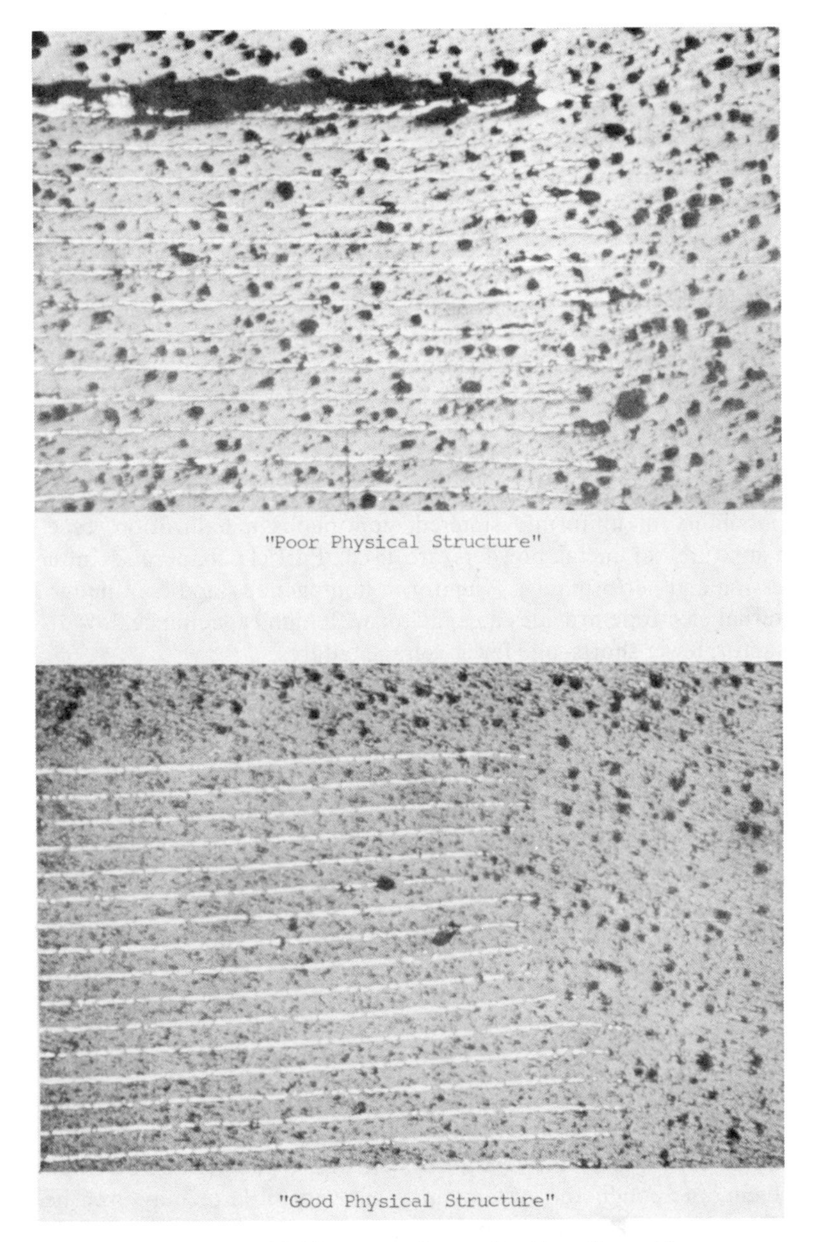

Figure 32.3. Multilayer capacitor quality. From Capozzi.[2]

sion. It may be necessary to adjust firing time to permit and compensate for optimization of powder characteristics.

Capacitors

Microstructure of the dielectric body and macrostructure of the capacitor have already been covered above. In the rest of this section the electrode inks for capacitors are discussed separately as internal and external or end-termination electrodes.

Internal electrodes consist of metallization between the dielectric layers in multilayer capacitors. These normally do not use any glass frit and may normally be fired at high temperatures simultaneously with the dielectric body. In this case the area covered by printed electrode paste on a dielectric tape is of major importance.

Fine particle size and adequate mixing and dispersion of powders increase the possibility of uniformly sintered, continuous metallization, especially when mixtures of metal powders are used. Particle shape also influences processing and performance. A uniform, homogenous, and continuous layer of internal electrode provides a capacitor with high capacitance, low dissipation factor, fewer shorts, and fewer voltage failures.

In the case of external electrodes or end terminations, glass frit is normally used with a mixture of metals or alloys; these systems are fired at lower temperature. A densely sintered electrode is desired to resist leaching and dewetting in the solder melt.

Blisters, cracks, and voids on the surface can be minimized by increased mixing, proper choice of vehicle, and higher degree of powder dispersion in the paste system. If excessive fusion and flow of glass frit at the firing temperature can be avoided, finer glass-frit size gives a denser electrode layer with improved adhesion, as well as improved temperature and voltage response of capacitance.

Thick Films

A significant part of this area has been covered under the heading of paste systems, in reference to the screenability and rheology of pastes.

In thick films where the resistor systems have glass frit as an important ingredient, the conductor systems may use very little or may even be fritless. For example, two developments in recent years are fritless gold and fritless platinum/silver conductor systems.

For quality testing of thick-film components, the performance characteristics listed above, such as macrostructure, leaching and dewetting, adhesion with the substrate, bondability, screenability, and area coverage, are common to both resistor and conductor systems. Therefore, their response to powder preparation and characterization will be similar.

Increased grinding and mixing to obtain better dispersion of solids in the paste system seem to improve macrostructure defects, such as pinholes and cavities on the surface of the film. Leaching and dewetting improve because of better sintering of the film and adhesion with the substrate. Electrical performance is also generally dependent on milling time.

For a given size range denser particles show better leaching and dewetting resistance, though they do not necessarily improve adhesion and bondability at the same time. It would appear, however, that adhesion and bondability should be improved by a suitable choice of glass frit or low-melting metal or alloy mixture, and that leaching and dewetting may be improved by a proper choice of metal, alloy, or oxide mixtures. For example, it may be expected that in a Pd/Ag system, leaching of Ag into a solder bath is reduced when the allow of Pd and Ag is formed.

REFERENCES

1. L. F. Miller, "Screenability and Rheology," *Solid State Technol.* **17**, 54–60 (Oct. 1974).
2. V. F. Capozzi, "Multilayer Ceramic Capacitor Materials and Manufacture," EMD, Sel-Rex Co., Santa Ana, Ca., 92705.

33

Future Directions in Processing Research

OPENING REMARKS BY PANEL LEADER—J. A. PASK

This timely conference reflects progress during the last 10 to 15 years on the science of ceramic processing. While we are far from providing all the answers, this conference has taken a long step in that direction.

Microstructure is a key toward understanding processing. Around 20 years ago we started talking about how it was necessary to characterize materials, characterize processes, and correlate character features with features of microstructure. To do that, the microstructure must be observed. The availability of exotic tools and microscopes makes the job possible. We have seen important applications of the scanning electron microscope and the transmission electron microscope, and see that the petrographic microscope is now being used again because features can be seen with this microscope that cannot be seen with the others. The major point is that we can now observe microstructure in great detail. Features can be seen on a microscale rather than only a macroscale. But we still have to quantify much more than we have been able to do so far.

This chapter summarizes the presentations given by panel members at the conference. The subject of the panel was processing research needs. The speakers represented different branches of industry, including traditional ceramics, technical ceramics, refractories, and materials for energy, and an academic perspective. Also, comments given earlier in the conference by A. G. Pincus on particle characterization have been added in this chapter.

POWDER CHARACTERIZATION—A. G. PINCUS

Why characterize particulates? One reason is that characterization is needed at the transfer point between powder supplier and the plant because the vendor and buyer have to be able to talk to each other. They have to agree on a set of specifications if they are to do business with each other. Also, the characterization of particulates has even more significance because, unless you can properly characterize a particulate, you will not have the subsequent forming and firing steps under control. The character of the particulate is going to affect the processing and certainly affect the product all the way through until it is shipped out of the factory, and maybe even in service. We think maybe some of this character of the particulates is wiped out in firing, but it has a way of hanging on until it shows itself.

Several years ago we talked of it being completely wiped out if you make your crystalline ceramics by way of melting or vapor-deposition routes. This wipes out any surface-history effects and you are no longer dependent upon fabrication of particulates. I even predicted one time that forming as a glass and then crystallizing to make a ceramic is so far superior to powder processing that it is going to displace powder processing entirely. We've seen now that after more than 15 years of glass-ceramics being around, it hasn't happened. The attendance at this conference is an example of the fact that powder processing is very much alive. We will be staying with powder processing and we have to learn how to do the characterization job well all the way through, from the preparation of the powders themselves on through to the shipping of the material out of the factory.

The problem that man and plant face is a bewildering array of tools, techniques, and instruments to do a characterization job. Particle-size measurement again is a good example of this. We talk about characterization as if it just got invented in this era of the new ceramics, this whole post World War II development, which has seen technical ceramics move into so many new fields. I think if you go back into the history of ceramics there has always had to be an intuitive or even quantitative understanding of characterization. The new feature we face now is the explosion of instrumentation. There are more and more instruments to do the job and more and more data piling out needing to be interpreted and applied. When you hook the computer up to the rapid electronic measuring device, you really have a problem. It is not only how to characterize, but how much data to accumulate. An incident quoted by a colleague relates to this question. He visited a ceramic whiteware plant and saw many file cabinets, and he asked the ceramist in charge, "What is all of this, what have you got in there?" "Particle-size measurements," was the response. "Do you measure particle size?" and he said every shipment that comes in he measures. So my colleague asked what he did with the information once he gets it, and the ceramist said, "I put it in my files!" This has always been a prime problem; how do you effectively use your characterization data.

Now if you shorten the time for making measurements from a day or more to a half an hour or minutes, and then have the computer able to absorb all this data and feed it back to you, a mammoth job of data selection must be faced. This leads us back to the kind of papers that have been presented at this meeting, where there is a real striving for understanding. Understanding the process and tying processing into the characterization well enough so that maybe just a few measurements will give you the controls that you need in large-scale manufacturing must be our prime objective.

TRADITIONAL CERAMICS—R. RUSSELL, JR.

A little boy rushed excitedly into the house and said, "Wow! Our new neighbors have a new lawnmower that doesn't require gasoline or oil or anything, all you do is push it!" I believe that story has some implications to ceramic processing in that it does say that there may be some good ideas in old technology and progress can come through better understanding of old technology and its application. During this panel I am to deal with traditional ceramics which have old technologies. First, I identify traditional ceramics as products such as triaxial porcelain and other whitewares for nonelectronic use; for example, dinnerware, sanitary ware, floor tiles, and various types of pottery. I also include structural clay products, enamels, abrasives, ceramic cements, etc.

It is a difficult challenge to say in a few minutes something relevant to such a wide product line. That is why I have chosen to talk, more or less, philosophically. My experience in production, research, education, and consulting has emphasized over and over the critical importance of ceramic processing. I'm not sure when I first fully recognized that a ceramic product represents a combination of many complex processes. But, the awakening came very early in my career and has been reconfirmed continually for some 40 years. I keep emphasizing this fact to my students, but I suspect they will accept only the thesis that "processing is vital" when ceramic production, development, quality, or characterization problems confront them in real-life situations.

When we set about some 12 years ago to define the criteria in ceramic engineering education, the significance of ceramic processing in justifying our various ceramic engineering curricula was a central theme, distinguishing us from the metallurgical or polymer engineering people, for example. The resulting objective criteria in the ceramic engineering education report of the American Society for Engineering Education included considerable emphasis on ceramic processing. However, I've wondered if any modern curriculum is doing justice to this area; and I doubt that any of our graduates, confronted with material-engineering problems, would claim overexposure to processing technology despite the sincere efforts of some of my peers. One has only to examine the ceramic literature in recent years, or for that matter, our Masters and Doctorate dissertations to gain some insight into the lack of emphasis on ceramic processing compared with basic and applied research on materials and their behavior. While processing is by no means ignored in our current literature, its coverage is hardly prolific or, in my opinion, adequate.

Now, rather than continuing to emphasize the negative, let us take the positive view. If we agree that ceramic processing is vital, why do we not understand it more fully, do it better, and publish more widely? I summarize a number of factors which I think are involved:

1. Institutional research and development is largely sponsored, with recent sponsorship being highly science and materials-properties oriented.
2. The processing research that is sponsored tends to relate overwhelmingly to modern or advanced materials needs, rather than traditional ceramics.
3. The processing of conventional ceramics tends to be traditionalized, and a natural resistance to change plus a resignation to the status quo is in effect all too often.
4. Lack of a general realization that a more thorough study of traditional ceramics and their processing has the potential rewards to justify an extensive effort.
5. Individual manufacturers, and in some cases groups of manufacturers, that study or develop processing technology often consider the information proprietary or just don't care to make it available publicly. (I've had some experience in that direction that I've thought was unfortunate.)
6. The demise of engineering experiment stations with ceramic-processing orientation and the greater science emphasis in such organizations as the National Bureau of Standards, the Bureau of Mines, and governmental and private research organizations tends to deemphasize traditional ceramic processing research.
7. Ceramic technical curricula are controlled by fewer and fewer engineering-oriented educators and reflect the character of the available

research funding as well as national trends in education, neither of which has favored an engineering or processing emphasis.

Now, while other factors are involved, the pertinent question is how to shift some of the research, development, and publication emphasis to processing, technology, and fundamentals, especially as regards to more traditional ceramics, which I assure you are an important part of our field, both volumewise and dollarwise. There are no easy answers, but some obvious changes based on part on my above-listed items are needed. First and foremost is the recognition of the great potential in ceramic-processing studies. Conferences, such as this, must increase. These must be developed to promote the processing studies needed for progress. Ceramic processing must be elevated in status, and those willing and able to contribute must be encouraged. There is great opportunity here for the talented and energetic ceramist who can lead the way and do much to encourage this direction. It is my hope that this conference has been one firm step towards better and more economical processing.

TECHNICAL CERAMICS—J. E. BURKE

One of the major goals of industrial ceramics is to obtain reproducible powders and to characterize them decently. For theoretical purposes we tend to think of powders as monodispersed, individual particles, and that their behavior can be characterized by measuring the surface area. In fact, surface area so frequently tells us very little about the activity (sinterability) of the powder.

We also must learn more about powder activity. Use of sintered density as a measure of powder activity is indirect and too often misleading. The activity of a powder, from the viewpoint of how it affects sinterability, is strongly dependent upon agglomeration. The small particles of the agglomerates sinter very rapidly, but big holes remain forever. A coarse powder may sinter more slowly but may go to a higher density because of the absence of large voids due to agglomerates.

Let us summarize what I think are some of the important points that have been made in this conference. Hard agglomerates do not break down during processing. The role of the strength of agglomerates is not yet clear. Some sintering-rate measurements are very helpful in indicating what role hard agglomerates are playing. Finally, a very good way to obtain information about the characteristics of powders is to look at the microstructure, because that indicates all of the factors of concern.

REFRACTORIES—R. A. ALLEIGRO

This conference did not cover the problems of the refractory industry in much detail. The papers mainly were associated with fine particles. Refractories are concerned with the other end of the spectrum. We start with rocks and try to find materials to put with them to get some degree of bonding. I would like to talk about the bonding of these large aggregates with fine particulates and clays. These aggregates are not micron- or submicron-size particles, but range up to 3000 μ in size (in some cases even larger). Yet when refractories are formed, the same consideration of particle size, shape, orientation, packing, and composition are important.

It has been mentioned (Chapter 23) that the examination of the fired product gives an indication of process character and control. In refractories we are continuously guided by those final results, with feedback to raw materials and processes involved in consolidation and firing. Examination of the fired product gives an indication of process character and control. This is essential before second-guessing the cause of rejects. Microstructural examination provides information on the amount and distribution of glassy phases bonding the coarse particles, the possible fractures in the bonding phases or in the large grain due to thermal shock, and the quality of adhesion.

The large particles, or aggregates, can have a wide range of physical and chemical variation. Some are dense, while others are quite open; others are rough or smooth. Some may be blocky or angular platelets. These characteristics greatly affect the bonding mechanism between the matrix and the aggregate. Chemical variations, either intentional by mixing of two types or through impurities often affect the final product.

What we strive for is to determine the desired product characteristics and feed this back to specifications which reflect a reasonable degree of constancy and an acceptable rejection level. Characterization of raw materials must be tied to those critical attributes upon which performance is based.

ENERGY-RELATED MATERIALS—W. B. CRANDALL

I would like to focus on the processing and characterization of oxynitride powders, which have been neglected so far in this conference, but offer outstanding potential for achieving superior high-temperature performance. An important factor in the processing of sialons, a solid solution of alumina and silicon nitride, is a large volume expansion that occurs as the solid solution is formed. Consequently, continuous application of pressure in hot pressing is largely compensated by the volume change. However, characterization of the alumina and silicon nitride powder end-members led to significant processing improvements. The petrographic microscope was found to be the most useful powder-characterization tool. Calcination of aluminum trihydroxide could be controlled to an optimum activity for the solid-solution-densification reaction by measuring the index of refraction.

Secondly, we found that the powder activity could be maintained during hot pressing by keeping the oxygen pressure below 10^{-18} atm and nitrogen about 10^{-3} atm. By use of a two-stage pressure and temperature application in the controlled atmosphere, powder activity is preserved and high-density compacts can be achieved.

ACADEMIC RESEARCH—D. W. FUERSTENAU

The title of this conference was the Science of Ceramic Processing before Firing, and I notice that at least two-thirds, if not three-fourths, of the topics talked about were "after firing." However, what you saw was that a lot of the problems that end up in your end products result from various factors in the preparation in the raw materials, in the preparation of the green bodies, and eventually the firing characteristics of the compacts. I thought there would be considerably more discussion on how do you actually characterize a particle and what kind of physical/chemical characteristics must be used for doing this. I once had a tour of Coors alumina plant and I noticed that a lot of their alumina products are made by filling molds with granules, yet only once did I hear some discussion about the effects of these granules. An area that certainly deserves study is what is the structure and size distribution of these granules, what do they do to the active behavior and flow characteristics in the die. I think one has to start right at the very beginning of particle preparation and work your way all the way along to sintering. I can see that some of the items briefly mentioned must tie very directly to the factors that control particle activity. One problem restricting progress in this field, as I see it, is that not one single professor of ceramics in this country devotes his entire professional interest to processing—someone who studies the preparation of particles. I've done a lot of work on grinding, but my interest in grinding is on the scale of boulders compared to what you people are dealing with. The kinds of products that are involved in the mineral processing industry in something like 100 to 300 microns in size and the phenomenon that exists in grinding at this relatively coarse size is far different from that encountered when you are down in the micron size. I think there is a whole world of study in particle preparation followed by particle characterization or mixing or formation of agglomerates, packing, and stresses that exist within packed powders. I imagine this is what contributes considerably to the nonuniformity and is worthy of basic studies without worrying about the final product. I think this is a good opportunity for some person to set as his life's goal. This is really what I've done in the world of mineral processing. I'm not personally concerned with whether I am going to be able to make a slurry of such and such a grade of ore and produce a concentrate or metal from that grade of ore; but I've really devoted many years to understanding basics and intend to continue this effort.

Index